The Journal of The American Association of Variable Star Observers

Volume 47
Number 1
2019

AAVSO
49 Bay State Road
Cambridge, MA 02138
USA

ISSN 0271-9053 (print)
ISSN 2380-3606 (online)

Publication Schedule

The Journal of the American Association of Variable Star Observers is published twice a year, June 15 (Number 1 of the volume) and December 15 (Number 2 of the volume). The submission window for inclusion in the next issue of JAAVSO closes six weeks before the publication date. A manuscript will be added to the table of contents for an issue when it has been fully accepted for publication upon successful completion of the referee process; these articles will be available online prior to the publication date. An author may not specify in which issue of JAAVSO a manuscript is to be published; accepted manuscripts will be published in the next available issue, except under extraordinary circumstances.

Page Charges

Page charges are waived for Members of the AAVSO. Publication of unsolicited manuscripts in JAAVSO requires a page charge of US $100/page for the final printed manuscript. Page charge waivers may be provided under certain circumstances.

Publication in *JAAVSO*

With the exception of abstracts of papers presented at AAVSO meetings, papers submitted to JAAVSO are peer-reviewed by individuals knowledgeable about the topic being discussed. We cannot guarantee that all submissions to JAAVSO will be published, but we encourage authors of all experience levels and in all fields related to variable star astronomy and the AAVSO to submit manuscripts. We especially encourage students and other mentees of researchers affiliated with the AAVSO to submit results of their completed research.

Subscriptions

Institutions and Libraries may subscribe to JAAVSO as part of the Complete Publications Package or as an individual subscription. Individuals may purchase printed copies of recent JAAVSO issues via Createspace. Paper copies of JAAVSO issues prior to volume 36 are available in limited quantities directly from AAVSO Headquarters; please contact the AAVSO for available issues.

Instructions for Submissions

The *Journal of the AAVSO* welcomes papers from all persons concerned with the study of variable stars and topics specifically related to variability. All manuscripts should be written in a style designed to provide clear expositions of the topic. Contributors are encouraged to submit digitized text in MS WORD, LATEX+POSTSCRIPT, or plain-text format. Manuscripts may be mailed electronically to journal@aavso.org or submitted by postal mail to JAAVSO, 49 Bay State Road, Cambridge, MA 02138, USA.

Manuscripts must be submitted according to the following guidelines, or they will be returned to the author for correction:

Manuscripts must be:
1) original, unpublished material;
2) written in English;
3) accompanied by an abstract of no more than 100 words.
4) not more than 2,500–3,000 words in length (10–12 pages double-spaced).

Figures for publication must:
1) be camera-ready or in a high-contrast, high-resolution, standard digitized image format;
2) have all coordinates labeled with division marks on all four sides;
3) be accompanied by a caption that clearly explains all symbols and significance, so that the reader can understand the figure without reference to the text.

Maximum published figure space is 4.5" by 7". When submitting original figures, be sure to allow for reduction in size by making all symbols, letters, and division marks sufficiently large.

Photographs and halftone images will be considered for publication if they directly illustrate the text.

Tables should be:
1) provided separate from the main body of the text;
2) numbered sequentially and referred to by Arabic number in the text, e.g., Table 1.

References:
1) References should relate directly to the text.
2) References should be keyed into the text with the author's last name and the year of publication, e.g., (Smith 1974; Jones 1974) or Smith (1974) and Jones (1974).
3) In the case of three or more joint authors, the text reference should be written as follows: (Smith et al. 1976).
4) All references must be listed at the end of the text in alphabetical order by the author's last name and the year of publication, according to the following format: Brown, J., and Green, E. B. 1974, *Astrophys. J.*, **200**, 765.
Thomas, K. 1982, *Phys. Rep.*, **33**, 96.
5) Abbreviations used in references should be based on recent issues of JAAVSO or the listing provided at the beginning of *Astronomy and Astrophysics Abstracts* (Springer-Verlag).

Miscellaneous:
1) Equations should be written on a separate line and given a sequential Arabic number in parentheses near the right-hand margin. Equations should be referred to in the text as, e.g., equation (1).
2) Magnitude will be assumed to be visual unless otherwise specified.
3) Manuscripts may be submitted to referees for review without obligation of publication.

Online Access

Articles published in JAAVSO, and information for authors and referees may be found online at: https://www.aavso.org/apps/jaavso/

The Journal of the American Association of Variable Star Observers
Volume 47, Number 1, 2019

Editorial

Variable Star Research

Instruments, Methods, and Techniques

Table of Contents continued on following pages

Variable Star Data

History and Biography

Abstracts of Papers and Posters Presented at the 107th Annual Meeting of the AAVSO, Held in Flagstaff, Arizona, November 15–17, 2018

Table of Contents continued on next page

Errata

Editorial

Citizen Science

John R. Percy
Editor-in-Chief, *Journal of the AAVSO*

Department of Astronomy and Astrophysics, and Dunlap Institute for Astronomy and Astrophysics, University of Toronto, 50 St. George Street, Toronto, ON M5S 3H4, Canada; john.percy@utoronto.ca

Received May 23, 2019

Citizen science, according to the Oxford English Dictionary, is: "Scientific work undertaken by members of the general public, often in collaboration with or under the direction of professional scientists and scientific institutions." It's a new name for an old activity, but it's exactly what AAVSO observer/researchers do.

Long ago, there wasn't much distinction between professionals and amateurs. Much of science was done by well-educated, well-to-do generalists, but that began to change in the 19th century, as science became more professionalized. AAVSO citizen science goes back to the words of John Herschel who, in 1833 said "this (variable star observation) is a branch of practical astronomy which has been too little followed up, and it is precisely that in which amateurs of the science, provided with only good eyes, or moderate instruments, might employ their time to excellent advantage." Friedrich Argelander in 1844, said "Could we be aided in this matter (variable star observing) by the cooperation of a goodly number of amateurs, we would perhaps in a few years be able to discover laws in these apparent irregularities, and then in a short time accomplish more than in all the 60 years which have passed since their discovery." The Variable Star Section of the British Astronomical Association appeared in 1890, and the AAVSO in 1911—building on the work of previous US amateurs.

The growth and maturation of citizen science

A generation or two ago, articles on citizen science would highlight ornithology (especially the work of the Audubon Society and the Cornell Lab of Ornithology), the work of the AAVSO, and perhaps a few more examples. Now, projects listed in *wikipedia* are so numerous that the AAVSO is mentioned only in passing. The well-known *Zooniverse* site (zooniverse.org/projects) contains 91 projects in the arts and sciences, including 18 in astronomy/space. The US Government website (citizenscience.gov) lists 439 projects, including 22 in astronomy/space. According to studies mentioned in *wikipedia*, the monetary value of the voluntary work in some of these projects is estimated to exceed an average of $200,000 each. I wonder what is the monetary value of AAVSO observers' work?

Technology has helped. AAVSO data are freely available on-line. There is software to analyze the data: think VSTAR. Our website provides charts, manuals, and connections to mentors and courses. There is email, and social media. Many observers have photometers and CCDs. On cloudy nights, our idle computers can be used by SETI@home (though I don't consider that as true citizen science, because the human brain is not engaged).

Technology can increase our opportunities for citizen science in other ways. My student Lucas Fenaux and I have just published (Percy and Fenaux 2019) a critique of the automated analysis and classification of tens of thousands of pulsating red giants in the massive All-Sky Automated Survey for Supernovae (ASAS-SN: www.astronomy.ohio-state.edu/~assassin/index.shtml), and showed that the majority of the automated analyses and classifications are incorrect or incomplete. There is still a place for the power and experience of the human brain! ASAS-SN provides great opportunities for analysis and research by students and knowledgeable amateur astronomers, as well as by professionals.

One characteristic of a mature field of endeavor is that it grows infrastructure. I remember amateur astronomy's photelectric photometry "revolution" in the 1980s. Previously, photoelectric photometry was done by electronics hobbyists, generally with unique equipment that they had built themselves. With the availability of off-the-shelf photometers came an organization (International Amateur-Professional Photoelectric Photometry, IAPPP) and its conferences, journal, and books.

Citizen science has spawned organizations in many countries; the US-based Citizen Science Association (citizenscience.org) has conferences, and a journal. Its interests lean towards nature and the environment. These organizations stress the need to establish goals and objectives, to adopt "best practices" which avoid bias and error, and to educate their members, evaluate their work, and provide feedback so their members' work is even more effective. Could the AAVSO do that more effectively?

The European Citizen Science Association has drawn up a statement of ethics—"Ten Principles of Citizen Science." Such statements of ethics and values have been drawn up for *professional* scientists by their institutions and organizations. A link to AAVSO's policy statements is clearly presented at the bottom of its home page, but the addition of an explicit values statement would be useful and appropriate.

And what about diversity? In North America, amateur astronomy seems to be the preserve of well-to-do graying white males like me. Is this generally true of citizen science? What can be done to attract a more diverse population?

Citizen science and the classroom

Modern school science curricula encourage students to *do* science, not just hear and read about it. Numerous citizen science projects can appeal to students, including astronomy projects and especially environmental projects. The AAVSO's *Hands-On Astrophysics* project (now called *Variable Star Astronomy*: www.aavso.org/education/vsa), was designed to enable students to develop and integrate their skills in science, math, and computing through variable star observation and analysis, motivated by the excitement of doing real science, with real data. At the post-secondary level, there's an increasing emphasis on "Work-Integrated Learning." This can be achieved in various ways including by volunteering in citizen-science projects. Perhaps by making citizen science part of the formal education system, we can start to increase diversity among citizen scientists in general.

Alternate definitions of citizen science

Two things have expanded my concept of citizen science. One was supervising the senior thesis of a very creative undergraduate student who was majoring in both astronomy and in the humanities and social sciences. She reminded me of the social dimensions of the term—it can help bridge the gap between science and society, scientists and non-scientists, and help the latter to have more "ownership" of science. She has decided to pursue this in her graduate studies.

The other thing was reading the entry on citizen science in *wikipedia*. It includes multiple definitions of the term, and a long and diverse list of citizen science projects. It emphasizes engagement with the *application* of science to society. Many citizen scientists (such as AAVSO observers) do science as a personal hobby or pastime but, in fields more directly connected to societal needs, it is important for citizen scientists to be concerned about the societal implications of their work—and often to take action.

The importance of citizen science today

Our planet is facing major environmental challenges. Climate is changing, and this and other human factors are leading to significant declines in thousands of species—among many other changes. A recent program on Ontario's excellent public TV channel (TVO 2019) dealt with the precipitous decline in the number and diversity of insects in our environment, and the ecological implications of this. One of the most important and widely-publicized studies showed that there had been a 78 percent drop-off in insect populations in dozens of *nature reserves* in Germany. The bulk of the work had been done by primarily-amateur members of the Entomological Society Krefeld. Similar studies of birds, butterflies, and other insects are being carried out. In April 2019, three one-hour episodes of *Nature*—"American Spring LIVE"—on the US Public Broadcasting System highlighted many opportunities for citizen scientists, both in school and among the general public.

More than ever, the world needs evidence-based policy making, in a wide variety of fields from environment, to health and medicine, to energy, population, and hunger. That, in turn, requires evidence and data. Citizens can help to collect that evidence and data. At the same time, the alternate definitions of citizen science must kick in. Non-scientists (and scientists) must engage with policy-makers in both evidence-gathering and decision-making, at the municipal, state/provincial, and national levels. Both must engage with the democratic process, while the democratic process still exists. The world can no longer afford politicians who govern by gut reaction, personal bias, and "fake news."

References

Percy, J. R., and Fenaux, L. 2019, arxiv.org/abs/1905.03279.

TVO. 2019, TV Ontario program (www.tvo.org/video/the-bug-apocalypse).

Observations of the Suspected Variable Star Ross 114 (NSV 13523)

Sriram Gollapudy
Yerkes Observatory, P.O. Box 0369, Williams Bay, WI (High school research student from Brookfield Academy, Brookfield, Wisconsin. Now at Duke University); sriram.gollapudy@duke.edu

Wayne Osborn
Yerkes Observatory, P.O. Box 0369, Williams Bay, WI; Wayne.Osborn@cmich.edu

Received December 3, 2018; revised January 28, 2019; accepted January 29, 2019

Abstract A study of the suspected variable star Ross 114 (NSV 13523) has been carried out. The star is confirmed to be a variable with an amplitude of about 2.5 magnitudes in B. Photometry and light curves from CCD images and archival photographic plates are presented. The observations show the star is a long period variable with a period of about 296 days.

1. Introduction

This paper reports the results of an educationally oriented research project. Yerkes Observatory, before its recent closing, offered activities for students at many grade levels designed to stimulate interest in science and engineering. As part of Yerkes' McQuown Scholars Program, high school students had the opportunity to work with a mentor on a project related to one of the observatory's professional activities. Projects have included such activities as constructing a scientific instrument, writing a software application, and engaging in astronomical research. The Scholars project reported here was to investigate a star announced as a possible variable in 1926 but still unstudied. Yerkes Observatory's collection of astronomical photographic plates would be used to seek brightness variations. If variability was found, new CCD observations would be obtained as time and equipment permitted in an attempt to determine the type of variability and produce a paper suitable for publication in a scientific journal.

F. Ross of Yerkes Observatory discovered 379 suspected variable stars from a comparison of photographs he took in the 1920s–1930s with plates taken a decade or more earlier by E. E. Barnard. Most of these stars have since been confirmed to be variable, but a few are still unstudied. One of these is Ross 114 (hereafter R114), more commonly known as NSV 13523. It was discovered when Ross noted a 12th magnitude star on a plate he exposed on 1925 November 5 was not visible on a similar plate of the same field taken by Barnard in 1909 (Ross 1926). R114 is located at R.A. = $21^{h}05^{m}14.0^{s}$, Dec. = +38° 37' 12" (2000).

Only two papers have dealt with R114 since its discovery, one showing that the 1925 observation was not of an asteroid (Marsden 2007) and another giving magnitudes improved over Ross's estimates (Osborn and Mills 2012). As this project was nearing completion we learned that R114 is among the stars monitored by the All-Sky Automated Survey for Supernovae (ASAS-SN) sky patrol (Shappee *et al.* 2014a, 2014b; Jayasinghe *et al.* 2018). The on-line light curve from those observations indicated large brightness variations with a period of 288.4 d.

2. Observations

Ross's original discovery plates are in the Yerkes plate archive. We were able to locate the two plates and confirmed the variability seen by Ross. We next identified 42 more plates showing the field. Eye estimates were made of the star's B magnitude relative to the comparison sequence shown in Figure 1 and are listed in Table 1, where the comparison B magnitudes are from the photometry of the AAVSO APASS survey (Henden *et al.* 2018). Each plate was estimated at least twice, but R114 was found to be blended with a nearby star (which we denote Star X) on almost all plates. The results of the plate observations are given in Table 2, where magnitudes with an uncertainty over 0.3 magnitude are marked with colons. The photographic observations show a large change in brightness.

After confirming the star's variability, we began obtaining CCD images of R114 field using the Skynet robotic observing system (Smith, Caton, and Hawkins 2016). The telescope employed was the Yerkes 1-m f/8 Cassegrain reflector. Imaging began in 2016 April. Unfortunately, equipment problems prevented CCD observations from June through August. Observations resumed in September and continued through December 2016.

CCD images were obtained on 16 nights, six nights in April and May and the remainder September through December. Multiple images were taken on all nights. A B filter was used to permit comparison with the plate results. However, R114 and Star X are mostly separated on the CCD images (and were individually measured) but are blended on the plates, making direct comparison of the CCD and plate results somewhat incompatible.

Aperture photometry was carried out on the CCD data to obtain B magnitudes using the Skynet AFTERGLOW program and employing the same local comparison stars as used for the plates. Table 3 shows the average magnitude and standard deviation (σ) of the measures for each comparison star, Star X and R114 for the April–May observations, then the September–December ones, and finally all measures. The comparison star results agree well with one another from the different

Figure 1. The field of R114 with the variable and comparison stars marked. North is at the top and east to the right.

Table 1. Identifications of comparison stars, Star X and R114.

Name	*R. A. (2000)* *h m s*	*Dec. (2000)* *° ′ ″*	*B*
A	21 04 43.4	+38 37 04	13.75
B	21 05 03.7	+38 37 24	14.03
C	21 05 18.2	+38 38 09	14.52
D	21 05 05.3	+38 35 54	15.71
E	21 05 00.0	+38 36 20	16.33
Star X	21 05 13.4	+38 37 24	16.82
Ross 114	21 05 14.0	+38 37 12	—

Table 2. Magnitudes of R114 from photographic plates.

Plate #	*Julian Date*	*B*	*Plate #*	*Julian Date*	*B*
L-33	2412741.79	15.47	6B-548	2418535.64	<14.03
L-83	2413096.78	14.63	6B-551	2418536.64	13.87
10B-103	2416703.75	14.78	10B-629	2418921.75	<14.52
6B-103	2416703.75	<14.52	6B-629	2418921.75	<14.52
10B-274	2417065.90	14.17	10B-700	2419207.70	14.78:
6B-274	2417065.90	14.30	6B-700	2419207.70	15.00
3B-274	2417065.90	14.28	10B-853	2419951.70	15.56
10B-276	2417067.93	14.27	6B-853	2419951.70	<14.52
6B-276	2417067.93	14.31	10B-910	2420387.64	15.31:
3B-276	2417067.93	14.35	6B-910	2420387.64	15.23:
10B-290	2417083.78	14.82	10B-911	2420388.62	14.78:
6B-290	2417083.78	14.82	6B-911	2420388.62	<14.03
6B-291	2417083.78	14.78	10B-1077	2421069.72	16.00
10B-353	2417448.8	<14.03	6B-1077	2421069.72	<14.52
10B-354	2417449.84	<14.52	10B-1822	2423552.80	<13.75
6B-354	2417449.84	<14.52	10R-73	2424460.52	13.85
10B-407	2417768.67	15.82	6R-73	2424460.52	13.83
6B-407	2417768.67	15.93	ILL-2071	2435988.72	<14.03
10B-521.5	2418447.82	15.90	ILL-2063a	2435994.65	<14.03
6B-521.5	2418447.82	15.79:	ILL-2065a	2436009.69	14.10
10B-547	2418534.74	13.86	ILL-2076	2436098.7	14.13:
6B-547	2418534.74	13.94	ILL-E-05	2432189.7	<13.5*

**Yellow magnitude.*

Table 3. Averages and standard deviations of the CCD measures by star.

Star	*Apr.–May 2016* *Avg. B*	*σ*	*Sep.–Dec. 2016* *Avg. B*	*σ*	*All Observations* *Avg. B*	*σ*
A	—	—	13.74	0.02	13.74	0.02
B	14.01	0.03	14.00	0.02	14.01	0.03
C	14.56	0.03	14.58	0.02	14.57	0.03
D	15.70	0.03	15.69	0.05	15.70	0.04
E	16.33	0.03	16.31	0.08	16.32	0.06
X	16.85	0.07	16.78	0.12	16.81	0.09
R114	16.63	0.12	15.35	0.73	15.96	0.84

Table 4. CCD photometry measures.

Julian Date *Apr.–May 2016*	*B*	*Julian Date* *Sep.–Dec. 2016*	*B*
2457492.886	16.66	2457633.685	14.18
2457492.892	16.55	2457633.686	14.19
2457492.893	16.60	2457636.703	14.09
2457516.818	16.72	2457636.704	14.09
2457516.821	16.61	2457680.583	14.99
2457516.823	16.69	2457680.585	15.03
2457524.767	16.86	2457683.588	15.07
2457524.780	16.65	2457683.590	15.09
2457524.787	16.63	2457696.635	15.55
2457528.769	16.58	2457696.636	15.57
2457528.770	16.40	2457702.650	15.70
2457528.771	16.50	2457702.651	15.69
2457532.769	16.89:	2457705.659	15.65
2457532.771	16.73:	2457705.660	15.68
2457532.771	16.62:	2457709.578	15.82
2457538.744	16.60	2457709.579	15.85
2457538.745	16.55	2457729.540	16.13
2457538.746	16.59	2457729.541	16.11
		2457736.519	16.35
		2457736.521	16.24

observing periods. The observational scatters (σ) are relatively small and provide indicators of the errors of our CCD observations at different magnitude levels. The CCD observations of R114 are given in Table 4, where the observations made in April–May 2016 are listed in the first column and those made September–December 2016 are in the second column.

3. Analysis

The light curve of R114 from the CCD observations is shown in Figure 2. The derived magnitudes in 2016 April–May fluctuate around 16.6. Although the fluctuations vary over about 0.5 magnitude, we believe that they are not real, due to the faintness of the star during this period and the effect on some measures of the nearby companion, Star X. When CCD observations resumed in the fall of 2016 R114 was found to be much brighter. In September, the variable was about B = 14, after which it decreased in brightness by more than two magnitudes over a three-month period. Visual comparison of R114 to its nearby companion, Star X, on the images easily confirmed the large change in brightness between April–May and September and the subsequent autumn decline. Figure 2 also shows the light curve of Star X; its brightness remained constant within the errors of the observations. The approximate two-magnitude drop of R114 from September to December demonstrates it is a fairly large amplitude variable, as found by ASAS-SN.

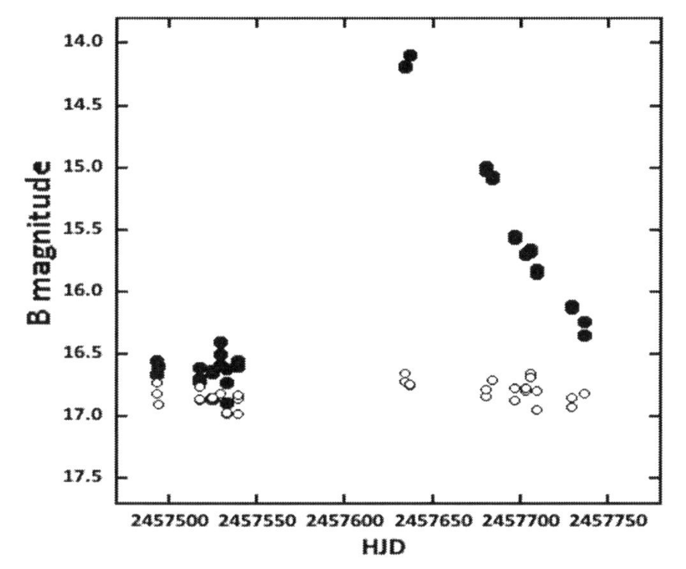

Figure 2. CCD-measured magnitudes of Ross 114 (solid circles) and of its close companion, Star X (open circles).

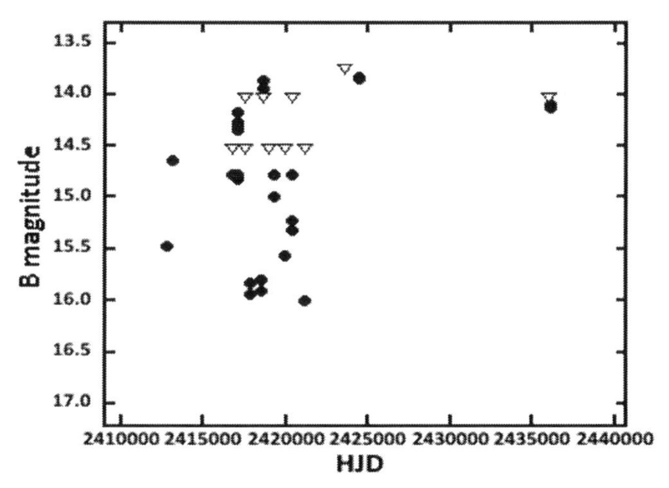

Figure 3. Magnitudes from eye estimates of the brightness of R114 on photographic plates. The dots show derived magnitudes from plates where the star was visible. The down-triangles show the magnitude below which the star's brightness must be for those cases when it was below the plate limit.

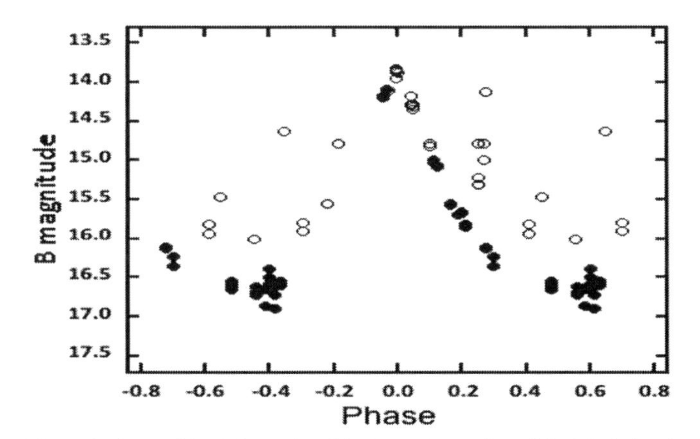

Figure 4. The combined phase plot of the CCD observations (filled circles) and photographic ones (open circles) with a period of 296.3 d.

The light curve from the plate observations is shown in Figure 3. The estimated magnitudes varied from as bright as B = 13.8 to as faint as about 16. Because many of the plates were taken simultaneously in pairs using co-mounted 10- and 6-inch telescopes (with corresponding plate numbers), we were able to confirm that the large magnitude differences on different dates were not due to plate defects. In general, the magnitudes from contemporaneous plates agreed within 0.12 of a magnitude, with the greatest discrepancy being 0.32 magnitude. The approximate two-magnitude variation shown by the plates is in agreement with the CCD results. The mid-magnitude was about B = 15, although this includes the light of both Star X and the variable.

The CCD observations suggested a period of 270 d or more. A period search of the plate data in the range 200–340 d using the AAVSO VSTAR tool (Benn 2012) yielded periodicities near 238 d and 297 d. The ASAS-SN observations cover three maxima that occur near HJD 2457340, 2457650, and 2457940, giving an average cycle time that varies slightly about 300 d and thus ruling out our shorter possibility.

Phased light curves with the combined photographic and CCD data using periods in the range 286–305 d showed no period fits all the data well, likely because the light variations are not strictly periodic. The best fit seemed to be with P = 296.3 d, and the phased light curve with this period using a modern reference epoch of 2457944.0 is shown in Figure 4. The CCD and photographic maximum are consistent, while the effect on the brightness of R114 from Star X is obvious for the photographic minimum. This ephemeris also fits the three ASAS-SN maxima, which occur at phases −0.04, +0.01, and 0.00.

4. Conclusions

Recent CCD observations and archival photographic plates both show that NSV 13523 = R114 varies over a fairly large range as found by the ASAS-SN survey. Our CCD photometry showed a range from B = 14.1 to 16.7. The magnitude estimates from plates are consistent, ranging from about 14 to 16 for the combined light of R114 and Star X. An ephemeris based on our observations and ASAS-SN times of maxima is: HJD (Max) = 2457944.0 + 296.3 E.

Finally, we offer the following comments on the educational aspects of this project. The lead author undertook this research as part of Yerkes Observatory's McQuown Scholars program for high school students. He feels a research project such this one can be of tremendous value for a student exploring an interest in astronomy. The opportunity to work one-on-one with a professional astronomer and to learn about and see first hand the need to collect and analyze data over an extended period of time as well as the effort that goes into writing and publishing a scientific paper clearly showed the nature of astronomical work. It was incredibly satisfying at the end to find our plate and CCD variations were in agreement. Being able to obtain CCD observations during the time period that was lost (2016 June–August) would have greatly helped in determining the light curve and its period, but this unexpected problem probably also illustrated a feature of observational astronomy work.

Overall, he sees this project as certainly one of the most valuable experiences he had in high school.

5. Acknowledgements

We express our appreciation to the anonymous referee for comments that led to a significant improvement in the paper. The lead author also thanks Yerkes Observatory for use of its resources and its McQuown Scholars Program for high school students without which this project would have been impossible. This research has made use of the SIMBAD database, operated at CDS, Strasbourg, France, the Skynet Robotic Telescope Network, and the AAVSO Photometric All-Sky Survey (APASS), funded by the Robert Martin Ayers Sciences Fund and NSF Grant AST-1412587.

References

Benn D. 2012, *J. Amer. Assoc. Var. Star Obs.*, **40**, 852.

Henden, A. A., Levine, S., Terrell, D., Welch, D., Munari, U., and Kloppenborg, B. K. 2018, American Astronomical Society Meeting #232, id. 223.06 (see APASS web page http://www.aavso.org/apass).

Jayasinghe, T., *et al.* 2018, arXiv 1809.07329.

Marsden, B. G. 2007, *Perem. Zvezdy*, **27**, 3.

Osborn, W. and Mills, O. F. 2012, *J. Amer. Assoc. Var. Star Obs.*, **40**, 929.

Ross, F. E. 1926, *Astron. J.*, **36**, 167.

Shappee, B. J., *et al.* 2014a, *Astrophys. J.*, **788**, 48.

Shappee, B. J., *et al.* 2014b, American Astronomical Society Meeting #223, id. 236.03 (see ASAPP-SN web page: http://www.astronomy.ohio-state.edu/~assassin/index.shtml).

Smith, A. B., Caton, D. B. and Hawkins, R. L. 2016, *Publ. Astron. Soc. Pacific*, **128**, 055002.

CCD Photometry, Light Curve Modeling, and Period Study of the Overcontact Binary Systems V647 Virginis and V948 Monocerotis

Kevin B. Alton
UnderOak Observatory, 70 Summit Avenue, Cedar Knolls, NJ 07927; mail@underoakobservatory.com

Received December 27, 2018; revised February 19, March 26, 2019; accepted May 2, 2019

Abstract Prior to this investigation, monochromatic CCD data for V647 Vir and V948 Mon had only been generated from automated surveys which employ sparse sampling strategies. In this study precise multi-color (B, V, and I_c) light curve data for V647 Vir (2018) and V948 Mon (2017–2018) were acquired at Desert Bloom Observatory (DBO). Both targets produced new times of minimum which were used along with other eclipse timings from the literature to update their corresponding ephemerides. Despite the limited amount of published data, preliminary evidence suggests a secular decrease in the orbital period of V948 Mon. Roche modeling to produce synthetic fits to the observed light curve data was accomplished using the Wilson-Devinney code. Since each system exhibits a total eclipse, a reliable value for the mass ratio (q) could be determined leading in turn to initial estimates for the physical and geometric elements of both variable systems.

1. Introduction

CCD-derived photometric data for V647 Vir (NSVS 13280611; GSC 00314-00388) were first acquired from the ROTSE-I survey between 1999–2000 (Akerlof *et al.* 2000; Wozniak *et al.* 2004; Gettel *et al.* 2006) and later from the Catalina Sky Survey (Drake *et al.* 2014). Its classification as a W UMa variable was assigned according to Hoffman *et al.* (2009). The variability of V948 Mon (GSC 04846-00809) was initially observed from data collected (1994–1996) in a calibration field for the Sloan Digital Sky Survey (Henden and Stone 1998) and later confirmed by Greaves and Wils (2003). Sparsely sampled photometric data for V948 Mon were also acquired from the ROTSE-I, ASAS (Pojmański *et al.* 2005), and Catalina surveys. Although other times of minimum light have been sporadically published, this paper marks the first detailed period analysis and multi-color Roche model assessment of light curves (LC) for V647 Vir and V948 Mon.

2. Observations and data reduction

Time-series images were acquired at Desert Bloom Observatory (DBO, USA—110.257 W, 31.941 N) with an SBIG STT-1603ME CCD camera mounted at the Cassegrain focus of a 0.4-m f/6.8 catadioptric telescope. This instrument produces an image scale of 1.36 arcsec/pixel (bin = 2 × 2) and a field of view (FOV) of 11.5' × 17.2'. Image acquisition (75-s) was performed using MAXIM DL v.6.13 (Diffraction Limited 2019) or THESKYX PRO v.10.5.0 (Software Bisque 2019). The CCD-camera is equipped with B, V, and I_c filters manufactured to match the Johnson-Cousins Bessell prescription. Dark subtraction, flat correction, and registration of all images collected at DBO were performed with AIP4WIN v.2.4.0 (Berry and Burnell 2005). Instrumental readings were reduced to catalog-based magnitudes using the APASS star fields (Henden *et al.* 2009, 2010, 2011 and Smith *et al.* 2011) built into MPO CANOPUS v.10.7.1.3 (Minor Planet Observer 2010). In order to minimize any potential error due to differential refraction and color extinction only data from images taken above 30° altitude (airmass < 2.0) were included.

3. Results and discussion

LCs for V647 Vir and V948 Mon were generated using an ensemble of five non-varying comparison stars in each FOV. The identities, J2000 coordinates, V-mags, and APASS color indices (B–V) for these stars are listed in Table 1. Uncertainty in comparison star measurements made in the same FOV with V647 Vir or V948 Mon typically stayed within ± 0.007 mag for V- and I_c- and ± 0.010 mag for B-passbands.

3.1. Photometry and ephemerides

Times of minimum were calculated using the method of Kwee and van Woerden (1956) featured in PERANSO v.2.5 (Paunzen and Vanmunster 2016; Vanmunster 2018). Long-term or secular changes in orbital period can sometimes be revealed by plotting the difference between the observed eclipse times and those predicted by a reference epoch against cycle number. Curve fitting all eclipse timing differences (ETD) was accomplished using scaled Levenberg-Marquardt algorithms. The results from these analyses are separately discussed for each binary system in the subsections below.

3.1.1. V647 Vir

A total of 334 photometric values in B-, 350 in V-, and 333 in I_c-passbands were acquired from V647 Vir between January 25, 2018 and March 19, 2018. Included in these determinations were seven new times of minimum (ToM) which are summarized in Table 2. Photometric data from the NSVS (1999–2000) and ASAS (2001–2009) surveys were folded together with V-mag data generated at DBO (2018). This was accomplished by applying periodic orthogonals (Schwarzenberg-Czerny 1996) to fit observations and analysis of variance to assess fit quality (PERANSO v.2.5; Paunzen and Vanmunster 2016; Vanmunster 2018). Despite significant scatter in the survey data, near congruence of the light curves was observed when P = 0.3478960 d (Figure 1). NSVS and ASAS timings contained within the uncertainty calculated (Kwee and van Woerden 1956) for the midpoint of the folded LCs during Min I and Min II were added to the list of ToM values summarized in Table 2. These results (n = 2) along with other published eclipse timings

Table 1. Astrometric coordinates (J2000), V-mags, and color indices (B–V) for V647 Vir, V948 Mon, and their corresponding five comparison stars used in this photometric study.

Star Identification	*R.A. (J2000)* h m s	*Dec. (J2000)* ° ' "	*V-mag*[a]	*(B–V)*[a]
V647 Vir	13 47 51.86	+07 00 45.79	12.603	0.725
GSC 00314-00627	13 47 24.50	+06 54 32.22	11.732	0.696
GSC 00314-00198	13 47 39.99	+06 52 37.88	12.101	0.743
GSC 00314-00530	13 47 55.13	+06 54 27.90	13.744	0.746
GSC 00314-00282	13 47 57.83	+07 04 50.99	12.814	0.854
GSC 00314-00009	13 47 44.60	+07 04 18.31	10.487	1.031
V948 Mon	08 01 51.19	-00 33 26.27	13.184	0.471
GSC 04846-00921	08 01 37.94	–00 38 31.38	12.653	0.506
GSC 04846-00463	08 01 19.32	–00 36 42.52	13.437	0.572
GSC 04846-02159	08 01 31.56	–00 37 38.89	13.528	0.611
GSC 04846-00795	08 01 30.12	–00 36 19.76	13.280	0.514
GSC 04846-01147	08 01 48.35	–00 28 24.17	13.179	0.509

a. V-mag and (B–V) for comparison stars derived from APASS database described by Henden et al. *(2009, 2010, 2011) and Smith* et al. *(2011), as well as on the AAVSO web site (https://www.aavso.org/apass).*

Table 2. V647 Vir times-of-minimum (February 2, 2000–March 19, 2018), cycle number, and residuals (ETD) between observed and predicted times derived from the updated linear ephemeris (Equation 2).

HJD 2400000+	*HJD Error*	*Cycle No.*	*ETD*[a]	*Reference*
51604.0046	0.0018	–18951	0.00023	NSVS[b]
54585.6480	0.0009	–10380.5	0.00069	ASAS[b]
54948.6773	0.0006	–9337	0.00048	Diethelm 2009
54948.8499	0.0001	–9336.5	–0.00087	Diethelm 2009
55634.9009	0.0001	–7364.5	–0.00084	Diethelm 2011
55687.7813	0.0002	–7212.5	–0.00064	Diethelm 2011
56000.8888	0.0002	–6312.5	0.00043	Diethelm 2012
58143.9284	0.0001	–152.5	0.00046	This study
58144.9712	0.0002	–149.5	–0.00039	This study
58146.0155	0.0001	–146.5	0.00021	This study
58156.9745	0.0002	–115	0.00049	This study
58184.9804	0.0001	–34.5	0.00072	This study
58196.8080	0.0002	–0.5	–0.00014	This study
58196.9812	0.0003	0	–0.00085	This study

a. ETD = Eclipse Time Difference between observed time-of-minimum and predicted values using the updated ephemeris (Equation 2).
b. Estimated following superimposition of NSVS, ASAS, and DBO (2018) lightcurves when folded at P = 0.3478960 d.

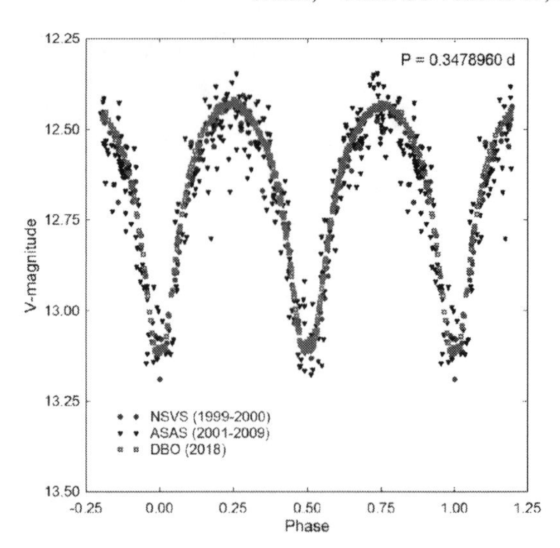

Figure 1. Period folded (P = 0.3478960 d) LCs for V647 Vir produced from NSVS, ASAS, and DBO photometric data. NSVS and ASAS LCs were offset to match the V-mag values determined from precise CCD photometry performed at DBO.

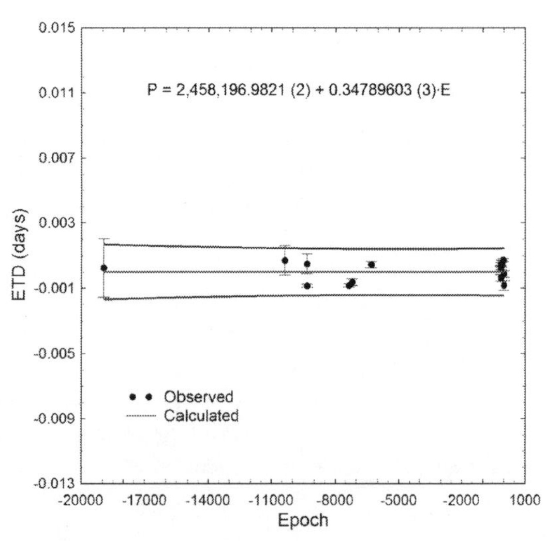

Figure 2. Eclipse timing differences (ETD) vs. epoch for V647 Vir calculated using the updated linear ephemeris (Equation 2). Measurement uncertainty is demarked by the hatched vertical lines. The solid red line indicates the linear fit while the blue lines represent the 95% confidence intervals which include the zero intercept.

from 1999 through 2017, were used to initially calculate ETD values with the reference epoch (Kreiner 2004) defined by the following linear ephemeris (Equation 1):

$$\text{Min I (HJD)} = 2454506.847 + 0.347896\ E. \quad (1)$$

An updated linear ephemeris (Equation 2) was thereafter derived as follows:

$$\text{Min I (HJD)} = 2458196.9821\ (2) + 0.34789603\ (3)\ E \quad (2)$$

It should be noted that eclipse timing data for V647 Vir are only available for the past 18 years with large time gaps between 2001–2008 and 2012–2018. The residuals (ETD) which are best described by a straight-line fit indicate that no substantive change in the orbital period has occurred since 2000 (Figure 2). Not surprisingly given the paucity of data, no other underlying variations in the orbital period stand out, such as those that might be caused by the magnetic cycles (Applegate 1992) or the presence of an additional gravitationally bound stellar-size body.

3.1.2. V948 Mon

A total of 604 photometric values in B-, 379 in V-, and 381 in I_c-passbands were acquired from V948 Mon between December 23, 2017 and January 5, 2018. Included in these determinations were seven new ToM values which are provided in Table 3. These data along with other published results were used to initially analyze eclipse timings according to the

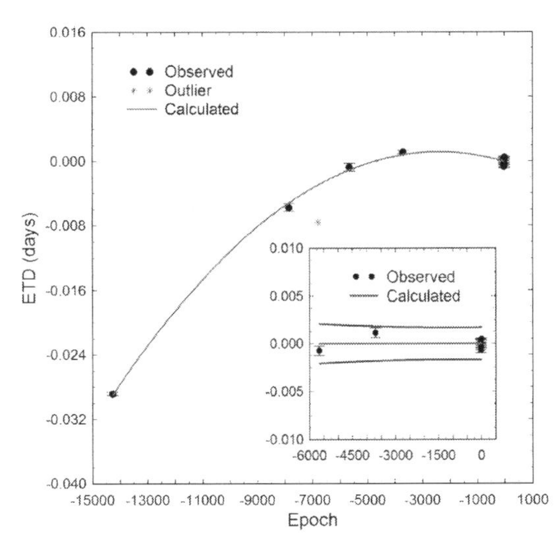

Figure 3. The downwardly directed quadratic fit to the ETD vs. epoch data (Equation 4) is shown with a solid red line and suggests the orbital period of V948 Mon is decreasing with time. The inset panel shows the near-term data which produced the updated linear ephemeris (Equation 5). Measurement uncertainty is demarked by the hatched vertical lines. The solid red line indicates the linear fit while the blue lines represent the 95% confidence intervals which include the zero intercept.

Table 3. V948 Mon times of minimum (April 13, 2003–January 5, 2018), cycle number, and residuals (ETD) between observed and predicted times derived from the updated linear ephemeris (Equation 5).

HJD 2400000+	*HJD Error*	*Cycle No.*	*ETD*[a]	*Reference*
52742.6100	—[c]	–14270	–0.02885	Greaves and Wils 2003
55164.7680	—[c]	–7847	–0.00574	Diethelm 2011
55564.8728[b]	0.0002	–6786	–0.00761	Diethelm 2011
55989.6866	0.0005	–5659.5	–0.00074	Diethelm 2012
56726.3599	0.0005	–3706	0.00114	Hübscher and Lehmann 2015
58110.8940	0.0002	–34.5	0.00023	This study
58115.7964	0.0002	–21.5	0.00030	This study
58115.9840	0.0003	–21	–0.00069	This study
58116.9276	0.0002	–18.5	0.00017	This study
58117.8696	0.0001	–16	–0.00056	This study
58118.8126	0.0001	–13.5	–0.00030	This study
58123.9043	0.0001	0	0.00047	This study

a. ETD = Eclipse Time Difference between observed time of minimum and those calculated using the updated ephemeris (Equation 5).
b. Outlier value shown as an asterisk in Figure 3 not included in period analyses.
c. Not reported.

Table 4. Estimation of effective temperature (T_{effl}) of V647 Vir based upon dereddened (B–V) data from five surveys and the present study.

	USNO-B1.0	*USNO-A2.0*	*2MASS*	*SDSS-DR8*	*UCAC4*	*Present Study*
$(B–V)_0$[a]	0.500	0.704	0.682	0.814	0.703	0.679
T_{effl}[b] (K)	6278	5590	5653	5283	5594	5663
Spectral Class[b]	F6V-F7V	G5V-G6V	G5V-G6V	G9V-K0V	G5V-G6V	G4V-G5V

a. Intrinsic $(B–V)_0$ determined using reddening value E(B–V) = 0.023 ± 0.001.
b. T_{effl} interpolated and spectral class range estimated from Pecaut and Mamajek (2013). Median value, $(B–V)_0$ = 0.693 ± 0.012, corresponds to a G5V-G6V primary star (T_{effl} = 5620 ± 102 K).

reference epoch (Kreiner 2004) defined by the following linear ephemeris (Equation 3):

$$\text{Min I (HJD)} = 2455164.7622 + 0.3771061\,E. \quad (3)$$

Plotting (Figure 3) the difference between the observed eclipse times and those predicted by the linear ephemeris against epoch (cycle number) uncovered what appears to be a quadratic relationship (Equation 4) where:

$$ETD = -4.8328 \cdot 10^{-5} - 1.0266 \cdot 10^{-6} E - 2.1387 \cdot 10^{-10} E^2. \quad (4)$$

In this case the ETD residuals vs. epoch can be best described by an expression with a negative quadratic coefficient ($-2.1387 \cdot 10^{-10}$) suggesting that the orbital period has been slowly decreasing over time at the rate of 0.0358 (13) $s \cdot y^{-1}$.

An updated linear ephemeris (Equation 5) based on near term ETD values (2012–2018) was calculated as follows:

$$\text{Min I (HJD)} = 2458123.9038\,(3) + 0.3771034\,(1)\,E. \quad (5)$$

These data are shown as a horizontal line within the inset for Figure 3. Nevertheless, since the orbital period appears to be decreasing linearly with time, ephemerides for V948 Mon will need to be updated on a regular basis.

3.2. Effective temperature estimation

Throughout this paper the primary star is defined as the hotter and more massive member of each binary system. No classification spectra are published for either W UMa-type variable so that the effective temperature (T_{effl}) of each primary star has been estimated using color index (B–V) data acquired at DBO and others determined from astrometric (USNO-A2.0, USNO-B1.0, and UCAC4) and photometric (2MASS, SDSS-DR8, and APASS) surveys. Interstellar extinction (A_V) was calculated (E(B–V) × 3.1) using the reddening value (E(B–V)) estimated from Galactic dust map models reported by Schlafly and Finkbeiner (2011).

Intrinsic color ($(B–V)_0$) for V647 Vir that was calculated from measurements made at DBO and those acquired from five other sources are listed in Table 4. The median value (0.693 ± 0.012) which was adopted for Roche modeling indicates a primary star with an effective temperature (5620 ± 102 K) that probably ranges in spectral class between G5V and G6V. This result is nearly identical to the Gaia DR2 release of stellar parameters (Andrae *et al.* 2018) in which the T_{eff} for V647 Vir is reported to be 5620^{+36}_{-240} K.

Table 5. Estimation of effective temperature (Teff1) of V948 Mon based upon dereddened (B-V) data from six surveys and the present study.

	USNO-B1.0	*USNO-A2.0*	*2MASS*	*SDSS-DR8*	*UCAC4*	*APASS*	*Present Study*
$(B–V)_0$ [a]	0.931	0.646	0.447	0.845	0.445	0.445	0.428
T_{eff1} [b] (K)	4962	5779	6477	5179	6483	6483	6560
Spectral Class[b]	K2V-K3V	G1V-G2V	F5V-F6V	K0V-K1V	F5V-F6V	F5V-F6V	F5V-F6V

a. Intrinsic $(B–V)_0$ determined using reddening value $E(B–V) = 0.028 \pm 0.001$.
b. T_{eff1} interpolated and spectral class range estimated from Pecaut and Mamajek (2013). Median value, $(B–V)_0 = 0.447 \pm 0.019$, corresponds to an F3V-F7V primary star ($T_{eff1} = 6480 \pm 274$ K).

Similarly, dereddened color indices ($(B–V)_0$) for V948 Mon from different sources are listed in Table 5. The median value (0.447 ± 0.019) adopted for Roche modeling corresponds to a primary star with an effective temperature (6480 ± 270 K) that likely ranges in spectral class between F3V and F7V. The median result is somewhat higher than the value reported in the Gaia DR2 release of stellar parameters (Andrae *et al.* 2018) but well within the documented confidence intervals ($T_{eff} = 6337^{+418}_{-168}$ K).

3.3. Roche modeling approach

Roche modeling of LC data from V647 Vir and V948 Mon was primarily accomplished using the programs PHOEBE 0.31a (Prša and Zwitter 2005) and WDWINT56A (Nelson 2009). Both feature an easy-to-use GUI interface to the Wilson-Devinney WD 2003 code (Wilson and Devinney 1971; Wilson 1979, 1990). WDWINT56A makes use of Kurucz's atmosphere models (Kurucz 2002) which are integrated over BVR_cI_c optical passbands. In both cases, the selected model was Mode 3 for an overcontact binary. Other modes (detached and semi-detached) were explored but never approached the goodness of fit achieved with Mode 3. Since the internal energy transfer to the surface of both variable systems is driven by convective (<7500 K) rather than radiative processes, the value for bolometric albedo ($A_{1,2} = 0.5$) was assigned according to Ruciński (1969) while the gravity darkening coefficient ($g_{1,2} = 0.32$) was adopted from Lucy (1967). Logarithmic limb darkening coefficients (x_1, x_2, y_1, y_2) were interpolated (Van Hamme 1993) following each change in the effective temperature (T_{eff2}) of the secondary star during model fit optimization using differential corrections (DC). All but the temperature of the more massive star (T_{eff1}), $A_{1,2}$, and $g_{1,2}$ were allowed to vary during DC iterations. In general, the best fits for T_{eff2}, i, q, and Roche potentials ($\Omega_1 = \Omega_2$) were collectively refined (method of multiple subsets) by DC using the multicolor LC data. LCs from V647 Vir (Figures 4 and 5) and V948 Mon (Figure 6) do not exhibit significant asymmetry during quadrature (Max. I ≅ Max. II) which is often attributed to the so-called O'Connell effect (O'Connell 1951). No spots were introduced during Roche modeling of V948 Mon, however, a cool spot was necessary to achieve the best fit of LC data for V647 Vir during Min II (Figure 5). Third light contribution (l_3) during DC optimization did not lead to any value significantly different from zero with either binary system. Since both systems clearly undergo a total eclipse during Min II, Roche model convergence to a unique value for q should be self-evident. To make this point and also to demonstrate that both systems are most likely A-type overcontact variables, a grid search was

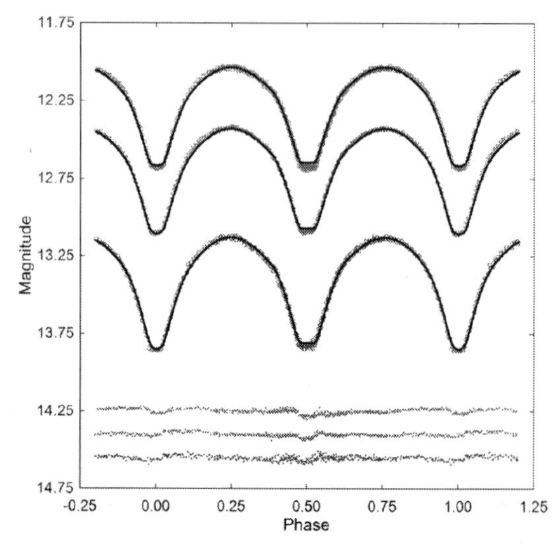

Figure 4. Folded CCD light curves for V647 Vir produced from photometric data obtained between January 25, 2018 and March 19, 2018. The top (I_c), middle (V), and bottom curve (B) shown above were reduced to APASS-based catalog magnitudes using MPO CANOPUS (Minor Planet Observer 2010). In this case, the Roche model assumed an A-type overcontact binary with no spots; residuals from the model fits are offset at the bottom of the plot to keep the values on scale.

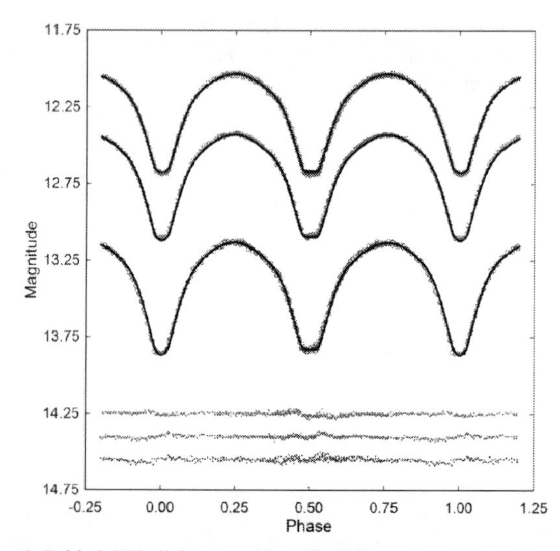

Figure 5. Folded CCD light curves for V647 Vir produced from photometric data obtained between January 25, 2018 and March 19, 2018. The top (I_c), middle (V), and bottom curve (B) shown above were reduced to APASS-based catalog magnitudes using MPO CANOPUS (Minor Planet Observer 2010). In this case, the Roche model assumed an A-type overcontact binary with a cool spot on the primary star; residuals from the model fits are offset at the bottom of the plot to keep the values on scale.

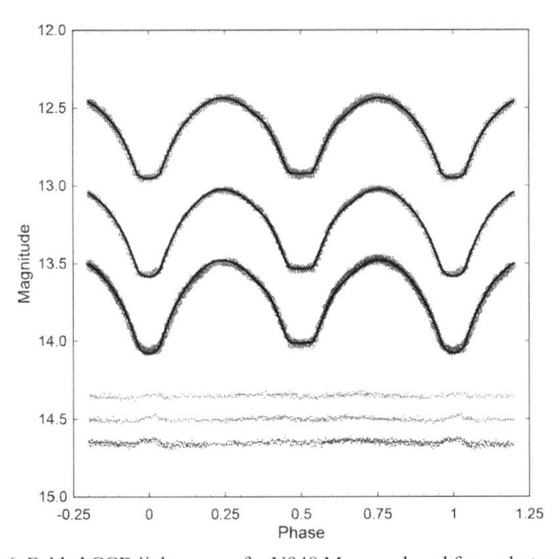

Figure 6. Folded CCD light curves for V948 Mon produced from photometric data obtained between December 23, 2017 and January 5, 2018. The top (I_c), middle (V), and bottom curve (B) shown above were reduced to APASS-based catalog magnitudes using MPO CANOPUS (Minor Planet Observer 2010). In this case, the Roche model assumed an A-type overcontact binary with no spots; residuals from the model fits are offset at the bottom of the plot to keep the values on scale.

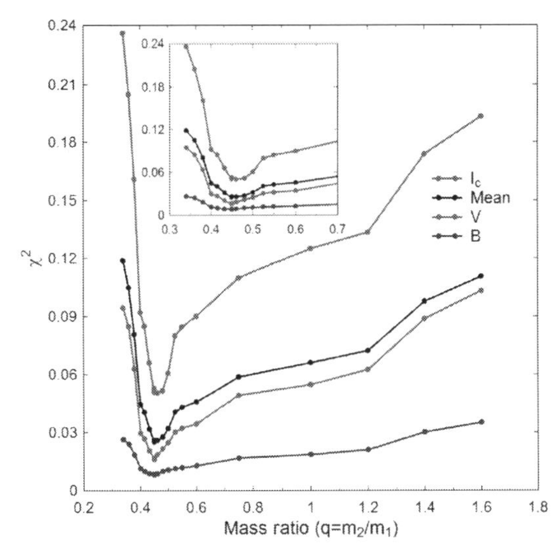

Figure 7. V647 Vir Roche model fit error minimization (χ^2) from PHOEBE 0.31a (Prša and Zwitter 2005) using the "q-search" approach. The figure inset zooms in where the best fit for q (~0.45) is observed.

performed on V647 Vir (Figure 7) and V948 Mon (Figure 8) by fixing the mass ratio at various intervals and finding a best fit for (T_{eff2}, i, and $\Omega_1 = \Omega_2$) using DC. These results are described in more detail within the subsections for each variable that follow.

3.4. Roche modeling results

Without radial velocity (RV) data, it is not possible to unambiguously determine the mass ratio or total mass. The total eclipse observed in the LCs from both systems greatly improves the chances of finding a unique mass ratio value for each star. Still, there is some risk at attempting to establish a mass ratio (q_{ptm}) with photometric data alone (Terrell and Wilson 2005). Standard errors reported in Tables 6–8 are computed from the DC covariance matrix and only reflect the model fit to the

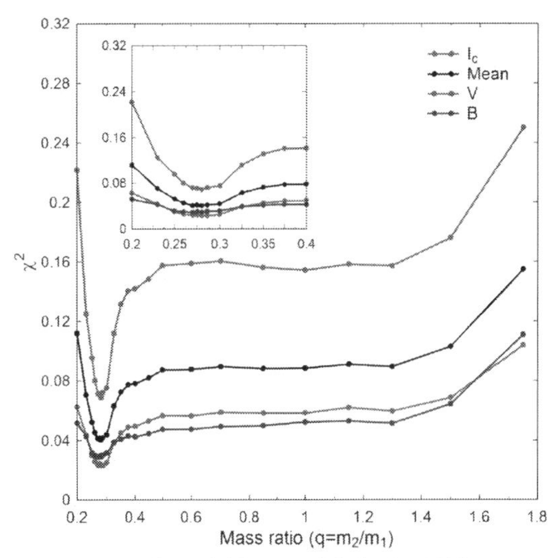

Figure 8. V948 Mon Roche model fit error minimization (χ^2) from PHOEBE 0.31a (Prša and Zwitter 2005) using the "q-search" approach. The figure inset zooms in where the best fit for q (~0.28) is observed.

observations which assume exact values for any fixed parameter. These intra-study errors may appear unrealistically small considering the estimated uncertainties associated with the mean adopted T_{eff1} values (Tables 6–8) along with basic assumptions about $A_{1,2}$, $g_{1,2}$, and the influence of spots added to the Roche model. Alternative approaches to locate the best Roche model fit in a multi-parameter space which have gained popularity include simplex optimization and heuristic scanning (also known as Monte Carlo simulation). Nonetheless, as discussed in more detail by Wilson and Van Hamme (2016), there is nothing inherently wrong with using DC for parameter estimation and determination of standard errors. Furthermore, Abubekerov *et al.* (2008, 2009) argue that for significantly nonlinear multi-parameter relationships the standard errors produced from DC or Monte-Carlo simulations are nearly equivalent. One normally fixes the value for T_{eff1} during modeling with the WD code despite acknowledging measurement uncertainty which can easily approach ±400 K. To address this concern, the effect that adjusting T_{eff1} would have on modeling estimates for q, i, $\Omega_{1,2}$, T_{eff2} along with the putative cool spot on V647 Vir was explored (Tables 6–8). In order to maximize the possibility of observing an effect, the worst case estimates for V647 Vir (5620±240 K) and V948 Mon (6480±418 K) obtained from Gaia DR2 were used for this analysis. Interestingly, with the obvious exception of T_{eff2}, varying T_{eff1} did not appreciably affect the model estimates (R.S.D. <2%) for i, q, or $\Omega_{1,2}$ (Tables 6–8). Said another way, assuming that the true T_{eff1} for V647 Vir falls within 5620±240 K and that for V948 Mon within 6480±418 K, the model fits for both systems were relatively insensitive to the T_{eff1}. These findings are consistent with similar results reported for AR CrB (Alton and Nelson 2018), a W-type overcontact binary in which T_{eff1} was tested over an even wider (±3σ) range.

The fill-out parameter (f) which corresponds to the outer surface shared by each star was calculated according to Equation 6 (Kallrath and Malone 1999; Bradstreet 2005) where:

$$f = (\Omega_{inner} - \Omega_{1,2}) / (\Omega_{inner} - \Omega_{outer}). \quad (6)$$

Table 6. V647 Vir lightcurve parameters evaluated by Roche modeling and the geometric elements derived for V647 Vir assuming it is an A-type W UMa variable with no spots. Modeling estimates also include those determined at the uncertainty boundaries (T_{eff1} = 5620 ± 240 K) for the primary star.

Parameter	*No spot*	*No spot*	*No spot*	*Mean*
T_{eff1} (K)[b]	5380	5620	5860	5620
T_{eff2} (K)	5499 (1)	5743 (2)	5974 (1)	5739 (238)
q (m_2 / m_1)	0.449 (1)	0.449 (1)	0.458 (1)	0.452 (5)
A[b]	0.5	0.5	0.5	0.5
g[b]	0.32	0.32	0.32	0.32
$\Omega_1 = \Omega_2$	2.742 (1)	2.738 (2)	2.754 (1)	2.745 (8)
i°	86.3 (2)	87.47 (8)	89.3 (8)	87.7 (1.5)
$L_1 / (L_1 + L_2)_B$[c]	0.6420 (2)	0.6435 (2)	0.6435 (2)	0.6430 (9)
$L_1 / (L_1 + L_2)_V$	0.6501 (1)	0.6515 (1)	0.6504 (1)	0.6507 (7)
$L_1 / (L_1 + L_2)_{Ic}$	0.6568 (2)	0.6576 (2)	0.6559 (2)	0.6568 (9)
r_1 (pole)	0.4292 (2)	0.4300 (3)	0.4286 (2)	0.4293 (7)
r_1 (side)	0.4582 (3)	0.4592 (4)	0.4577 (3)	0.4584 (8)
r_1 (back)	0.4880 (4)	0.4893 (5)	0.4880 (4)	0.4884 (8)
r_2 (pole)	0.2975 (3)	0.2983 (3)	0.3003 (3)	0.2987 (14)
r_2 (side)	0.3113 (3)	0.3122 (4)	0.3143 (3)	0.3126 (15)
r_2 (back)	0.3481 (5)	0.3497 (6)	0.3520 (5)	0.3499 (20)
Fill-out factor (%)	12.4	14.0	14.5	13.6 (1.1)
SSR (B)[d]	0.0280	0.0296	0.0272	0.0283 (12)
SSR (V)[d]	0.0146	0.0198	0.0156	0.0167 (28)
SSR (Ic)[d]	0.0342	0.0342	0.0321	0.0328 (12)

a. All error estimates for T_{eff2}, q, $\Omega_{1,2}$, i, $r_{1,2}$, and L_1 from WDWINT56A *(Nelson 2009).*
b. Fixed during DC.
c. L_1 and L_2 refer to scaled luminosities of the primary and secondary stars, respectively.
d. Monochromatic sum of squares residual fit from observed values.

Table 7. V647 Vir lightcurve parameters evaluated by Roche modeling and the geometric elements derived for V647 Vir assuming it is an A-type W UMa variable with a cool spot on the primary star. Modeling estimates also include those determined at the uncertainty boundaries (T_{eff1} = 5620 ± 240 K) for the primary star.

Parameter	*Spotted*	*Spotted*	*Spotted*	*Mean*
T_{eff1} (K)[b]	5380	5620	5860	5620
T_{eff2} (K)	5419 (1)	5607 (1)	5843 (2)	5623 (212)
q (m_2 / m_1)	0.460 (1)	0.466 (1)	0.468 (1)	0.465 (2)
A[b]	0.5	0.5	0.5	0.5
g[b]	0.32	0.32	0.32	0.32
$\Omega_1 = \Omega_2$	2.776 (3)	2.796 (1)	2.793 (1)	2.788 (11)
i°	86.8 (2)	86.2 (2)	88.24 (77)	87.1 (1.1)
$A_S = T_S / T_\star$[c]	0.80 (1)	0.78 (1)	0.77 (1)	0.78 (2)
Θ_S(spot co-latitude)[c]	90 (6)	90 (6)	90 (4)	90 (3)
φ_S (spot longitude)[c]	180 (1)	180 (1)	180 (1)	180 (1)
r_S (angular radius)[c]	10.0 (1)	12.0 (1)	12.0 (1)	11.3 (1.2)
$L_1 / (L_1 + L_2)_B$[d]	0.6590 (2)	0.6705 (2)	0.6702 (2)	0.6666 (65)
$L_1 / (L_1 + L_2)_V$	0.6620 (1)	0.6699 (1)	0.6693 (1)	0.6671 (44)
$L_1 / (L_1 + L_2)_{Ic}$	0.6643 (2)	0.6693 (2)	0.6686 (2)	0.6674 (27)
r_1 (pole)	0.4251 (2)	0.4226 (3)	0.4233 (2)	0.4237 (13)
r_1 (side)	0.4531 (4)	0.4500 (4)	0.4510 (3)	0.4514 (16)
r_1 (back)	0.4822 (5)	0.4786 (4)	0.4800 (4)	0.4803 (18)
r_2 (pole)	0.2973 (10)	0.2970 (9)	0.2983 (2)	0.2975 (7)
r_2 (side)	0.3108 (12)	0.3102 (12)	0.3118 (3)	0.3109 (8)
r_2 (back)	0.3460 (20)	0.3445 (19)	0.3468 (4)	0.3458 (12)
Fill-out factor (%)	8.1	5.2	7.3	6.9 (1.5)
SSR (B)[e]	0.0259	0.0265	0.0253	0.0259 (6)
SSR (V)[e]	0.0128	0.0131	0.0121	0.0127 (5)
SSR (I_c)[e]	0.0236	0.0222	0.0214	0.0224 (11)

a. All error estimates for T_{eff2}, q, i, $\Omega_{1,2}$, A_S, Θ_S, φ_S, r_S, $r_{1,2}$, and L_1 from WDWINT56A *(Nelson 2009).*
b. Fixed during DC.
c. Temperature factor (A_S); location (Θ_S, φ_S) and size (r_S) parameters in degrees.
d. L_1 and L_2 refer to scaled luminosities of the primary and secondary stars, respectively.
e. Monochromatic sum of squares residual fit from observed values.

Ω_{outer} is the outer critical Roche equipotential, Ω_{inner} is the value for the inner critical Roche equipotential, and $\Omega = \Omega_{1,2}$ denotes the common envelope surface potential for the binary system. In both cases the systems are considered overcontact since $0 < f < 1$.

3.4.1. V647 Vir

LC parameters and geometric elements derived from the WD code are summarized in Table 6 (no spot) and Table 7 (cool spot). According to Binnendijk (1970) the deepest minimum (Min I) of an A-type overcontact system occurs when the hotter and larger star is occulted by the cooler less massive member of the binary system. With V647 Vir, the flat-bottomed dip in brightness indicative of a total eclipse of the secondary occurs at Min II while the round-bottomed deeper minimum (Min I) results from a transit across the primary face. As expected, the "q-search" results (Figure 7) clearly illustrate that model error quickly reaches a minimum as the mass ratio approaches ~0.45. It is also evident that V647 Vir is most likely an A-type overcontact binary; consequently, WD modeling proceeded under this assumption. Min II from the I_c-band LC includes two data points that are slightly deeper (< 0.012 mag) than Min I, which could indicate that V647 Vir is a W-type system. Attempts to simultaneously model all LC data under this assumption (q ≃ 2.22 and fixed values for T_{eff2}) produced grossly misshaped fits and were thereafter abandoned. Instead, adding a cool spot to the WD model improved the light curve fits during Min II (Figure 5), which resulted in lower sum of squared residuals (SSR) compared to the unspotted fit (Tables 6 and 7). A three-dimensional image rendered (Figure 9) using BINARYMAKER3 (BM3; Bradstreet and Steelman 2004) illustrates the transit during Min I (φ = 0) and the cool spot location on the primary star (φ = 0.60).

It could be argued in some cases that an A-type system is a cool or hot spot away from being classified as a W-type overcontact binary (and vice-versa). Inspection of the sparsely sampled ASAS and NSVS survey data folded with high cadence V-mag data from DBO (Figure 1) suggests that there is significant variability in the depth of Min II. Also, it should be noted that contrary to expectations for an A-type system, the best fit of the unspotted LC data occurred when the effective temperature of the secondary star (T_{eff2}) was higher (114–119 K) than the primary (T_{eff1}) component (Table 6). Not without precedence, this phenomenon has also been observed for EK Com (Deb *et al.* 2010), HV Aqr (Gazeas *et al.* 2007), BO CVn (Zola *et al.* 2012), and TYC 1664-0110-1 (Alton and Stępień 2016). It is therefore not unreasonable to propose that V647 Vir has in the past or will at some future date give the appearance of a W-type overcontact system.

3.4.2. V948 Mon

The broad flattened bottom (Figure 6) observed during Min II is a diagnostic indicator for a total eclipse of the secondary star. It follows that minimum light (Min I) occurs when the smaller secondary transits the primary star. As shown

Table 8. V984 Mon lightcurve parameters evaluated by Roche modeling and the geometric elements derived for V948 Mon assuming it is an A-type W UMa variable. Modeling estimates also include those determined at the uncertainty boundaries (T_{eff1} = 6480 ± 418 K) for the primary star.

Parameter	No spot	No spot	No spot	Mean
T_{eff1} (K)[b]	6062	6480	6898	6480
T_{eff2} (K)	6100 (2)	6505 (4)	6926 (2)	6510 (413)
q (m_2 / m_1)	0.283 (1)	0.286 (1)	0.287 (1)	0.285 (2)
A[b]	0.5	0.5	0.5	0.5
g[b]	0.32	0.32	0.32	0.32
$\Omega_1 = \Omega_2$	2.342 (2)	2.347 (3)	2.342 (1)	2.344 (3)
i°	86.5 (4)	88.2 (8)	87.3 (3)	87.3 (9)
$L_1 / (L_1 + L_2)_B$[c]	0.7410 (2)	0.7414 (2)	0.7389 (2))	0.7404 (13)
$L_1 / (L_1 + L_2)_V$	0.7431 (1)	0.7427 (1)	0.7403(1)	0.7420 (15)
$L_1 / (L_1 + L_2)_{Ic}$	0.7448 (1)	0.7440 (2)	0.7420 (1)	0.7436 (14)
r_1 (pole)	0.4792 (3)	0.4787 (5)	0.4799 (2)	0.4793 (6)
r_1 (side)	0.5218 (4)	0.5212 (6)	0.5229 (3)	0.5220 (9)
r_1 (back)	0.5535 (6)	0.5531 (7)	0.5555 (4)	0.5540 (13)
r_2 (pole)	0.2773 (3)	0.2786 (15)	0.2803 (3)	0.2787 (15)
r_2 (side)	0.2919 (4)	0.2933 (18)	0.2954 (3)	0.2935 (18)
r_2 (back)	0.3458 (9)	0.3477 (43)	0.3521 (7)	0.3485 (32)
Fill-out factor (%)	52.2	49.2	51.0	51.1 (1.6)
SSR (B)[d]	0.0631	0.0625	0.0611	0.0623 (11)
SSR (V)[d]	0.0210	0.0208	0.0203	0.0207 (4)
SSR (I_c)[d]	0.0224	0.0227	0.0220	0.0224 (3)

a. All error estimates for T_{eff2}, q, i, $\Omega_{1,2}$, $r_{1,2}$, and L_1 from WDWINT56A (Nelson 2009).
b. Fixed during DC.
c. L_1 and L_2 refer to scaled luminosities of the primary and secondary stars, respectively.
d. Monochromatic sum of squares residual fit from observed values.

in Figure 8, model error quickly reaches a minimum as the mass ratio approaches ~0.28. In this regard V948 Vir behaves like an A-type overcontact binary and was therefore modeled according to this assumption. The Roche model for V948 Mon did not require the addition of a spot to improve the LC fits. LC parameters and geometric elements derived from the WD code are summarized in Table 8. Similar to V647 Vir, the best fit of V948 Mon LC data occurred when the effective temperature of the secondary star (T_{eff2}) was slightly higher (25–38 K) than the primary (T_{eff1}) component. In this regard, attempts to model V948 Mon as a W-type overcontact system also proved unsuccessful. A three-dimensional rendering produced using BM3 (Figure 10) shows the transit during Min I (φ=0) and the Roche lobe surface outline (φ=0.75).

3.5. Absolute parameters

Fundamental stellar parameters were estimated for both binary stars using results from the best fit simulations of the 2018 LCs. However, without the benefit of RV data and classification spectra, these results should be more accurately described as "relative" rather than "absolute" parameters and considered preliminary in that regard.

3.5.1. V647 Vir

Gazeas and Stępień (2008) noted that primary (defined as more the massive component) stars in cool contact binaries obey the mass-radius relation associated with main-sequence (MS) stars. Power-law fits for the primary radii correspond very closely to those determined from single MS stars with masses lower than 1.8 $M_\odot$ (Giménez and Zamorano 1985). Therefore,

Table 9. Fundamental stellar parameters for V647 Vir using the mean photometric mass ratio (q_{ptm} = m_2 / m_1) from the spotted Roche model fits of LC data (2018) and the estimated mass for a putative G5V-G6V primary star in a W UMa variable.

Parameter	Primary	Secondary
Mass ($M_\odot$)	1.13 ± 0.05	0.53 ± 0.02
Radius ($R_\odot$)	1.10 ± 0.01	0.78 ± 0.01
a ($R_\odot$)	2.46 ± 0.03	2.46 ± 0.03
Luminosity ($L_\odot$)	1.09 ± 0.03	0.54 ± 0.01
M_{bol}	4.66 ± 0.03	5.43 ± 0.03
Log (g)	4.41 ± 0.02	4.38 ± 0.02

Table 10. Fundamental stellar parameters for V948 Mon using the mean photometric mass ratio (q_{ptm} = m_2 / m_1) from the spotted Roche model fits of LC data (2017–2018) and the estimated mass for a putative F3V-F7V primary star in a W UMa variable.

Parameter	Primary	Secondary
Mass ($M_\odot$)	1.32 ± 0.07	0.38 ± 0.02
Radius ($R_\odot$)	1.28 ± 0.02	0.73 ± 0.01
a ($R_\odot$)	2.62 ± 0.04	2.62 ± 0.04
Luminosity ($L_\odot$)	2.61 ± 0.07	0.85 ± 0.02
M_{bol}	3.71 ± 0.03	4.93 ± 0.03
Log (g)	4.34 ± 0.03	4.29 ± 0.03

reasonable estimates for the mass and radius of a primary star (most often populated by a MS star) in a W UMa-type binary system can be derived using data published for MS stars. These tabulations cover a wide range of spectral types typically attributed to primary stars in an overcontact binary system. For a putative G5V-G6V system (T_{eff1} ~ 5620 K) like V647 Vir, this includes a value (M_1 = 1.07 ± 0.05 $M_\odot$) interpolated from Harmanec (1988) and another (M_1 = 0.98 ± 0.02 $M_\odot$) from Pecaut and Mamajek (2013). A final relationship reported by Torres *et al.* (2010) for main sequence stars above 0.6 $M_\odot$ predicts a mass of 1.03 $M_\odot$ for the primary component. Importantly, three different empirically derived mass-period relationships for W UMa binaries have been published by Qian (2003) then later by Gazeas and Stępień (2008) and Gazeas (2009). According to Qian (2003) when the primary star is less than 1.35 $M_\odot$ its mass can be determined from Equation 7:

$$\log M_1 = 0.391\ (59) \log P + 1.96\ (17), \quad (7)$$

or alternatively when M_1 > 1.35 $M_\odot$ then Equation 8:

$$\log M_1 = 0.761\ (150) \log P + 1.82\ (28), \quad (8)$$

where P is the orbital period in days. Using Equation 7 leads to M_1 = 1.07 ± 0.08 $M_\odot$ for the primary. The mass-period relationship (Equation 9) derived by Gazeas and Stępień (2008):

$$\log M_1 = 0.755\ (59) \log P + 0.416\ (24), \quad (9)$$

corresponds to a W UMa system where M_1 = 1.17 ± 0.10 $M_\odot$. Gazeas (2009) reported another empirical relationship (Equation 10) for the more massive (M_1) star of a contact binary such that:

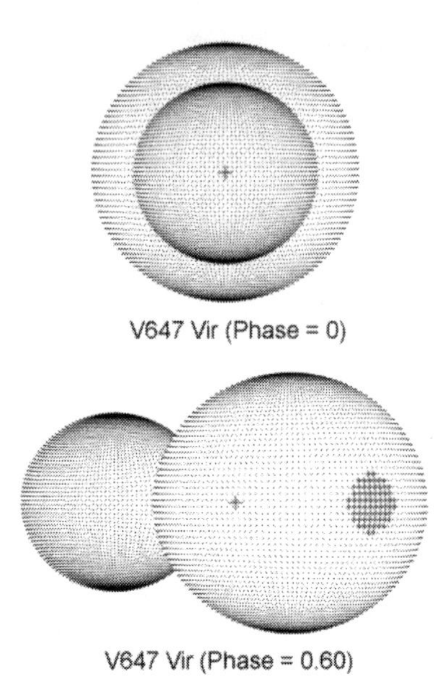

Figure 9. Three-dimensional spatial model of V647 Vir illustrating the transit of the secondary star across the primary star face at Min I ($\varphi = 0$) and cool spot location ($\varphi = 0.60$) on the primary star.

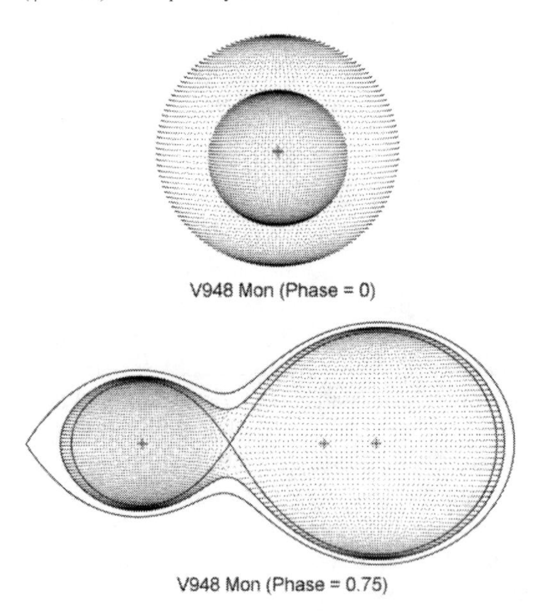

Figure 10. Three-dimensional spatial model of V948 Mon showing the transit at Min I ($\varphi = 0$) and the Roche lobe surface outline ($\varphi = 0.75$).

$$\log M_1 = 0.725\,(59) \log P - 0.076\,(32) \log q + 0.365\,(32), \quad (10)$$

in which $M_1 = 1.14 \pm 0.08\ M_\odot$. The mean of three values ($M_1 = 1.13 \pm 0.05\ M_\odot$) estimated from empirical models (Equations 7, 9, and 10) for W UMa binaries was used for subsequent determinations of M_2, semi-major axis a, volume-radii r_L, and bolometric magnitudes (M_{bol}) for V647 Vir. The mass estimates from Harmanec (1988), Pecaut and Mamajek (2013), and Torres (2010) are interesting in that they reflect values expected from single stars. In this case a single G5V-G6V star was estimated to be less massive ($1.02 \pm 0.05\ M_\odot$) than the primary star in V647 Vir. The secondary mass ($0.53 \pm 0.02\ M_\odot$) and total mass ($1.66 \pm 0.06\ M_\odot$) of the system were subsequently determined using the mean photometric mass ratio (0.465 ± 0.002) from the spotted Roche model. By comparison, a single main sequence star with a mass similar to the secondary (late K-type) would likely be much smaller ($R_\odot \sim 0.55$), cooler ($T_{eff} \sim 4000$), and far less luminous ($L_\odot \sim 0.07$). The semi-major axis, $a(R_\odot) = 2.46 \pm 0.03$, was calculated from Newton's version (Equation 11) of Kepler's third law where:

$$a^3 = (G \times P^2 (M_1 + M_2)) / (4\pi^2). \quad (11)$$

The effective radius of each Roche lobe (r_L) can be calculated over the entire range of mass ratios ($0 < q < \infty$) according to an expression (Equation 12) derived by Eggleton (1983):

$$r_L = (0.49q^{2/3}) / (0.6q^{2/3} + ln(1 + q^{1/3})), \quad (12)$$

from which values for r_1 (0.4462 ± 0.0002) and r_2 (0.3152 ± 0.0002) were determined for the primary and secondary stars, respectively. Since the semi-major axis and the volume radii are known, the radii in solar units for both binary components can be calculated where $R_1 = a \cdot r_1 = 1.10 \pm 0.01\ R_\odot$ and $R_2 = a \cdot r_2 = 0.78 \pm 0.01\ R_\odot$.

Luminosity in solar units ($L_\odot$) for the primary (L_1) and secondary stars (L_2) was calculated from the well-known relationship (Equation 13) where:

$$L_{1,2} = (R_{1,2} / R_\odot)2\ (T_{1,2} / T_\odot)^4. \quad (13)$$

Assuming that $T_{eff1} = 5620$ K, $T_{eff2} = 5607$ K, and $T_\odot = 5772$ K, then the solar luminosities ($L_\odot$) for the primary and secondary are $L_1 = 1.09 \pm 0.02$ and $L_2 = 0.54 \pm 0.01$, respectively. According to the Gaia DR2 release of stellar parameters (Andrae *et al.* 2018), the reported T_{eff} (5620^{+36}_{-240} K) is nearly identical to the adopted T_{eff1} (5620 K) value while the size ($R_\odot = 1.21$) and luminosity ($L_\odot = 1.32$) of the primary star in V647 Vir are greater than the corresponding values generated by the study herein. Based on the Bailer-Jones (2015) correction for parallax data in Gaia DR2 (Gaia *et al.* 2016, 2018) this system can be found at a distance of $444.5^{+8.2}_{-7.9}$ pc. By comparison, a value derived using the distance modulus equation corrected for interstellar extinction ($A_V = 0.070 \pm 0.003$) places V647 Vir slightly farther (471 ± 7 pc) away. Other values derived herein and necessary to perform this calculation include $V_{avg} = 12.77 \pm 0.01$, bolometric correction ($BC = -0.14$), $A_V = 0.070 \pm 0.002$, and the absolute V-magnitude ($M_V = 4.33 \pm 0.03$) from the combined luminosity (4.19 ± 0.03).

3.5.2. V948 Mon

The same approach described above for V647 Vir was used to estimate the primary star mass for V948 Mon (Table 8) but this time for a putative F3V-F7V system ($T_{eff1} \sim 6480$ K). The mass-period empirical relationships (Equations 8–10) lead to a mean value of $M_1 = 1.32 \pm 0.07\ M_\odot$ for the primary star. Interestingly, this was nearly identical ($M_1 = 1.31 \pm 0.04\ M_\odot$) to that obtained from single star estimates for an F3V-F7V system. The secondary mass $= 0.38 \pm 0.02\ M_\odot$ and total mass ($1.69 \pm 0.07\ M_\odot$) of the system were derived from the mean photometric mass ratio (0.285 ± 0.002). If the secondary was a single main sequence star with a similar mass (early M-type) it would probably be much smaller ($R_\odot \sim 0.42$), cooler ($T_{eff} \sim 3600$), and far less luminous ($L_\odot \sim 0.03$). The semi-major axis, $a(R_\odot) = 2.62 \pm 0.04$, was calculated from Equation 11 while the effective radius of each Roche lobe (r_L) was calculated according to Equation 12 from which values for r_1 (0.4898 ± 0.0006) and r_2 (0.2773 ± 0.0005) were determined for the primary and secondary stars, respectively. The radii in solar units for both binary components were calculated such that $R_1 = 1.28 \pm 0.02$ $R_\odot$ and $R_2 = 0.73 \pm 0.01$ $R_\odot$. Luminosity in solar units ($L_\odot$) for the primary (L_1) and secondary stars (L_2) was calculated according to Equation 13. Assuming that $T_{eff1} = 6480$ K, $T_{eff2} = 6505$ K, and $T_\odot = 5772$ K, then the solar luminosities for the primary and secondary are $L_1 = 2.61 \pm 0.07$ and $L_2 = 0.85 \pm 0.02$, respectively. According to the Gaia DR2 release of stellar parameters (Andrae *et al.* 2018), the reported T_{eff} (6337^{+418}_{-168} K) is not meaningfully different from the adopted value ($T_{eff1} = 6480$ K) used herein which was based on intrinsic color. However, the size ($R_\odot = 1.44$) and luminosity ($L_\odot = 3.03$) of the primary star in V948 Mon are greater than the values estimated by the study herein. This system is estimated to be $879.3^{+36.1}_{-33.4}$ pc away using the Bailer-Jones (2015) correction for parallax-derived distances reported in Gaia DR2 (Gaia *et al.* 2016, 2018). A value independently derived from the distance modulus equation using data generated herein ($V_{avg} = 13.30 \pm 0.01$, $A_V = 0.079 \pm 0.003$, BC $= -0.042$, and $M_V = 3.41 \pm 0.03$) places V948 Mon a similarly distant 901 ± 14 pc away.

4. Conclusions

Seven new times of minimum were observed for both V647 Vir and V948 Mon based on recent (2017–2018) CCD-derived LC data collected with B, V, and I_c filters. These along with other published values led to an updated linear ephemeris for each system. Potential changes in orbital period were assessed using differences between observed and predicted eclipse timings. A quadratic relationship was established with ETD values determined from V948 Mon, suggesting that the orbital period has been slowly decreasing at a rate of 0.0358 s $\cdot$ y^{-1}. Both systems will require many more years of eclipse timing data to further substantiate any potential change(s) in orbital period. The adopted effective temperatures (T_{eff1}) for V647 Vir (5620 K) and V948 Mon (6480 K) based on intrinsic color indices ($(B-V)_0$) were well within the confidence intervals reported from the Gaia DR2 release of stellar characteristics (Andrae *et al.* 2018). Estimates for the primary star luminosity ($L_\odot$) and radii ($R_\odot$) in both systems were lower (10–20%) than those reported in Gaia DR2. It is not known at this time whether this finding is coincidental or the result of a systematic bias in either method of determination. Both A-type overcontact systems clearly exhibit a total eclipse which is most evident as a flattened bottom during Min II. Therefore the photometric mass ratios for V647 Mon ($q = 0.465$) and V948 Mon ($q = 0.285$) determined by Roche modeling should prove to be a reliable substitute for mass ratios derived from RV data. Nonetheless, spectroscopic studies (RV and classification spectra) will be required to unequivocally determine a mass ratio, total mass, and spectral class for both systems.

5. Acknowledgements

This research has made use of the SIMBAD database operated at Centre de Données astronomiques de Strasbourg, France. Time of minima data tabulated in the Variable Star Section of Czech Astronomical Society (B.R.N.O.) website proved invaluable to the assessment of potential period changes experienced by this variable star. In addition, the Northern Sky Variability Survey hosted by the Los Alamos National Laboratory, the International Variable Star Index maintained by the AAVSO, the Catalina Sky Survey Data Release 2 maintained at Caltech, and the ASAS Catalogue of Variable Stars were mined for photometric data. The diligence and dedication shown by all associated with these organizations is very much appreciated. This work also presents results from the European Space Agency (ESA) space mission Gaia. Gaia data are being processed by the Gaia Data Processing and Analysis Consortium (DPAC). Funding for the DPAC is provided by national institutions, in particular the institutions participating in the Gaia MultiLateral Agreement (MLA). The Gaia mission website is https://www.cosmos.esa.int/gaia. The Gaia archive website is https://archives.esac.esa.int/gaia. Last, but certainly not least, the selfless support of the editorial staff at *JAAVSO* and the anonymous referee who provided critical review of this paper is greatly appreciated.

References

Abubekerov, M. K., Gostev, N. Yu. and Cherepashchuk, A. M. 2008, *Astron. Rep.*, **52**, 99.

Abubekerov, M. K., Gostev, N. Yu. and Cherepashchuk, A. M. 2009, *Astron. Rep.*, **53**, 722.

Akerlof, C., *et al.* 2000, *Astron. J.*, **119**, 1901.

Alton, K. B., and Nelson, R. H. 2018, *Mon. Not. Roy. Astron. Soc.*, **479**, 3197.

Alton, K. B., and Stępień, K 2016, *Acta Astron.*, **66**, 357.

Andrae, R., *et al.* 2018, *Astron. Astrophys.*, **616A**, 8.

Applegate, J. H. 1992, *Astrophys. J.*, **385**, 621.

Bailer-Jones, C. A. L. 2015, *Publ. Astron. Soc. Pacific*, **127**, 994.

Binnendijk, L. 1970, *Vistas Astron.*, **12**, 217.

Berry, R., and Burnell, J. 2005, *The Handbook of Astronomical Image Processing*, 2nd ed., Willmann-Bell, Richmond, VA.

Bradstreet, D. H. 2005, in *The Society for Astronomical Sciences 24th Annual Symposium on Telescope Science*, The Society for Astronomical Sciences, Rancho Cucamonga, CA, 23.

Bradstreet, D. H., and Steelman, D. P. 2004, BINARY MAKER 3, Contact Software (http://www.binarymaker.com).

Deb, S., Singh, H. P., Seshadri, T. R., and Gupta, R. 2010, *New Astron.*, **15**, 662.

Diethelm, R. 2009, *Inf. Bull. Var. Stars*, No. 5894, 1.

Diethelm, R. 2011, *Inf. Bull. Var. Stars*, No. 5992, 1.

Diethelm, R. 2012, *Inf. Bull. Var. Stars*, No. 6029. 1.

Diffraction Limited. 2019, MAXIMDL v.6.13, image processing software (http://www.cyanogen.com).

Drake, A. J., *et al.* 2014, *Astrophys. J., Suppl. Ser.*, **213**, 9.

Eggleton, P. P. 1983, *Astrophys. J.*, **268**, 368.

Gaia Collaboration, *et al.* 2016, *Astron. Astrophys.*, **595A**, 1.

Gaia Collaboration, *et al.* 2018, *Astron. Astrophys.*, **616A**, 1.

Gazeas, K. D. 2009, *Commun. Asteroseismology*, **159**, 129.
Gazeas, K. D., Niarchos, P.G., and Zola, S. 2007, in *Solar and Stellar Physics Through Eclipses*, eds. O. Demircan, S. O. Selam, B. Albayrak, ASP Conf. Ser. 370, 279.
Gazeas, K., and Stępień, K. 2008, *Mon. Not. Roy. Astron. Soc.*, **390**, 1577.
Gettel, S. J., Geske, M. T., and McKay, T. A. 2006, *Astron. J.*, **131**, 621.
Giménez, A., and Zamorano, J. 1985, *Astrophys. Space Sci.*, **114**, 259.
Greaves, J., and Wils, P. 2003, *Inf. Bull. Var. Stars*, No. 5458, 1.
Harmanec, P. 1988, *Bull. Astron. Inst. Czechoslovakia*, **39**, 329.
Henden, A. A., Levine, S. E., Terrell, D., Smith, T. C., and Welch, D. L. 2011, *Bull. Amer. Astron. Soc.*, **43**, 2011.
Henden, A. A., and Stone, R. C. 1998, *Astron. J.*, **115**, 296.
Henden, A. A., Terrell, D., Welch, D., and Smith, T. C. 2010, *Bull. Amer. Astron. Soc.*, **42**, 515.
Henden, A. A., Welch, D. L., Terrell, D., and Levine, S. E. 2009, *Bull. Amer. Astron. Soc.*, **41**, 669.
Hoffman, D. I., Harrison, T. E., and McNamara, B. J. 2009, *Astron. J.*, **138**, 466.
Hübscher, J., and Lehmann, P. B. 2015, *Inf. Bull. Var. Stars*, No. 6149, 1.
Kallrath, J., and Milone, E. F. 1999, *Eclipsing Binary Stars: Modeling and Analysis*, Springer-New York.
Kreiner, J. M. 2004, *Acta Astron.*, **54**, 207.
Kurucz, R. L. 2002, *Baltic Astron.*, **11**, 101.
Kwee, K. K., and van Woerden, H. 1956, *Bull. Astron. Inst. Netherlands*, **12**, 327.
Lucy, L. B. 1967, *Z. Astrophys.*, **65**, 89.
Minor Planet Observer. 2010, MPO Software Suite (http://www.minorplanetobserver.com), BDW Publishing, Colorado Springs.
Nelson, R. H. 2009, WDWINT v.56a astronomy software (https://www.variablestarssouth.org/bob-nelson).
O'Connell, D. J. K. 1951, *Publ. Riverview Coll. Obs.*, **2**, 85.
Paunzen, E., and Vanmunster, T. 2016, *Astron. Nachr.*, **337**, 239.
Pecaut, M. J., and Mamajek, E. E. 2013, *Astrophys. J. Suppl. Ser.*, **208**, 9.
Pojmański, G., Pilecki, B., and Szczygiel, D. 2005, *Acta Astron.*, **55**, 275.
Prša, A., and Zwitter, T. 2005, *Astrophys. J.*, **628**, 426.
Qian, S.-B. 2003, *Mon. Not. Roy. Astron. Soc.*, **342**, 1260.
Ruciński, S. M. 1969, *Acta Astron.*, **19**, 245.
Schlafly, E. F., and Finkbeiner, D. P. 2011, *Astrophys. J.*, **737**, 103.
Schwarzenberg-Czerny, A. 1996, *Astrophys. J., Lett.*, **460**, L107.
Smith, T. C., Henden, A. A., and Starkey, D. R. 2011, in *The Society for Astronomical Sciences 30th Annual Symposium on Telescope Science*, The Society for Astronomical Sciences, Rancho Cucamonga, CA, 121.
Software Bisque. 2019, THESKYX PRO v.10.5.0 software (http://www.bisque.com).
Terrell, D., and Wilson, R. E. 2005, *Astrophys. Space Sci.*, **296**, 221.
Torres, G., Andersen, J., and Giménez, A. 2010, *Astron. Astrophys. Rev.*, **18**, 67.
Van Hamme, W. 1993, *Astron. J.*, **106**, 2096.
Vanmunster, T. 2018, PERANSO v.2.50, light curve and period analysis software, (http://www.cbabelgium.com/peranso).
Wilson, R. E. 1979, *Astrophys. J.*, **234**, 1054.
Wilson, R. E. 1990, *Astrophys. J.*, **356**, 613.
Wilson, R. E., and Devinney, E. J. 1971, *Astron. J.*, **143**, 1.
Wilson, R. E., and Van Hamme, W. 2016, "Computing Binary Star Observables" (ftp://ftp.astro.ufl.edu/pub/wilson/lcdc2015/ebdoc.6jun2016.pdf).
Wozniak, P. R., *et al.* 2004, *Astron. J.*, **127**, 2436.
Zola, S., Nelson, R. H., Senavci, H. V., Szymanski, T., Kuźmicz, A., Winiarski, M., and Jableka, D. 2012, *New Astron.*, **17**, 673.

SSA Analysis and Significance Tests for Periodicity in S, RS, SU, AD, BU, KK, and PR Persei

Geoff B. Chaplin
Hokkaido, Kamikawa-gun, Biei-cho, Aza-omura Okubo-kyosei, 071-0216, Japan; geoff@geoffgallery.net

Received January 4, 2019; revised March 19, 2019; accepted March 25, 2019

Abstract Visual observations made by experienced observers are adjusted for individual observer bias. We examine the time series using signal processing methods to identify periodicities and test for the significance of the results finding a reliable periods in S Per, and to a limited extent in RS, SU, BU, and KK Per. Recommendations for future visual and electronic observation are made.

1. Introduction

We look at seven semiregular and irregular variables in Perseus. All but S Per are narrow range variables. The stars studied are shown in Table 1 together with previously quoted periodicities and references to the source.

SRc variables are massive, young, Population I stars with a magnitude range of under 5 and irregular periodicity typically in the 250–1000 day range (Percy 2011). The best known such stars are alpha Orionis (Betelgeuse) and mu Cephei ("the garnet star"). Lc variables are also red supergiant stars but with irregular periods. The variables apart from S Per studied here have a narrow range of variation (less than 2 magnitudes) and as such pose a severe test for visual observers because of this, the individual's eyes' color sensitivity, the Purkinje effect (Purkinje 1825; Sigismondi 2011; AAVSO (2013) and references therein), and other observational factors such as local light pollution. We anticipate the "extrinsic" noise (related to the observational process) as opposed to the "intrinsic" noise (related to random events within the star or environment—if any) to be large, and attempt to reduce this as much as possible prior to analysis. For example, even experienced observers—defined as those with over 100 observations of the star—may differ by as much as a magnitude when observing the same star at roughly the same time. In this paper we restrict our attention to observations made by experienced observers and analyze these for consistent "bias," adjusting the data before further analysis. Adjustment of observer data is described in detail below and is atypical of standard procedures which generally reject outliers only.

A variety of analytical techniques for period identification are used in the literature: discrete Fourier transform (DFT) (Kendall 1984; Shumway and Stoffer 2017) with or without adjustment for the observational window, for example, using the CLEAN (Roberts *et al.* 1987) or CLEANEST algorithms (Foster 1995); autoregressive analysis and in particular the simple and efficient implementation by Percy and Sato (2009); wavelet analysis—see Foster (1996) or Sundararajan (2015) for theory and, for example, Percy and Kastrukoff (2001) for an application to pulsating variables, and Sabin and Zijlstra (2006) when analyzing instability in long-period variables. A general review of these techniques is given by Templeton (2004). The difficulties of using standard Fourier methods and obtaining reliable results should not be underestimated (see Thomson 1990). More recently, non-linear techniques have also been used by Kollath (1990) and Kollath *et al.* (1998), and others in the context of giant variable stars. The methods of Empirical Mode Decomposition (Huang *et al.* 1998; overview by Lambert *et al.* 2019) are also geared particularly towards non-linear series.

We apply methods from the field of singular spectrum analysis (SSA), explained in the context of astronomical data analysis by Chaplin (2018) and references therein, and derive the underlying signal (removing noise and trends). The aim here is to identify underlying patterns of behavior, summarizing them by periodicities where appropriate, although the techniques of

Table 1. Stars analyzed in this study, with previously quoted periodicities and references to the source.

Star	*GSC Designation*	*Spectral Type (Wenger 2000)*	*Class (Kiss et al. 2006)*	*Period(s)*	*Magnitude Range (BAAVSS 2019)*
S Per	03698-03073	M4.5-7Iae C	SRc	813 ± 60 (Kiss *et al.* 2006) 822 (Samus *et al.* 2017) 745, 797, 952, 2857 (Chipps *et al.* 2004)	7.9–12.8
RS Per	03694-01293	M3.5IabFe-1 C	SRc	244.5 (BAAVSS 2019) 4200 ± 1500 (Kiss *et al.* 2006)	7.8–9.0
SU Per	03694-01652	M3-M4Iab C	SRc	533 (BAAVSSS 2019) 430 ± 70 and 3050 ± 1200 (Kiss *et al.* 2006) 500 (Stothers and Leung 1971)	7.2–8.7
AD Per	03694-01613	M3Iab C	SRc	No discernable peak, rise to lowest frequencies (Kiss *et al.* 2006) 362.5 (Samus *et al.* 2017)	7.7–8.4
BU Per	03694-01247	M4Ib C	SRc	381 ± 30 and 3600 ± 1000 (Kiss *et al.* 2006) 367 (Samus *et al.* 2017)	9.0–10.0
KK Per	03693-01951	M2Iab-Ib B	Lc	No significant frequency (Kiss *et al.* 2006)	7.5–8.0
PR Per	03694-00152	M1-Iab-Ib B	Lc	No significant frequency (Kiss *et al.* 2006)	7.7–8.2

SSA do not relate any such periodicities to harmonic patterns of behavior. We then proceed to test whether these periodicities are likely to have arisen by chance from noisy data (in which case we reject such periodicity as not intrinsic to the star) or not (in which case we accept it as an intrinsic periodicity).

2. Data, observer bias, and adjustments

Data are taken from the BAA (2019) and the AAVSO (2010) databases, and from the VSOLJ (2018) database prior to 2000. The list of experienced observers for which a bias adjustment is made is given in Appendix A.

For each star other than S Per we proceed as follows. The mean magnitude of visual observations is calculated for each experienced observer separately and for all experienced observers for that star. The individual mean less the overall mean is called the observer bias and is deducted from each observation made by that observer to get the adjusted magnitude. This process generally leads to a substantial reduction in the overall variance. Results are shown for each star in Table 2 (but we intentionally do not wish to state the bias for each observer since this might lead to a change in the observer's methods—consistency is preferred to accuracy). Table 2 also gives the timespan of data but in all cases there were a number of isolated or widely separated observations at the beginning of the time series which were ignored. It should be noted that the stars for which the bias adjustment was made are all narrow range variables, so preferential observing (for example, when the variable is bright) should not be a significant source of bias. On the other hand, preferential observing is a factor for S Per so a bias adjustment is not made.

One would not expect bias for a given observer to be constant across different stars because different reference stars may be used and the group of observers being compared against is different. Nevertheless it was noted that observers' magnitude estimates tended to be consistently high or consistently low although the amount differed from star to star.

For S Per, which has a much greater range of variation, we take the data from experienced observers without further adjustment.

3. Analytical methods

3.1. Singular Spectrum Analysis (SSA)

SSA is used to extract a series from observations and is a method used widely in meteorology, medical science, economics, the sciences and industry, and appears to be becoming the method of choice for time series data analysis. In this section we very briefly outline the methods, and introduce the terminology of, singular spectrum analysis (SSA), explained more fully in the paper by Chaplin 2018, books by Golyandina and Zhigljavsky 2013, Golyandina *et al.* 2001, and Huffaker *et al.* 2017.

Table 2. Timespan and variance reduction through observer bias removal.

Star	*Start (2440000+)*	*End (2440000+)*	*Length (years)*	*Variance Reduction*
S	–21576	18115	108.7	n/a
RS	2744	18096	42.0	43.5%
SU	–5347	18115	64.2	49.0%
AD	1978	18115	44.2	65.1%
BU	1636	18082	45.0	40.8%
KK	3112	18115	41.0	36.3%
PR	3112	18115	41.1	54.4%

From an autocorrelation matrix calculated from the time series of magnitude observations the eigenvectors and eigenvalues are calculated. These eigenvectors are sorted in order from the strongest to the weakest according to the relative magnitudes of the associated eigenvalues. The related time series are then compared with each other to find correlations between them and to determine if the general patterns of behavior are similar. The original time series is "projected" along each of these eigenvectors to derive an EV-time series (which we subsequently refer to as the EV). We then group the series together into "trends" (long-term slow patterns), "cyclical" (possibly several different groups of series with different periods), and noise.

It should be noted that observations are required at equally spaced intervals in order to perform the above analysis—so we have to put data into equal time intervals (buckets), averaging values within the bucket. In the stars covered here data have been put into 20-day buckets. Also, reconstructed signals, although they may look periodic, do not necessarily have a constant period nor do they have a constant amplitude, and are not derived in any way from harmonic series—the EV time series are merely complicated averages of the original data. "Periods" indicated below represent an approximation to the actual behavior.

In this paper we use the R (2018a) statistical programming language and CRAN (2018b) libraries and in particular the function "ssa" in the R library "Rssa," and use the code as explained in detail in section 3.7 and Appendix B.

3.2. Significance tests of discovered signals

A white noise (uncorrelated random noise) is generally regarded as an insufficient test for the presence of signals in data, and Monte Carlo methods (MCSSA) have been devised to test significance (for example, Allan and Smith 1996; Ghil *et al.* 2002). We use the R implementation of MCSSA developed by Gudmundsson (2017) and in particular the functions decompSSA and MCSSA. Code is given in Appendix B5.

We also use a somewhat different approach inspired by analysis of variance methods and also by the following intuitive idea. If we see a signal in a period of data, then if the signal is a permanent feature of the underlying process we expect it to continue, but if it is an artefact arising from noisy data we expect it to cease to be present in the future.

In order to perform significance analysis we compare two different time intervals of the same series (first and second halves, H1 and H2), looking for common signals, and we need to do this (for reasons which will become apparent below) in an automated way. Any series typically contains trends—long term changes—which may be quite different in two sub-intervals of the series, together with potential signals and noise. The impact of this can be that a periodic signal manifests itself as one set of eigenvectors in one subset and a different set in another.

We therefore begin by finding trend components (defined as having too long a period or no period), taking the first signal which is not a trend and whose period is not too short as defining a potential signal, and look for remaining signals whose period matches the first to within a defined amount (the "acceptance criterion"). A potential signal is required to have two or more component signals. Code "XYZgetSignals_udf" performs this analysis.

First, an analysis of H1 in the observational data is made to determine the EV groupings that correspond to a signal. We then do the same for H2. If the period in H1 is *P1* and the period in H2 is ˆ and if

$$A > \mathrm{abs}((P1 - P2) \times 400 / (P1 + P2)) \quad (1)$$

where A is the acceptance criterion, we accept the two periods as belonging to a signal. If we find no correspondence between signals in H1 and H2 then we conclude there is no consistent period in the data. Secondly, (assuming we have found a potential signal) we then model the original entire series (i.e. before trend removal) as a "red" noise (AR(1)) process (see section 3.4 below). Code "XYZ actual data tests.R" performs this analysis.

The AR(1) model is then used to generate simulated data ("surrogates") over the same time period as the actual data which are then analyzed as above as if they were the real data. In cases where a signal is found in both H1 and H2 of the simulated series with a difference less than the acceptance criterion, it is then counted as a (simulated) signal. Note that the simulated signal is not required to be of the same frequency as that identified in the actual data. The process is then repeated over 1,000 simulations and the proportion generating simulated real signals for the wide and the narrow acceptance criterion is calculated (together with an estimate of the accuracy of this figure). This then gives an indication of the confidence that the real signal did not arise by chance. Code "XYZ significance tests.R" performs this analysis.

Finally it is important to test variation in the parameters used to perform the analysis, in particular by changing the bucketing length, the start date by one, two, or more days (which changes the bucket contents), and the SSA window length. We test using bucket sizes such as 17, 20, 23, 30, 34, 40, and 46 days (depending on the length of data available and the suspected period—aiming to keep within about one tenth of the period) and require that the signal is found in all the decompositions. We then reduce the acceptance period subject to the signal continuing to be discovered.

The process is described more fully in the case of SU Per, which is presented first in section 4 below.

3.3. Fourier analysis

Fourier analysis is a traditional method for analyzing time-series where there is underlying periodicity and where the underlying series is stationary. For general references on traditional time series analysis including Fourier and autoregressive techniques, see Kendall (1984) and Shumway and Stoffer (2017; the latter includes R examples and code).

In this paper we use the "spectrum" function in the R stats library to perform the Fourier analysis and smoothing.

Error bars on the spectral power can be calculated from surrogate data. However, a plot of the spectrum together with the percentiles of the surrogate distribution can be misleading and can overstate the significance of peaks—underlying AR(1) noise can exaggerate the height of peaks in the spectrum (Allen and Smith 1996). Code in Appendix B4 plots the spectrum and surrogate percentiles.

3.4. Autoregressive AR(1) model

Random noise is generated from a zero mean "red" noise (AR(1)) process according to the following formula:

$$x_t = alpha \times x_{t-1} + sigma \times epsilon_t \quad (2)$$

where *alpha* and *sigma* are constants and *epsilon* is generated from an independent random normal (zero mean, unit variance) process.

The parameters of the zero mean AR model are chosen by fitting such a model to the actual data series using the "ar" function in the R stats library.

3.5. Wavelet analysis

Where periodicity is known not to be strict or the time series non-stationary, Fourier methods are theoretically incorrect—although they may be a reasonable approximation. Instead a technique known as wavelet analysis (or more simply a moving window on the data as in Howarth and Greaves 2001) is often used. Here we use code based on the wavelet analysis code from the AAVSO (2017). For comparison with the SSA results we analyze the data using two window sizes determined by the "decay" factor—a factor of 0.0001 cycles per day (the "slow" window, roughly corresponding to a slow 10,000-day window) identifying periodicities which change slowly, and a factor of 0.003 (a "fast" 333-day window) identifying more rapid changes. In such analysis we identify the strongest period, then the next strongest, etc. It is the case that generally the second strongest period is virtually the same as the first, so when looking for a different period we require that the period is at least 20% different from its predecessor. In each case only periods significant at a certain level on an F-test (dependent on the star) are shown.

3.6. Missing data

Three methods for filling missing data were used. The first was simple linear interpolation between the last known data value and the immediately following known data value. The second followed the method of Kondrashov and Ghil (2006) by filling missing values from the first eigenseries, recentering and refitting until convergence of the eigenvalue was achieved, then potentially going on to the next eigenseries. A final method was to randomize the linearly interpolated values, the impact of which is to slightly lower the value of the autoregressive parameter in the fitted AR model.

3.7. Code

R code intended for the Rstudio environment for the analysis described in sections 3.2 and 3.3 is provided in Appendix B. Two main codes are used—one to analyze the real data ("first part" above) and another, if needed, to simulate and analyze the simulated data ("second part" above). Each part uses (directly or indirectly) some of the following helper functions (ending in _udf—user defined function) given in Appendix B1.

XYZspectrum_udf performs a spectral analysis using ar smoothing or no smoothing, producing a chart if required and returning a list of the periods discovered in declining order of strength.

XYZgetSignals_udf code performs 1d-ssa on the data, finding trends and signals meeting certain criteria.

XYZbucketData_udf takes the observational data and times of observation and collects values into the specified length of bucket, taking the average of all values in the bucket. Where gaps in the data occur linearly interpolated values are calculated, returning the bucketed data, a flag indicating whether in interpolated value is used, and other summary data.

Appendix B2 contains the **"first part"** code and loads the above functions and the data. The user sets various parameters and the code buckets data and performs a 1d-ssa analysis of the entire series and the first and second half separately, producing results for inspection. Additionally, the code fits an AR(1) model producing parameters for simulation use.

Appendix B3 contains the **"second part"** code and includes a function **matchTest2** to decide whether two signals are close (the user inputs the diffPeriodpercent figure, and other parameters, into the code where indicated), loads other helper functions, and simulates 1000 data series using user input AR(1) parameters, performing the analysis described in section 3.2 and outputting the proportion of simulations producing signals of the same period in each half of the data.

Appendix B4 contains code to plot the Fourier spectrum of the signal derived from the entire data series, together with upper and lower 2 and 10 percentiles calculated from the surrogate data and signals.

Appendix B5 contains the code to perform MCSSA on the actual data, producing a chart with error bars and identifying outlying frequencies.

4. The stars

4.1. SU Per

SU Per is covered in more detail than the following stars hence is presented first.

Prior to September 1974 data were sparse—even after bucketing into 20-day buckets more than half the buckets were empty and with long gaps prior to 1974. Attempts to fill the data using linear interpolation or the Khondrashov and Ghil method failed to give a satisfactory data series in this earlier period. Post-1974, 6,550 observations were bucketed into 803 20-day buckets. Less than 8% of the buckets were empty, with no long empty runs, and tests using linear interpolation versus the Khondrashev and Ghil method showed no material difference in the resulting signals; the following results are based on the linear interpolation gap filling method. Bucketing tests were run using 17, 20, 23, 30, 34, 40, and 46 days together with shifts in the start date by 1 or 2 days, and showed a consistent set of results with an acceptance criterion of 7% across all the following analysis. We describe the results in detail for the 20-day buckets.

An AR(1) model was fitted to the data (after removing the mean) and—after randomizing the linearly interpolated values, which reduces the alpha—showed an alpha of 0.71 and sigma of 0.145. If we assume bias adjusted observations have a standard deviation of 0.2 magnitude then the bucketed data (approximately 8 observations per bucket) should have a residual standard deviation of about 0.1. The AR model is therefore not inconsistent with observational error being by far the largest part of the noise in the data.

The following discussion and figures are based on a window length of 400 for the entire series. Tests with a window length of 200 show similar results but going much shorter than that starts to produce inconsistent results. Figure 1a shows the EVs and Figure 1b the correlation analysis for the entire series, with signals 5 and 6 meeting the criteria and showing a period of 475 days with the spectrum illustrated in Figure 1c. The data, trend, and signal are shown in Figure 1f.

The EVs for the first half are shown in Figure 1d, with signals 3 and 4 meeting the criteria and having a period of 464 days. Signals 5 and 6 are approximately the second harmonic.

EVs for the second half are shown in Figure 1e, with signals 6 and 7 meeting the criteria and giving a period of 475 days. EVs 9 and 10 are also approximately the second harmonic.

The AR model was then used to produce 1000 simulated sets of observational magnitudes, each of which was analyzed as described in section 3.2. If the period of the identified signal in the first half was within 7% of the period from the second half then this was counted as a "hit." It should be noted that there was no requirement that the spurious signal periods matched the signal period in the actual data—simply that there are closely similar signals in both intervals. Signals corresponding to periods of 1,000 days (50 buckets) or longer, or 100 days (5 buckets) or shorter were ignored in this test. (Very few spurious signals had periods outside this range and many signals had no identified period.)

Simulation results showed 3.7% (with a standard deviation (sd) of 2.0%) of simulations led to spurious signals of approximately the same frequency in both halves of the data.

As an independent test we use the Monte Carla SSA methods in the MCSSA algorithm from the "simsalabim" library to produce Figure 1g. Note that periods are 40 days/frequency. The figure identifies (as well as early trends) the signals at 475 days lying just below the 95% confidence level together with significant signals around the second harmonic.

We conclude that SU Per exhibits a periodicity of 475 ± 33 days with approximately 95% confidence.

4.1.1. Fourier analysis

We use the simulations generated above, together with the reconstructed signals and their spectra, to generate 10- and 2-percentile power levels. These are plotted in Figure 1h along with the (unsmoothed) spectrum of the signal in the actual data. Note the following points. The autoregressive process generates the typical "1/f" rise in power at lower frequencies widely seen in Fourier spectra of magnitude time series. Also, the figure misleadingly suggests the signal is significant at the 98% level—the noise process exaggerates the power in the actual signal, thereby overestimating its significance.

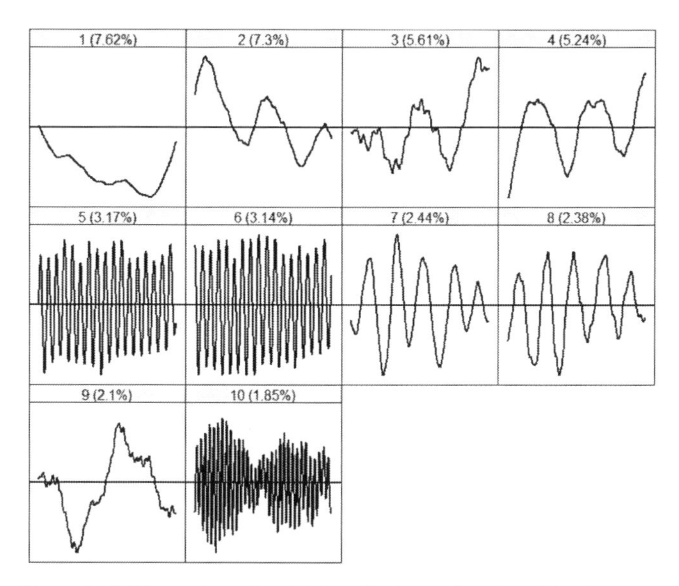

Figure 1a. SU Per entire series EVs (amplitude as a function of time).

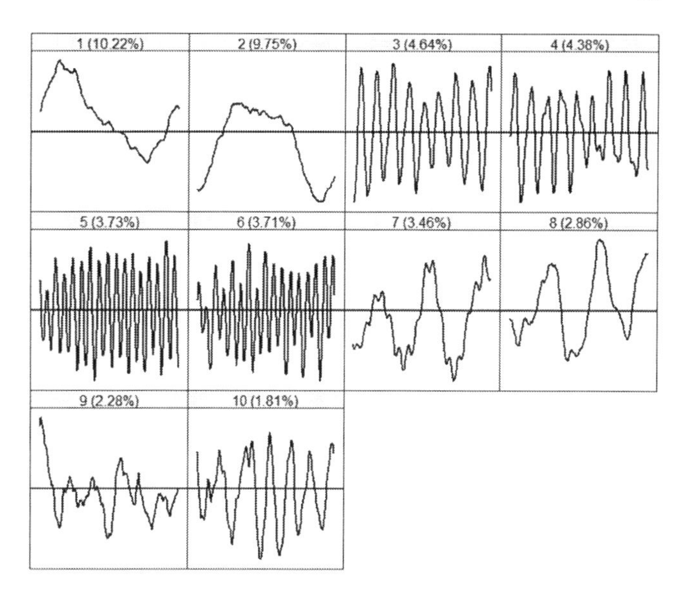

Figure 1d. SU Per first half EVs (amplitude as a function of time).

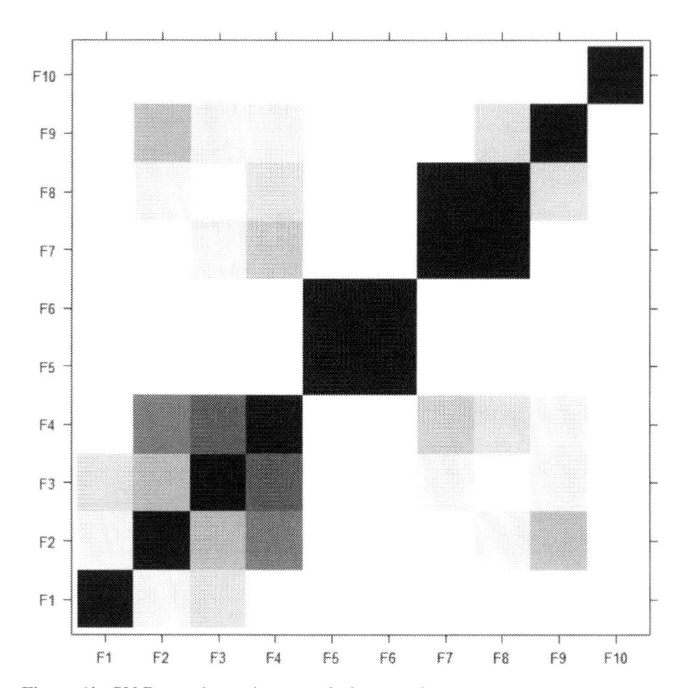

Figure 1b. SU Per entire series, correlation matrix.

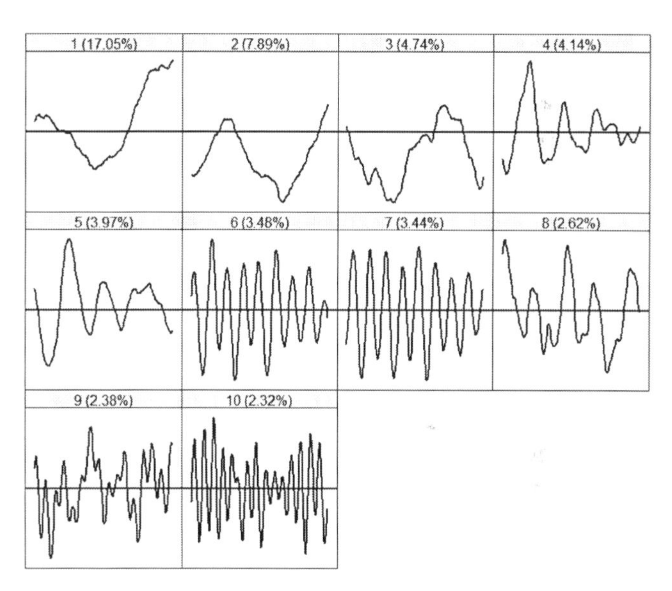

Figure 1e. SU Per second half EVs (amplitude as a function of time).

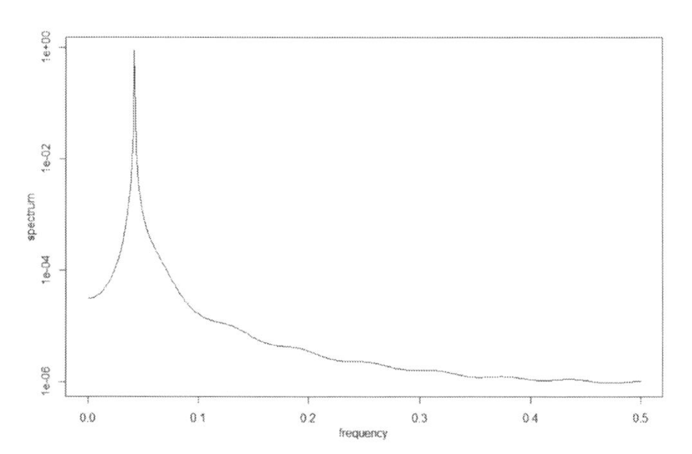

Figure 1c. SU Per spectrum derived from signals 5 and 6 in the entire series.

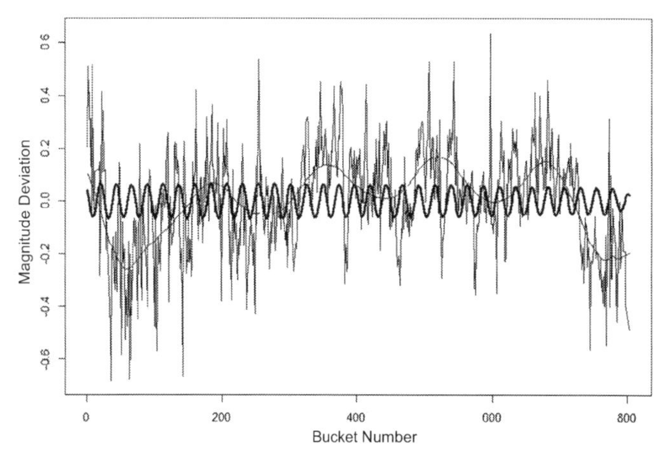

Figure 1f. SU Per entire data series with recovered trend components (EVs 1 to 4) and signal.

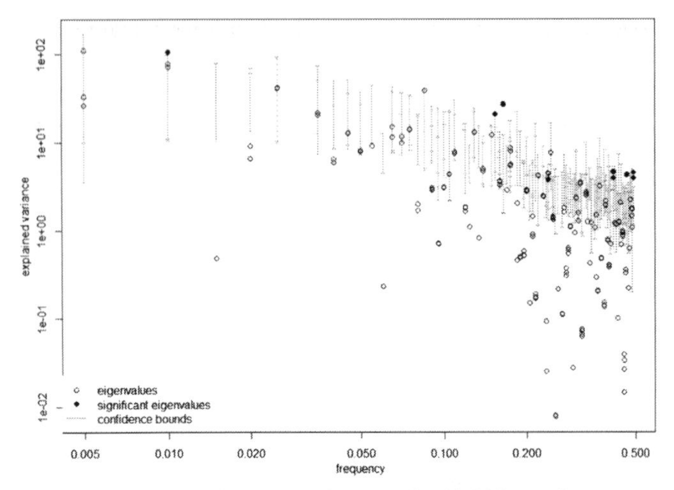

Figure 1g. SU Per significance test of EV signals with 95% error bars.

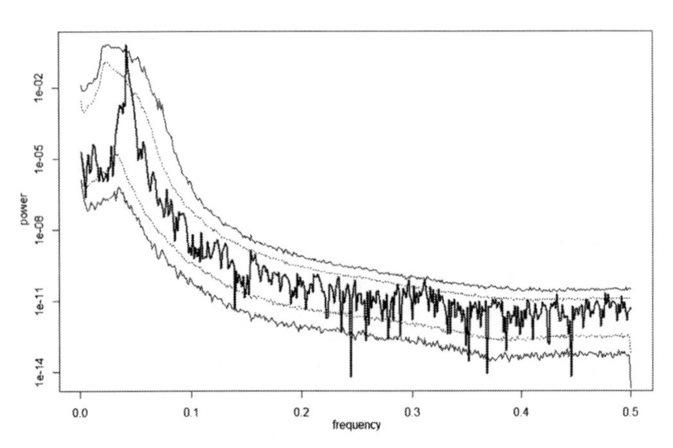

Figure 1h. Fourier spectrum of data signal together with simulation based 10 and 2 percentile envelopes.

4.1.2. Wavelet analysis

The slow wavelet identifies a 3,225-day period and many periods longer than 25% of the time series with 99% confidence—we reject longer periods as trends in our SSA analysis—together with a period at 476 days significant at the 97.5% level and briefly a period of approximately four times this. The fast wavelet also identifies the very long waves and identifies a period rising from about 1,600 days to 1,900 days.

We note however that simulated data regularly also show signals persisting over large fractions of the data span but, while these are significant in the context of that specific series, in the context of a series which may be generated by a random process, wavelet analysis carries little meaning and is therefore not covered further for the following stars.

4.2. S Per

S Per is analyzed in some detail in Chaplin (2018). We simply summarize the data and state the simulation results here.

The data are well populated from January 1920. From 25,860 observations 1,789 20-day buckets were constructed with less than 2% being empty. A fitted AR(1) model gave an alpha of 0.96 and sigma of 0.20, the higher sigma possibly arising because of unadjusted bias in the observations and the high alpha because of the large amplitude of variation relative to the noise.

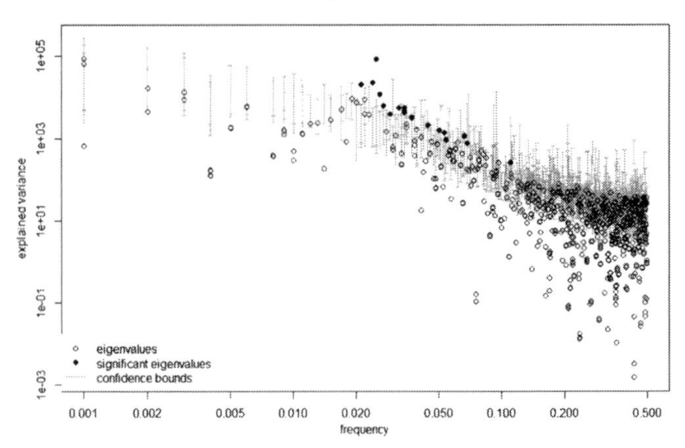

Figure 2. S Per significance test of EV signals with 99% error bars.

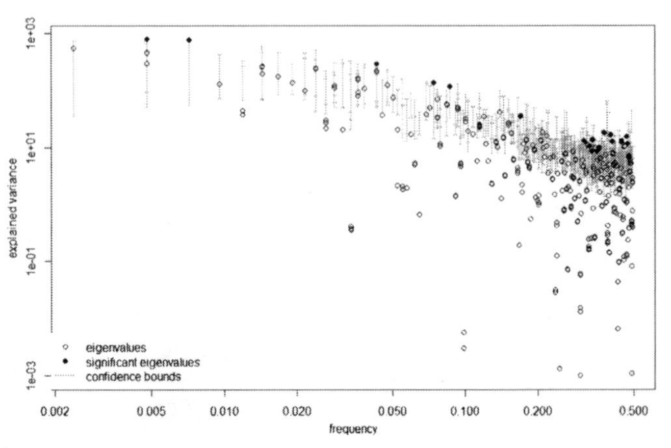

Figure 3. RS Per significance test of EV signals with 95% error bars.

In simulations the high alpha tends to generate very few signals with a period as short as that analyzed for S Per, and simulations resulted in only 0.1% generating signals within 5% of each other, hence we accept that S Per has a period of 815 ± 40 days with over 99% confidence.

Using the MCSSA significance testing methods produces the results shown in Figure 2. The 815-day signal lies well outside the error bars, with neighboring and many harmonics also outside the error bars consistent with amplitude and frequency modulation of the signal.

4.3. RS Per

Prior to November 1972 data were sparse. Post-1972 4,820 observations were bucketed into 838 20-day buckets. Less than 9% of the buckets were empty, with no long empty runs. A fitted AR(1) model gave an alpha of 0.78 and sigma of 0.165.

SSA consistently revealed periods in the 445–495 day range with a 5% acceptance criterion and simulations resulted in 4.6% generating signals.

Using the MCSSA significance testing methods produces the results shown in Figure 3. The signal lies outside the error bars, with neighboring and some harmonics also outside the error bars.

We conclude that S Per has a period of 475 ± 25 days with 95% confidence.

4.4. AD Per

Prior to September 1974 data were sparse. Post-1974 3,945 observations were bucketed into 805 20-day buckets.

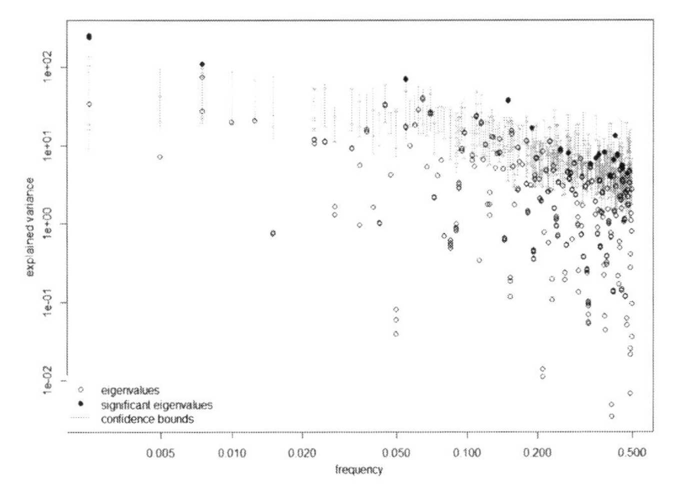

Figure 4.AD Per significance test of EV signals with 90% error bars.

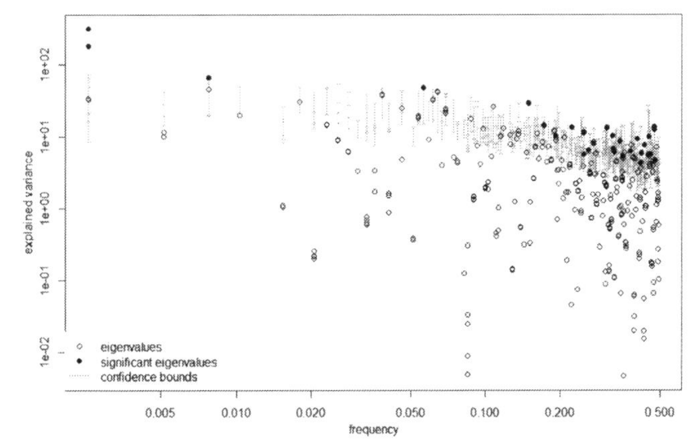

Figure 5. BU Per significance test of 20 day bucketing and signal from EVs 4 and 5 with 80% error bars.

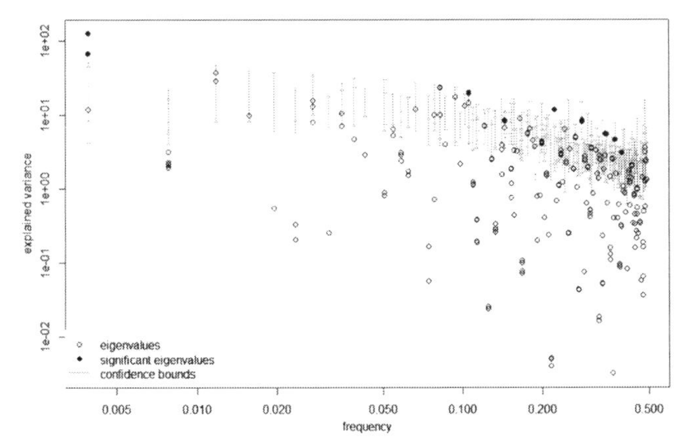

Figure 6. KK Per significance test of EV 5, 6, 9, 10 and 30 day bucketing with 90% error bars.

Approximately 8% were empty, with no long empty runs. A fitted AR(1) model gave an alpha of 0.61 and sigma of 0.13.

Testing with 17- to 35-day bucketing with a 15% acceptance criterion nevertheless revealed periods between 320 and 450 days, with many spectral peaks being very broadly defined. The entire series gave a signal period of 360 days. Simulation with a 15% acceptance criterion gave 13% generating signals. On the other hand, MCSSA using 20-day bucketing and signals 5–8 detected significance at the 90% level at periods of 350 days together with a second harmonic and an intermediate period depicted in Figure 4.

Because of the instability of the signal detected with changing bucketing we conclude AD Per has no clear intrinsic period.

4.5. BU Per

Prior to December 1975 data were sparse. Post-1975 3,562 observations were bucketed into 782 20-day buckets. Approximately 7% of the buckets were empty, with no long empty runs. A fitted AR(1) model gave an alpha of 0.56 and sigma of 0.13.

SSA gave periods in the range 300–360 days although there were exceptions with a 17-day bucketing and in one case the second harmonic gained preference. Simulations resulted in 14% of signals lying within a 15% acceptance criterion. MCSSA significance testing produces the results shown in Figure 5. The signal lies just outside the error bars with neighboring and some harmonics also outside the error bars.

We conclude BU Per has a period of 330 ± 50 days with 80% confidence.

4.6. KK Per

Prior to July 1976 data were sparse. Post-1976 3,391 observations were bucketed into 771 20-day buckets. Less than 7% of the buckets were empty, with no long empty runs. A fitted AR(1) model gave an alpha of 0.57 and sigma of 0.13 (virtually the same as BU Per).

SSA with 17- to 35-day buckets consistently gave well-defined periods in the range 330–360 days using a 7% acceptance criterion, with the entire series showing 348 days, and simulations resulted in 3% false signals. However, selection of EVs to form the signal was sensitive to whether or not linearly interpolated values were randomized. MCSSA significance testing produces the results shown in Figure 6. The signal lies just on the error bar with neighboring and some harmonics outside the error bars.

We tentatively conclude KK Per has a period of 345 ± 25 days with approximately 90% confidence.

4.7. PR Per

Prior to August 1982 data were sparse. Post-1982 2,826 observations were bucketed into 659 20-day buckets. Less than 9% of the buckets were empty, with no long empty runs. A fitted AR(1) model gave an alpha of 0.59 and sigma of 0.11.

SSA with 17- to 35-day gave no identified period in many cases and when signals were identified they tended to be 460 and 300 days.

We conclude PR Per has no clear period.

5. Conclusions and observer recommendations

SSA provides a means of exploring the signals within the data and separating trends and noise from cyclical patterns, but needs separate analysis to gain confidence that these signals are meaningful and not randomly generated by noise in the observations. It is clear from the above analysis that narrow range late spectral type stars are problematic for visual

Table 3. Results of the analysis.

Star	Period (days)	Uncertainty	Confidence
S Per	815	40	>99%
RS Per	475	25	95%
SU Per	475	33	95%
AD Per	No clear period		
BU Per	330	50	80%
KK Per	345	25	90%
PR Per	No clear period		

observation. Nevertheless a long run of data can help overcome the noise, but a better and available solution is to reduce the noise. We strongly recommend the use of CCD/DSLR equipment by amateurs as outlined further below to overcome the problem of the substantial component of extrinsic noise in future data.

We reject the use of wavelet analysis in the context of noisy data such as these.

Results of the analysis are summarized in Table 3.

S Per makes it clear that a long history and large range of magnitude variation lead to a period determination with high confidence. For the other stars, where a period is determined, the confidence is in the 80-95% region.

It is unfortunate that observations of SRc variables have reduced in recent years. These stars are not well understood and a rich long database of observations is essential for future study. Visual observation is helpful in order to relate visual and future electronic observations and in any event is likely to be more plentiful than electronic observations. Visual observers are encouraged to build up a series of over 100 observations, making observations no more frequently than once a week.

The narrow range of variability and the strong color make these objects ideal for CCD observation with a V filter, or DSLR observation. A good consumer digital camera and 200mm lens on an equatorial mount is sufficient to produce high quality data for these objects. Variable sky conditions can mean any single observation may be accurate to only 0.1 magnitude (even though the software stated reduction accuracy is much better), so electronic observations should ideally be a set of 30 to 100 observations to reduce the error in the mean to 0.01 magnitude or less. It should be noted that with short focal length instruments (500mm or less) six or more SRc variables in Perseus will fit on a 35mm frame sensor, making data collection efficient. A long history of accurate magnitudes derived from electronic data is essential to apply some of the analysis in this paper with a high level of confidence and is essential for a better understanding of pulsating variables.

6. Acknowledgements

The author is grateful to the AAVSO, the BAA, and the VSOLJ for providing the observational data used in this analysis and, in particular, to the dedicated observers listed in Appendix A who followed one or more of these stars regularly for a period of years or decades. The author would also like to thank J. Howarth and R. Pickard for helpful suggestions during an early version of this paper. The author is also grateful to the AAVSO for the code for the wavelet analysis performed in this paper. This paper benefited greatly from the comments of anonymous reviewers for which the author is very grateful.

References

AAVSO. 2010, Data access (https://www.aavso.org/data-access).

AAVSO. 2013, Purkinje and v faint estimates (https://www.aavso.org/purkinje-and-v-faint-estimates).

AAVSO. 2017, Software directory WWZ Fortran (https://www.aavso.org/software-directory).

Allen, M. R., and Smith, L. A. 1996, *J. Climate*, **9**, 3373.

British Astronomical Association. 2019, BAA Photometry Database (https://www.britastro.org/vssdb). **[baa2]**

British Astronomical Association Variable Star Section. 2019, BAA/VSS website (http://www.britastro.org/vss). **[baa1]**

Chaplin, G. B. 2018, *J. Amer. Assoc. Var. Star Obs.*, **46**, 157.

Chipps, K. A., Stencel, R. E., and Mattei, J. A. 2004, *J. Amer. Assoc. Var. Star Obs.*, **32**, 1.

Foster, G. 1995, *Astron. J.*, **109**, 1889.

Foster, G. 1996, *Astron. J.*, **112**, 1709.

Ghil, M., *et al.* 2002, *Rev. Geophys.*, **40**, 1003.

Golyandina, N., Nekrutkin, V., and Zhigljavsky, A. 2001, *Analysis of Time Series Structure*, Chapman and Hall, CRC Press, Boca Raton, FL.

Golyandina, N., and Zhigljavsky, A. 2013, *Singular Spectrum Analysis for Time Series*, Springer-Verlag, Berlin.

Gudmundsson , L. 2017, Singular System/Spectrum Analysis (SSA) (https://r-forge.r-project.org/projects/simsalabim).

Howarth, J., and Greaves, J. 2001, *Mon. Not. Roy. Astron. Soc.*, **325**, 1383.

Huang, N. E., *et al.* 1998, *Roy. Soc. London Proc, Ser. A*, **454**, 903.

Huffaker, R., Bittelli, M., and Rosa, R. 2017, *Non-linear Time Series Analysis with R*, Oxford Univ. Press, Oxford.

Kendall, M. 1984, *Time Series*, 2nd ed., Charles Griffin and Co., Ltd., London and New York, ch. 8.

Kiss, L. L., Szabo, Gy. M., and Bedding, T. R. 2006 *Mon. Not. Roy. Astron. Soc.*, **372**, 1721.

Kollath, Z. 1990, *Mon. Not. Roy. Astron. Soc.*, **247**, 377.

Kollath, Z., Buchler, J. R., Serre, T., and Mattei, J. A. 1998, *Astron. Astrophys.*, **329**, 147.

Kondrashov, D., and Ghil, M. 2006, *Nonlinear Process Geophys.*, **13**, 151.

Lambert, M., Engroff, A., Dyer, M. and Byer, B. 2019, Empirical Mode Decomposition (https://www.clear.rice.edu/elec301/Projects02/empiricalMode).

Percy, J. R. 2011, *Understanding Variable Stars*, Cambridge Univ. Press, Cambridge, 217.

Percy, J. R., and Kastrukoff, R. 2001, *J. Amer. Assoc. Var. Star Obs.*, **30**, 16.

Percy, J. H., and Sato, H. 2009, *J. Roy. Astron. Soc. Canada*, **103**, 11.

Purkinje, J. E. 1825, *Neue Beiträge zur Kenntniss des Sehens in Subjectiver Hinsicht*, Reimer, Berlin, 109.

The R Foundation for Statistical Computing. 2018a, R: A language and environment for statistical computing (https://www.R-project.org).

The R Foundation for Statistical Computing. 2018b, CRAN: The Comprehensive R Archive Network (https://cran.r-project.org).

Roberts, D. H., Lehar, J., and Dreher, J. W. 1987, *Astron. J.*, **93**, 968.

RStudio. 2018, RStudio software (https://www.rstudio.com).

Sabin, L., and Zijlstra, A. A. 2006, *Mem. Soc. Astron. Ital.*, **77**, 933.

Samus N. N., Kazarovets E. V., Durlevich O. V., Kireeva N. N., and Pastukhova E. N. 2017, *General Catalogue of Variable Stars*, version GCVS 5.1 (http://www.sai.msu.su/gcvs/gcvs/index.htm).

Shumway, R. H., and Stoffer, D. S. 2017, *Time Series Analysis and its Applications*, 4th ed., Springer, New York, ch. 4.

Sigismondi, C. 2011, arXiv:1106.6356v1.

Stothers R., and Leung K. C. 1971, *Astron. Astrophys.*, **10**, 290.

Sundararajan, D. 2015, *Discrete Wavelet Transform*, Wiley, Hoboken, NJ.

Templeton, M. 2004, *J. Amer. Assoc. Var. Star Obs.*, **32**, 4.

Thompson, D. J. 1990, *Philos. Trans. Roy. Soc. London, Ser. A*, **330**, 601.

Variable Star Observers League in Japan. 2018, VSOLJ variable star observation database (http://vsolj.cetus-net.org/database.html).

Wegner, M., *et al.* 2000, *Astron. Astrophys., Suppl. Ser.*, **143**, 9.

Appendix A

List of the observers for which a bias adjustment is made, and the number of visual observation for the stars analyzed.

Observer	Number
D. Stott	1153
G. Poyner	238
G. Ramsay	104
C. Hadhazi	1498
S. Hoeydalsvik	224
I. A. Middleton	1391
J. D. Shanklin	100
T. Kato	750
J. Krticka	221
A. Kosa-Kiss	1139
R. S. Kolman	352
L. K. Brundle	2239
W. Lowder	112
M. J. Nicholson	313
O. J. Knox	337
E. Oravec	588
P. J. Wheeler	467
S. Papp	202
R. C. Dryden	629
S. W. Albrighton	4211
S. Sharpe	1048
A. Sajtz	746
T. Markham	1781
P. Vedrenne	2926
W. J. Worraker	203
Y. Watanabe	503

Appendix B: R code

Notes:

1. We recommend the use of RStudio (2018) which provides a simple and highly efficient way of handling R code and results including the production of graphics.

2. The user needs to set the path according to where the R system has been installed—see the code comments below—and also define certain input parameters.

3. Comments are in italics, code in bold, headings in larger type italics.

B.1. Helper functions

```
# function to get periods corresponding to peak intensities
XYZspectrum_udf <-
function(x, drawPlot, graphText, smoothing) {
   if (drawPlot) spec.out←spectrum(x, main=graphText,
      method=smoothing)
   else spec.out←spectrum(x, plot=FALSE, method=smoothing)
   #Power Spectrum Plots
   power<-spec.out$spec # vertical axis values in spectral plot
   frequency<-spec.out$freq # all the frequencies on the x-axis
   cycle<-1/frequency # corresponding wavelengths
   #Sort cycles in order of magnitude of power spikes
   hold<-matrix(0,(length(power)-2),1)
   for(i in 1:(length(power)-2)){
      max1<-if(power[i+1]>power[i]&&power[i+1]>power[i+2])1 else (0)
       hold[i,]<-max1
   }
   max<-which(hold==1)+1
   if (length(max) == 0) {
      max = 1
   } else {
      if (power[1] == Inf) {
         max = 1
   } else {
      if(power[1]>power[max]) max = 1
      }
   }
   power.max<-power[max]
   cycle.max<-cycle[max]
   o←order(power.max, decreasing=TRUE)
   cycle.max.o<-cycle.max[o]
   peakFrequencies<-1/cycle.max[o]
   results<-list(cycle.max.o)
   return(results)
}

# function to identify trends and primary periodic signal
XYZgetSignals_udf <-
function(y, s, longestPeriod, shortestPeriod, bucketSize,
periodDiffpercent,outputVecCount){
   # find trends
   trendSignals = seq(0, 0, length.out=outputVecCount)
   EVPeaks = seq(0, 0, length.out=outputVecCount)
   for (i in 1:outputVecCount){
      r <- reconstruct(s, groups = list(EV = c(i:i)))
      recon = unlist(r[1])
      spec.out = XYZspectrum_udf(recon, drawPlot=FALSE, "",
         smoothing="ar")
      specPeaks = unlist(spec.out[1])*bucketSize
      if (length(specPeaks) == 0) specPeaks = 0
      EVPeaks[i] = specPeaks[1]
      if (EVPeaks[i]>longestPeriod) trendSignals[i] = i
   }
   trendSignals = trendSignals[trendSignals != 0]
   #determine first periodic signal neither too long nor too short a period
   periodSignals = seq(0, 0, length.out=outputVecCount)
   pStart = 0
   for (i in 1:(outputVecCount-1)) {
```

```
      if (pStart == 0) {
         if (EVPeaks[i] >= shortestPeriod & EVPeaks[i] <= longestPeriod) {
            itmp = 1
            periodSignals[itmp] = i
            pStart = i
         }
      }
   }
#determine subsequent periodic signals matching first
   if (pStart>0){
      for (i in (pStart+1):outputVecCount) {
         if (EVPeaks[i] >= shortestPeriod & EVPeaks[i] <= longestPeriod &
            abs(EVPeaks[i]-EVPeaks[pStart]) < periodDiffpercent*EVPe
               aks[pStart]/100) {
            itmp = itmp + 1
            periodSignals[itmp] = i
         }
      }
   }
   periodSignals = periodSignals[periodSignals != 0]
   if (length(periodSignals) > 1) { # NB single signal not allowed
      r2 <- reconstruct(s, groups = list(EV = periodSignals))
      signal = unlist(r2[1])
      spec.out2 = XYZspectrum_udf(signal, drawPlot=FALSE, "",
         smoothing="ar")
      signalPeaks = unlist(spec.out2[1])*bucketSize
   } else signalPeaks = NULL
   return(list(trendSignals, EVPeaks, periodSignals, signalPeaks))
}

# function to collect irregularly timed data into constant size buckets
XYZbucketData_udf <-
function(bucketSize, data, time, ndata){
   n = 1
   bucketSum = data[1] # sum within a bucket
   count = 1 # number of obs within bucket
   sumCount = 0 # calculates the average number of data points in non-
     empty buckets
   nbucket = 1 # number of buckets
   Tstart = time[1]
   maxBuckets = floor((time[ndata] - Tstart) / bucketSize) + 1
   bucketData = seq(0, 0, length.out=maxBuckets)
   EMPTYBUCKETFLAG = seq(0, 0, length.out=maxBuckets)
   while (n < ndata) {
      if (time[n+1]>=Tstart+nbucket*bucketSize){
         bucketData[nbucket] = bucketSum / count
         sumCount = sumCount + count
         count = 0
         bucketSum = 0
         while (time[n+1]>=Tstart+(nbucket+1)*bucketSize) {
            nbucket = nbucket + 1
            EMPTYBUCKETFLAG[nbucket] = 1
         }
         nbucket = nbucket + 1
         count = 1
         n = n + 1
         bucketSum = data[n]
      } else {
         n = n + 1
         count = count + 1
         bucketSum = bucketSum + data[n]
      }
   } #end while
   if (count > 0) { # final bucket (incomplete)
      bucketData[maxBuckets] = bucketSum / count
   } else emptBucketCount = emptBucketCount + 1
   totalEmpty = sum(EMPTYBUCKETFLAG)
   avgNoInNonemptyBuckets = ndata / (maxBuckets - totalEmpty)
   #now fill empty buckets by by linear interpolation
   LIbucketData = seq(0, 0, length.out=maxBuckets)
   iLast = 1
   LIbucketData[1] = bucketData[1]
   for (i in 2:maxBuckets){
      if (EMPTYBUCKETFLAG[i] == 0 & EMPTYBUCKETFLAG[i-1]
         == 1 ) {
         LIbucketData[iLast] = bucketData[iLast]
         LIbucketData[i] = bucketData[i]
         for (j in iLast+1:i-1) LIbucketData[j] =
            bucketData[iLast] + (bucketData[i] - bucketData[iLast])*
               (j-iLast)/(i-iLast)
         iLast = i
      }
      else if (EMPTYBUCKETFLAG[i] == 0) {
         iLast = i
         LIbucketData[i] = bucketData[i]
      }
   }
   ntmp = length(LIbucketData)
   bucketData = LibucketData[-(maxBuckets+1:ntmp)]
   result = list(bucketData, maxBuckets,totalEmpty,
      avgNoInNonemptyBuckets,
   EMPTYBUCKETFLAG, ntmp)
   return(result)
}
```

B.2. "first part" analysis in section 3.2

```
rm(list=ls(all=TRUE))
#Load User-Defined Functions
setwd("C:/Users/Geoff/Documents/R/GBC Defined Functions")
dump("XYZgetSignals_udf", file="XYZgetSignals_udf.R")
source("XYZgetSignals_udf.R")
dump("XYZspectrum_udf", file="XYZspectrum_udf.R")
source("XYZspectrum_udf.R")
dump("XYZbucketData_udf", file="XYZbucketData_udf.R")
source("XYZbucketData_udf.R")
#load Rssa R library from Install Packages
library(Rssa)
# end user defined functions

# USER INPUT# USER INPUT# USER INPUT# USER INPUT# USER INPUT
longestPeriod = 1000 # maximum acceptable period in days
shortestPeriod = 100 # shortest
periodDiffpercent = 10.0 # % of frequency or supposed period, used as
acceptance criterion
randomiseLinterp = TRUE
#NB user can set up a loop over the following variables and write output if
desired
Xfactor = 1 # change to adjust bucket size
dataStart = 1
baseBucketSize = 20
# USER INPUT# USER INPUT# USER INPUT# USER INPUT# USER INPUT

# STEP 1: Read in and select data
fileIn = "SU Per" # data is 3 col CSV file headers JD, mag and adjMag
setwd(paste0("C:/Users/Geoff/Documents/ASTRO/data analysis/", fileIn,
"/raw data"))
#D:/ or your own path here
tsIn←read.csv("biasAdjusted.csv") # data is 3 col CSV file headers JD,
mag and adjMag
plot(tsIn$adjMag,xlim=c(1,length(tsIn$adjMag)), xlab="", ylab="",
type="l", col="black",
lwd=2, main="complete series actual data")
if (dataStart > 1) ts = tsIn[-c(1:dataStart-1),] else ts = tsIn
ndata = nrow(ts)
mag<-ts$adjMag
timeJD = ts$JD

# STEP 2a: bucket data
bucketSize = baseBucketSize * Xfactor
tmp = XYZbucketData_udf(bucketSize,mag,timeJD,ndata)
bucketDates = seq(timeJD[1]+bucketSize/2,timeJD[ndata],by=bucketSize)
maxBuckets = unlist(tmp[2])
emptyBuckets = unlist(tmp[3])
avgFilledBucketCount = unlist(tmp[4])
bucketMag = unlist(tmp[1])
emptyFlag = unlist(tmp[5])
```

```
L = floor(maxBuckets/4)*2
magMean = mean(bucketMag)
bucketMag = bucketMag - magMean
outputVecCount = 10

# STEP 2b:
# calculate mean average change from bucket to bucket and randomise linterp values
if (randomiseLinterp) {
  bucketMagLagged = bucketMag[2:maxBuckets]
  delta = sum(abs(bucketMag-bucketMagLagged))/(maxBuckets-1)
  set.seed(0)
  randomNormal <- rnorm(maxBuckets)
  bucketMagRand = bucketMag
  for (i in 2:maxBuckets) {
    if (emptyFlag[i] == 1) bucketMagRand[i] = bucketMagRand[i] +
      delta*randomNormal[i]
    }
    bucketMag = bucketMagRand - mean(bucketMagRand)
  }

# STEP 3: automated SSA of actual data
x = bucketMag
x1 = x[1:L]
x2 = x[-(1:L)]
for (kk in 1:3) {
  if (kk == 1) { x = x - mean(x); Lx = L
  } else if (kk == 2) { x = x1 - mean(x1); Lx = L/2
  } else if (kk == 3) { x = x2 - mean(x2); Lx = L/2 }
  s←ssa(x, Lx, kind="1d-ssa")
  plot(s, type="vectors", idx=1:outputVecCount, xlim=c(1,Lx),
col="black", lwd=2)
  w←wcor(s, groups=c(1:outputVecCount))
  plot(w, title="correlation matrix")
  results = XYZgetSignals_udf(x, s, longestPeriod, shortestPeriod,
    bucketSize, periodDiffpercent, outputVecCount)
  actualTrendSignals = results[1]
  actualEVPeaks = results[2]
  actualPeriodSignals = results[3]
  actualSignalPeaks = results[4]
  count = length(actualPeriodSignals[[1]])
  signal = reconstruct(s, groups = list(EV = unlist(actualPeriodSignals)))
  XYZspectrum_udf(unlist(signal[1]), drawPlot=TRUE, "signal
    spectrum", smoothing="ar")
  }
# STEP 4: fit AR(1) model for later simulation use
autoAR1 = ar(bucketMag, aic=FALSE, order.max=1)
alphaLI = autoAR1$ar
errors = autoAR1$resid
sigmaLI = sqrt(var(errors[2:maxBuckets], y=NULL, na.rm=TRUE))
write("alphaAR, sigmaAR", file = "actualDataAnalysis.csv", ncolumns =
1, append = TRUE,
sep = ",")
write(paste(alphaLI, sigmaLI, sep=","), file = "actualDataAnalysis.csv",
ncolumns = 2,
append = TRUE, sep = ",")
write(" ", file = "actualDataAnalysis.csv", ncolumns = 1, append = TRUE,
sep = ",")
```

B.3. "second part" analysis in section 3.2

```
rm(list=ls(all=TRUE))
#Load User-Defined Functions
setwd("C:/Users/Geoff/Documents/R/GBC Defined Functions")
dump("XYZgetSignals_udf", file="XYZgetSignals_udf.R")
source("XYZgetSignals_udf.R")
dump("XYZspectrum_udf", file="XYZspectrum_udf.R")
source("XYZspectrum_udf.R")

# this helper function tests H1 signal frequency against H2
matchTest2 <- function(peakS1H1, peakS1H2, periodDiffpercent){
  hit = 0
    if ((peakS1H1>shortestPeriod) & (peakS1H1<longestPeriod)){
      if ((peakS1H2>shortestPeriod) & (peakS1H2<longestPeriod)){
        if (abs(peakS1H1-peakS1H2)<periodDiffpercent*(peakS1H1+
          peakS1H2)/200){
          hit = 1
        }
      }
    }
    return(hit)
  }
  # end user defined functions
  library(Rssa)

# USER INPUT# USER INPUT# USER INPUT# USER INPUT# USER INPUT
periodDiffpercent = 10.0 # % of frequency or supposed period, used as acceptance criterion
longestPeriod = 1000 # maximum acceptable period in days
shortestPeriod = 100 # shortest
bucketSize = 20 # used to calculate spectral peak in days
maxBuckets = 803
alpha = 0.71
sigma = 0.145
# USER INPUT# USER INPUT# USER INPUT# USER INPUT# USER INPUT

# simulate data, and perform analysis looking for a periodic signal
L = floor(maxBuckets/4)*2
LH = L/2
outputVecCount = 10
nsims = 1000
hitSimsCount = 1
set.seed(0)
hits = seq(0, 0, length.out=nsims) # number of hits within periofDiffPercent
for (j in 1:nsims){
  simulatedSeries <- arima.sim(list(ar=c(alpha,0,0)), sd=sigma,
  n=maxBuckets)
  y = simulatedSeries[1:L]
  y = y - mean(y)
  s<-ssa(y,LH,kind="1d-ssa")
  results = XYZgetSignals_udf(y, s, longestPeriod, shortestPeriod,
    bucketSize, periodDiffpercent, outputVecCount)
  signalPeaks = unlist(results[4])
  if(length(signalPeaks)==0)peakS1H1=0elsepeakS1H1=signalPeaks[[1]]
  y = simulatedSeries[-(1:L)]
  y = y - mean(y)
  s<-ssa(y,LH,kind="1d-ssa")
  results = XYZgetSignals_udf(y, s, longestPeriod, shortestPeriod,
    bucketSize, periodDiffpercent, outputVecCount)
  signalPeaks = unlist(results[4])
  if (length(signalPeaks) == 0) peakS1H2 = 0 else peakS1H2 =
    signalPeaks[[1]]
  # compare the strongest signal in H1 with first or second strongest in H2
  hits[j] = matchTest2(peakS1H1, peakS1H2, periodDiffpercent)
}
cat(sum(hits)*100/nsims, sum(hitsHalf)*100/nsims, "\n")
```

B.4. Fourier spectrum and percentiles analysis in section 3.3

```
# plots spectrum of signal in the actual data together with envelopes derived from the spectra of
# signals in surrogate series

rm(list=ls(all=TRUE))
#Load User-Defined Functions
setwd("C:/Users/Geoff/Documents/R/GBC Defined Functions")
dump("XYZgetSignals_udf", file="XYZgetSignals_udf.R")
source("XYZgetSignals_udf.R")
dump("XYZspectrum_udf", file="XYZspectrum_udf.R")
source("XYZspectrum_udf.R")
dump("XYZbucketData_udf", file="XYZbucketData_udf.R")
source("XYZbucketData_udf.R")
library(Rssa)

# USER INPUT# USER INPUT# USER INPUT# USER INPUT# USER INPUT
periodDiffpercent = 10.0 # % of frequency or supposed period, used as acceptance criterion
```

```
longestPeriod = 1000 # maximum acceptable period in days
shortestPeriod = 100 # shortest
bucketSize = 20 # used to calculate spectral peak in days
fileIn = "SU Per"
setwd(paste0("C:/Users/Geoff/Documents/ASTRO/data analysis/", fileIn,
"/raw data"))
alpha = 0.71
sigma = 0.145
# USER INPUT# USER INPUT# USER INPUT# USER INPUT# USER INPUT

# STEP 1: Read in data, bucket, find signal and perform spectral analysis for
the chart
ts<-read.csv("biasAdjusted.csv")
ndata = nrow(ts)
mag<-ts$adjMag
timeJD = ts$JD
tmp = XYZbucketData_udf(bucketSize, mag, timeJD, ndata)
bucketDates = seq(timeJD[1]+bucketSize/2, timeJD[ndata], by=bucketSize)
bucketMag = unlist(tmp[1])
maxBuckets = unlist(tmp[2])
bucketMag = bucketMag - mean(bucketMag)
outputVecCount = 10
L = floor(maxBuckets/4)*2
s←ssa(bucketMag, L, kind="1d-ssa")
results = XYZgetSignals_udf(x, s, longestPeriod, shortestPeriod, bucketSize,
periodDiffpercent, outputVecCount)
actualPeriodSignals = results[3]
signal = reconstruct(s, groups = list(EV = unlist(actualPeriodSignals)))
spec.out←spectrum(unlist(signal[1]), plot=FALSE, method="pgram")
x<-spec.out$freq # all the frequencies on the x-axis
actual = spec.out$spec
nnn = length(x)

# STEP 2: simulate data, and perform analysis looking for a periodic signal
nsims = 1000
L = floor(maxBuckets/4)*2
outputVecCount = 10
set.seed(0)
power2 = matrix(0, nsims, length(x))
for (j in 1:nsims){
   y <- arima.sim(list(ar=c(alpha,0,0)), sd=sigma, n=maxBuckets)
   y = y - mean(y)
   s<-ssa(y,L,kind="1d-ssa")
   results = XYZgetSignals_udf(y, s, longestPeriod, shortestPeriod,
      bucketSize, periodDiffpercent, outputVecCount)
   actualPeriodSignals = results[3]
   signal = reconstruct(s, groups = list(EV = unlist(actualPeriodSignals)))
   spec.out = spectrum(unlist(signal[1]), plot=FALSE, method="pgram")
   if (j==1) frequency<-spec.out$freq # all the frequencies on the x-axis;
      standard intervals
   power2[j,] = spec.out$spec
   }
# find 10% and 2% envelopes
lower10 = c(nnn)
upper10 = c(nnn)
lower2 = c(nnn)
upper2 = c(nnn)
for (ifreq in 1:nnn) {
   datax = power2[,ifreq]
   lower10[ifreq] = quantile(datax,0.1)
   upper10[ifreq] = quantile(datax,0.9)
   lower2[ifreq] = quantile(datax,0.02)
   upper2[ifreq] = quantile(datax,0.98)
}
plot(x,actual, log="y", xlab="frequency", ylab="power", type="l",
   col="black", lwd=2, main=paste0(fileIn, ": signal and 10 and 2
   percentiles"))
lines(x, y=upper10, col="black", lty=3, lwd=1)
lines(x, y=lower10, col="black", lty=3, lwd=1)
lines(x, y=upper2, col="black", lty=1, lwd=1)
lines(x, y=lower2, col="black", lty=1, lwd=1)
```

B.5. MCSSA code

```
rm(list=ls(all=TRUE))
#Load User-Defined Functions
setwd("C:/Users/Geoff/Documents/R/GBC Defined Functions")
dump("XYZspectrum_udf", file="XYZspectrum_udf.R")
source("XYZspectrum_udf.R")
dump("XYZbucketData_udf", file="XYZbucketData_udf.R")
source("XYZbucketData_udf.R")
library(Rssa)
#install.packages("simsalabim", repos="http://R-Forge.R-project.org")
library(simsalabim)

# USER INPUT# USER INPUT# USER INPUT# USER INPUT# USER INPUT
longestPeriod = 1000 # maximum acceptable period in days
shortestPeriod = 100 # shortest
periodDiffpercent = 10.0 # % of frequency or supposed period, used as
acceptance criterion
bucketSize = 20
fileIn = "SU Per"
setwd(paste0("C:/Users/Geoff/Documents/ASTRO/data analysis/", fileIn,
"/raw data"))
# USER INPUT# USER INPUT# USER INPUT# USER INPUT# USER INPUT

# STEP 1: Read in data and bucket
ts<-read.csv("biasAdjusted.csv")
ndata = nrow(ts)
mag<-ts$adjMag
timeJD = ts$JD
tmp = XYZbucketData_udf(bucketSize, mag, timeJD, ndata)
bucketDates = seq(timeJD[1]+bucketSize/2, timeJD[ndata], by=bucketSize)
maxBuckets = unlist(tmp[2])
bucketMag = unlist(tmp[1])
L = floor(maxBuckets/4)*2
x = bucketMag - mean(bucketMag)
outputVecCount = 10

# STEP 3: MCSSA analysis
s←decompSSA(x, L, toeplitz = FALSE, getFreq = TRUE)
x.rc1 <- reconSSA(s, x, list(5:6)) # the signal
signalFreq = XYZspectrum_udf(unlist(x.rc1), drawPlot=TRUE,
"signalspectrum",
smoothing="ar")
x.rc2 <- reconSSA(s, x, list(1:4)) # trend
plot(x,type="l")
lines(x.rc1,col="red",lwd=2)
points(x.rc2,col="blue")
y = MCSSA(s, x, n=1000, conf = 0.9, keepSurr = FALSE, ar.method="mle")
plot(y, by = "freq", normalize = FALSE, asFreq = TRUE,
lam.pch = 1, lam.col = "black", lam.cex = 1, sig.col = "black",
sig.pch = 19, sig.cex = 1, conf.col = "darkgray", log = "xy",
ann = TRUE, legend = TRUE, axes = TRUE)
```

The Southern Solar-type, Totally Eclipsing Binary PY Aquarii

Ronald G. Samec
Faculty Research Associate, Pisgah Astronomical Research Institute, 1 PARI Drive, Rosman, NC 28772; ronaldsamec@gmail.com

Heather A. Chamberlain
Pisgah Astronomical Research Institute, 1 PARI Drive, Rosman, NC 28772

Walter Van Hamme
Department of Physics, Florida International University, Miami, FL 33199

Received December 12, 2018; revised January 21, 2019; accepted January 24, 2019

Abstract We obtained new *BVRI* (Bessell) observations of the solar-type eclipsing binary PY Aqr in 2017 with the 0.6-m SARA South reflector located at Cerro Tololo Inter-American Observatory. A simultaneous Wilson-Devinney solution of the new 2017 light curves, the 2003 discovery curve, and the 2001–2009 ASAS light curve reveals a system configuration with a modest degree of over-contact (fill-out ≈ 18%) and total eclipses (duration ≈ 23 minutes). The photometrically determined mass ratio is ≈ 0.32. The temperature difference between the components is ≈ 130K, indicating two stars in reasonably good thermal contact. Light curve asymmetries are modeled with a cool spot region on the primary, more massive star. Spanning a 16-year time base, the light curves indicate a 0.049 ± 0.005 s/yr steady increase of the orbit period. This dp/dt is not unusual as compared to the unpublished poster paper at the 2018 IAU GA study of over 200 solar type binaries. Two methods were used in conducting the period study, the p and dp/dt parameters in the Wilson program and a Wilson program means of generating eclipse timings from discovery and patrol based observations.

1. Introduction

Studies of contact binaries have led to very exciting results. This was recently highlighted by the discovery of Red Novae, characterized by a violent event which appears to be the final coalescence of the components of an over-contact binary into a fast rotating, blue straggler-like single star. The recovery of archived observations of a contact binary with high fill-out at the site of the red nova V1309 Sco (Tylenda *et al.* 2011; Tylenda and Kamiński 2016) has underscored the need for the characterization and continued patrol of such binaries in transition. The color of these objects distinguishes them from the usually blue, high temperature novae and supernovae. Archival data indicate that other similar events have happened in the past, with V838 Mon (Bond *et al.* 2003) and M31-RV (Boschi and Munari 2004) as examples.

Other interesting results have been determined in the past years. For instance, many contact binaries are found to be a part of triple and multiple star systems. Chambliss noticed this fact (1992). This may give insight into their origins. Kinematics, and the high abundance of contact binaries, gives hints about their old age (Guinan and Bradstreet 1988). Oscillations of all amplitudes are common and are not only attributed to their orbits in triple star systems, but their magnetic cycles (Han *et al.* 2019). Also, continuous positive or negative period changes about near contact configurations may be due to Thermal Relaxations Oscillations (TRO; Lucy 1976; Flannery 1976; Robertson and Eggleton 1977). The TRO model explains that the binary configuration undergoes periodic oscillations between semidetached and contact configurations about a state of marginal contact. In the broken contact phase, the mass ratio (q) increases and the period decreases. In the contact phase q decreases and period increases. All of these results point to the importance of observations of contact and near contact systems. PY Aquarii is another of these interesting eclipsing binaries whose photometric study is summarized in the next few paragraphs.

PY Aqr (GSC 05191-00853, 2MASS J20535602−0632016, V = 12.7–13.3 mag) was discovered in 2003 by observers C. Demeautis, D. Matter, and V. Cotrez (Demeautis *et al.* 2005). The system is listed as Object No. 77 in "Reports of New Discoveries No. 17" (Olah and Jurcsik 2005), which gives an EW type and contains links to a finding chart, a figure of the light curve, and a light curve data file with ephemeris:

$$\mathrm{HJD(min)} = 2452877.558\mathrm{d} + 0.40210 \times \mathrm{E}. \quad (1)$$

The variable was also observed by the All Sky Automated Survey (Pojmański 2002) and is listed as ASAS J205356−0632.1 in the ASAS-3 data file. The binary received the name PY Aqr in the "80th Name List of Variable Stars" (Kazarovets *et al.* 2013). The new GAIA DR2 (Riello *et al.* 2018) results give a distance of 605 ±14 pc.

PY Aqr is a solar-type contact binary and since it is moderately bright and totally eclipsing, it is easily monitored with small telescopes. A preliminary study of PY Aqr was presented at the AAS meeting #231 (Chamberlain *et al.* 2018). A more complete photometric study and period analysis are presented here.

2. New photometry and data reduction

Light curves in *B*, *V*, *R*, and *I* were obtained with the 0.6-m SARA South reflector at Cerro Tololo Inter-American Observatory in remote mode on 17 July, 17 August, 23 September, and 17 October, 2017. The telescope was equipped

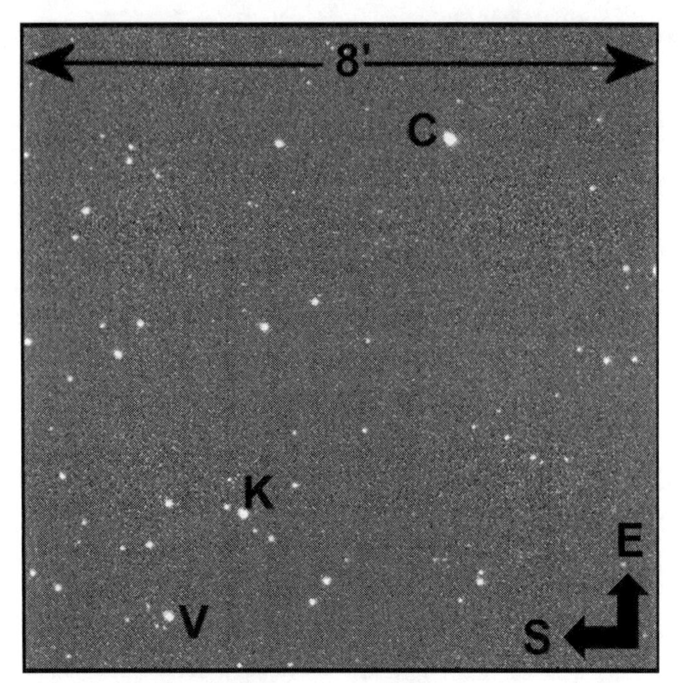

Figure 1. PY Aqr (V), comparison star C (2MASS J2054027−0630586), and check star K (2MASS J205356024−0632016).

with a thermoelectrically cooled (−38° C) 1 K × 1 K pixel FLI camera and Bessell *BVRI* filters. We obtained 111 individual observations in *B*, 136 in *V*, 131 in *R*, and 128 in *I*. The standard error of a single observation was 10 mmag in *B*, *R*, and *I*, and 12 mmag in *V*. The finding chart, given here for future observers, is shown in Figure 1. Characteristics of the variable, comparison, and check star are listed in Table 1.

The C–K magnitude differences remained constant throughout the observing run to better than 1%. Exposure times varied from 200–250 s in *B*, 100–140 s in *V*, and 30–75 s in *R* and *I*. Nightly images were calibrated with 25 bias frames, at least five flat frames in each filter, and ten 350-second dark frames. The light curve data are listed in Table 2. Light curve amplitudes and the differences in magnitudes at various quadratures are given in Table 3. Curve-dependent σs used in the Wilson program are given in Table 4. The new curves are of good precision, about 1% photometric precision.

The amplitude of the light curve varies from 0.61 to 0.53 magnitude in *B* to *I*. The O'Connell effect, an indicator of spot activity, averages several times the noise level, 0.02–0.04 mag, indicating magnetic activity. The differences in minima are small, 0.02–0.04 mag, indicating over-contact light curves in good thermal contact. A time of constant light appears to occur at minima and lasts some 23 minutes as measured by the light curve solution about phase 0.5.

3. Light curve solution

The new *B*, *V*, R, and I light curves were pre-modeled with BINARY MAKER 3.0 (Bradsteet and Steelman 2002), with each curve fitted separately. Each yielded an over-contact binary configuration. Averaged parameters were then used as starting values for a solution by the method of differential corrections (DC) using the Wilson-Devinney (WD) binary star program (Wilson and Devinney 1971; Wilson 1979) and revised several times, most recently as described in Wilson and Van Hamme (2014). To increase the time baseline and improve the determination of ephemeris parameters, including a period rate of change, the new multiband light curves were combined with the 2003 discovery light curve in Olah and Jurcsik (2005) and the ASAS light curve in the ASAS-3 database.

The 2MASS catalog lists a color *J–K* color of 0.384 ± 0.033, which is consistent with a primary component of solar spectral type (Houdashelt, Bell, and Sweigart 2000; Cox 2000). Accordingly, a surface temperature of 5750 K was adopted for the primary component mean surface temperature. Limb darkening coefficients were interpolated locally in terms of surface temperature and gravity in the tables of Van Hamme (1993) for a logarithmic law. The detailed reflection effect treatment with one reflection (Wilson 1990) was selected. The solution was run in mode 3 (over-contact) with convective values for the gravity brightening parameters ($g1 = g2 = 0.32$) and albedos ($A1 = A2 = 0.5$).

Essential information on light curve weighting is in Wilson (1979), including a discussion of level-dependent, curve-dependent, and individual data point weights. Level-dependent weights for the light curves were generated within the DC program assuming photon counting statistics. For individual data point weights, only weight ratios matter among the points of a given data subset. Accordingly, the scaling factor for individual weights can be set arbitrarily. Here, individual light curve points were given unit weights. Curve-dependent weights (Table 3) were based on fixed σs computed by the DC program.

Solution parameters are listed in Table 5. Solution 1 includes a period rate of change as one of the adjusted parameters, whereas Solution 2 does not include a dP/dt. Note that the orbit semi-major axis (a) is not a solution parameter since no radial velocities for the system exist. The Table 5 value of a is the adopted value that produces a primary mass close to that of the Sun. Light and Solution 1 curves vs. orbit phase are shown in Figures 2 and 3, and, for selected nights, vs. time in Figures 4 to 7. Figure 8 is the V plot with Solution 1 less the dark spot to show the effects of it. The spot affects the curve from phase 0.6 to phase 0.1.

Table 1. Photometric targets.

Role	*Label*	*Name*	*V*	*J–K*
Variable	V	PY Aqr	12.72–13.37	0.384 ± 0.033
Comparison	C	TYC 5191-971-1	11.51	0.313 ± 0.035
Check	K	2MASS J20540271−0630486	12.73	0.632 ± 0.033

Note: The C−K magnitude differences remained constant throughout the observing run to better than 1%. Exposure times varied: 200–250 s in B, 100–140 s in V, and 30–75 s in R and I. Nightly images were calibrated with 25 bias frames, at least five flat frames in each filter, and ten 350-second dark frames. The light curve data are listed in Table 2.

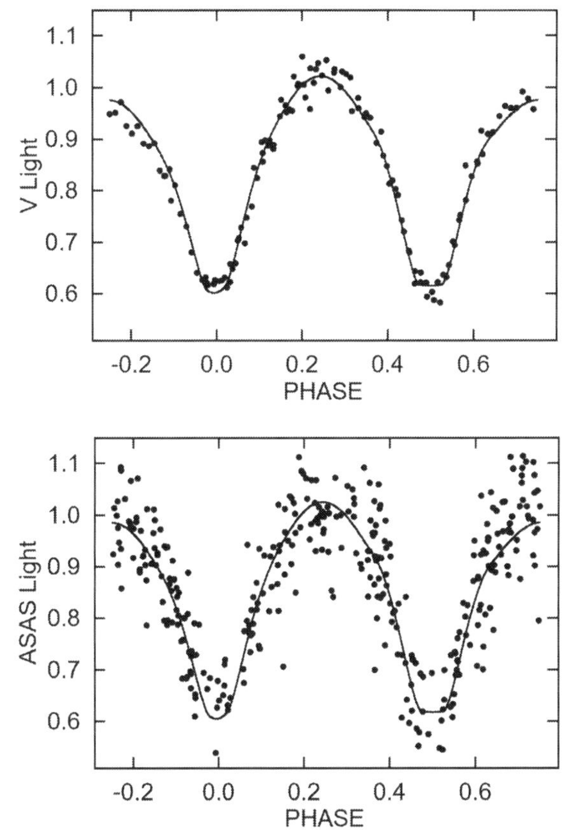

Figure 2. The PY Aqr 2003 (upper panel) and ASAS (lower panel) observed and Solution 1 computed light curves phased with the orbit period. Light units are normalized flux.

Light curve asymmetries were modeled with spots as described in Wilson (2012). The WD program allows for different spot configurations at different epochs, and this feature was exploited here. Times of onset and end of each of the intervals of spot growth, maximum, and decay are included in Table 5.

An eclipse duration of ~ 22 minutes was determined for the secondary eclipse (phase 0.5) from the light curve solution. The fill-out is ~ 18%, indicating a modest degree of over-contact. Fill-out is defined as:

$$\text{fill-out} = \frac{\Omega_1 - \Omega_{ph}}{\Omega_1 - \Omega_2} \qquad (2)$$

where Ω_1 is the inner critical potential where the Roche Lobe surfaces reach contact at L_1, and Ω_2 is the outer critical potential where the surface reaches L_2.

The more massive component has a lower temperature, characteristic of a W-type (smaller component is hotter) W UMa system. This conclusion is not very firm, however. We will have to await spectroscopic observations and the determination of radial velocities before the W-type nature of the binary can be confirmed. Although photometric mass ratios for totally eclipsing over-contact binaries are reliable (see e.g. Terrell and Wilson 2005), spot effects in PY Aqr are not fully modeled, as indicated by night-to-night variations in light curve shapes (Figures 4 to 7). Radial velocities will be needed to pin down the mass ratio and determine absolute dimensions.

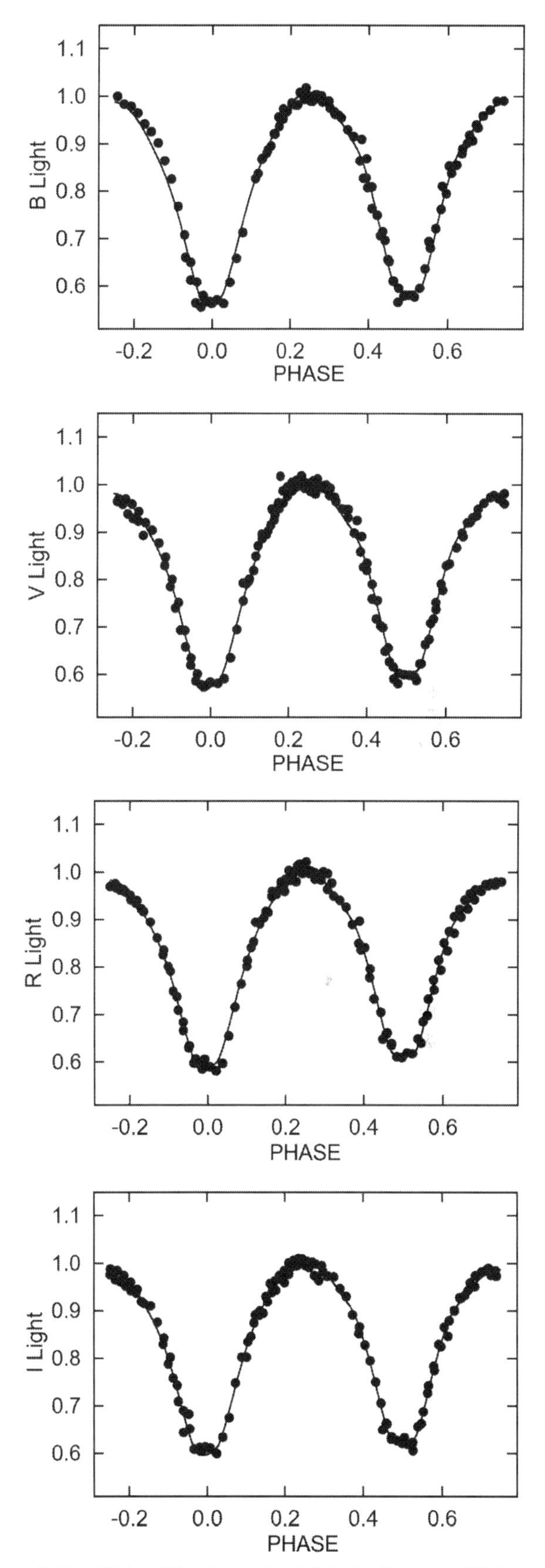

Figure 3. The PY Aqr 2017 observed and Solution 1 computed light curves phased with the orbit period. Light units are normalized flux.

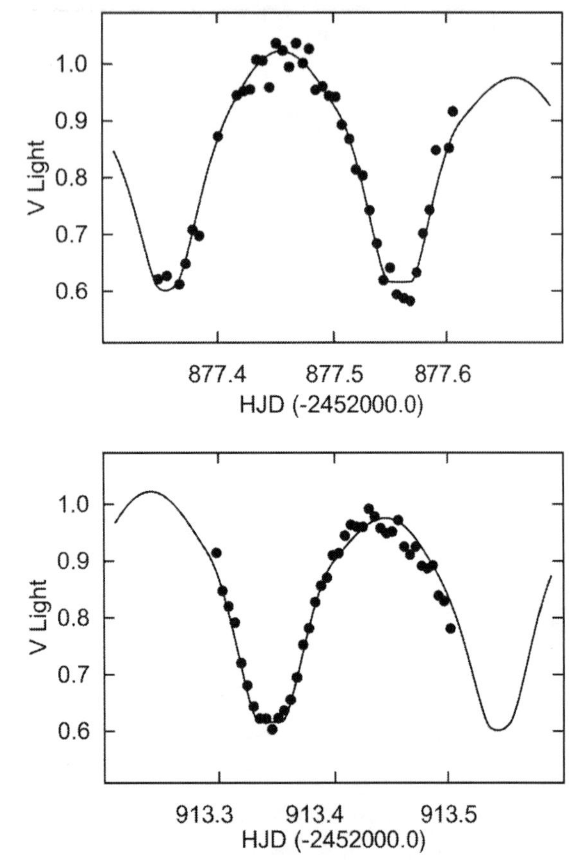

Figure 4. The PY Aqr 2003 August 25 (upper panel) and September 30 (lower panel) observed and Solution 1 computed light curves. Light units are normalized flux.

4. Orbit period and ephemerides

Solving the light curves with time (and not phase) as the independent variable allows adjustment of ephemeris parameters, which are then determined from whole light curves and not just timing minima (see e.g. Van Hamme and Wilson 2007). Solutions 2 and 1, respectively, yield linear and quadratic ephemerides:

$$\text{HJD(min)} = 2455460.00294 \pm 0.00034 + 0.402093519 \pm 0.000000051 \times E, \quad (3)$$

and

$$\text{HJD(min)} = 2455459.9909 \pm 0.0013 + 0.402093472 \pm 0.000000048 \times E + 3.10 \pm 0.32 \times 10^{-10} \times E^2, \quad (4)$$

with the coefficient of the E^2 term derived from the Solution 1 dP/dt value. The photometric data span an interval of about 16 years and show an orbital period that is increasing at a rate dP/dt $= +1.54 \pm 0.16 \times 10^{-9}$. Formally, the quadratic term is significant, having a value of 9 times its standard deviation. However, we should be cautious and not over-interpret this result. Information on dP/dt comes predominantly from six eclipses, four which occur in 2003 at the beginning of the 16-year span and two in 2017 at the end. Because of the low frequency of ASAS observations (one data point every one or two days), that light curve contains no eclipses with full phase coverage. However, there are a fair number of ingress and egress points in the ASAS

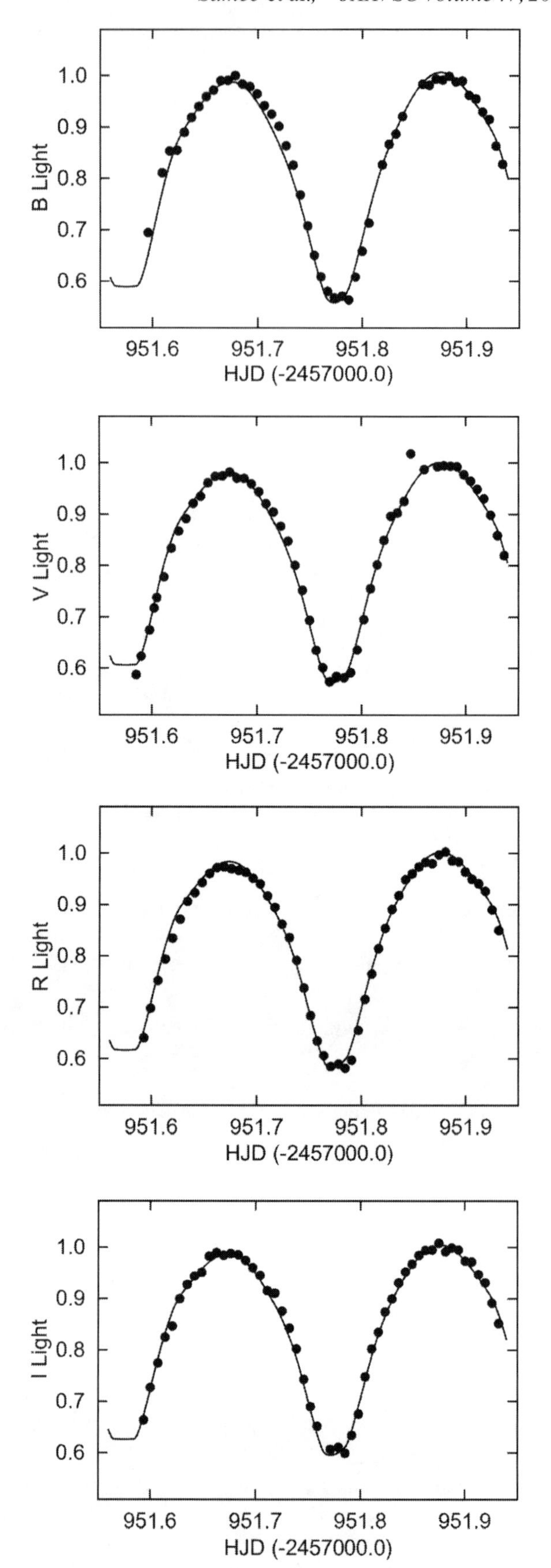

Figure 5. The PY Aqr July 17, 2017 observed and Solution 1 computed light curves. Light units are normalized flux.

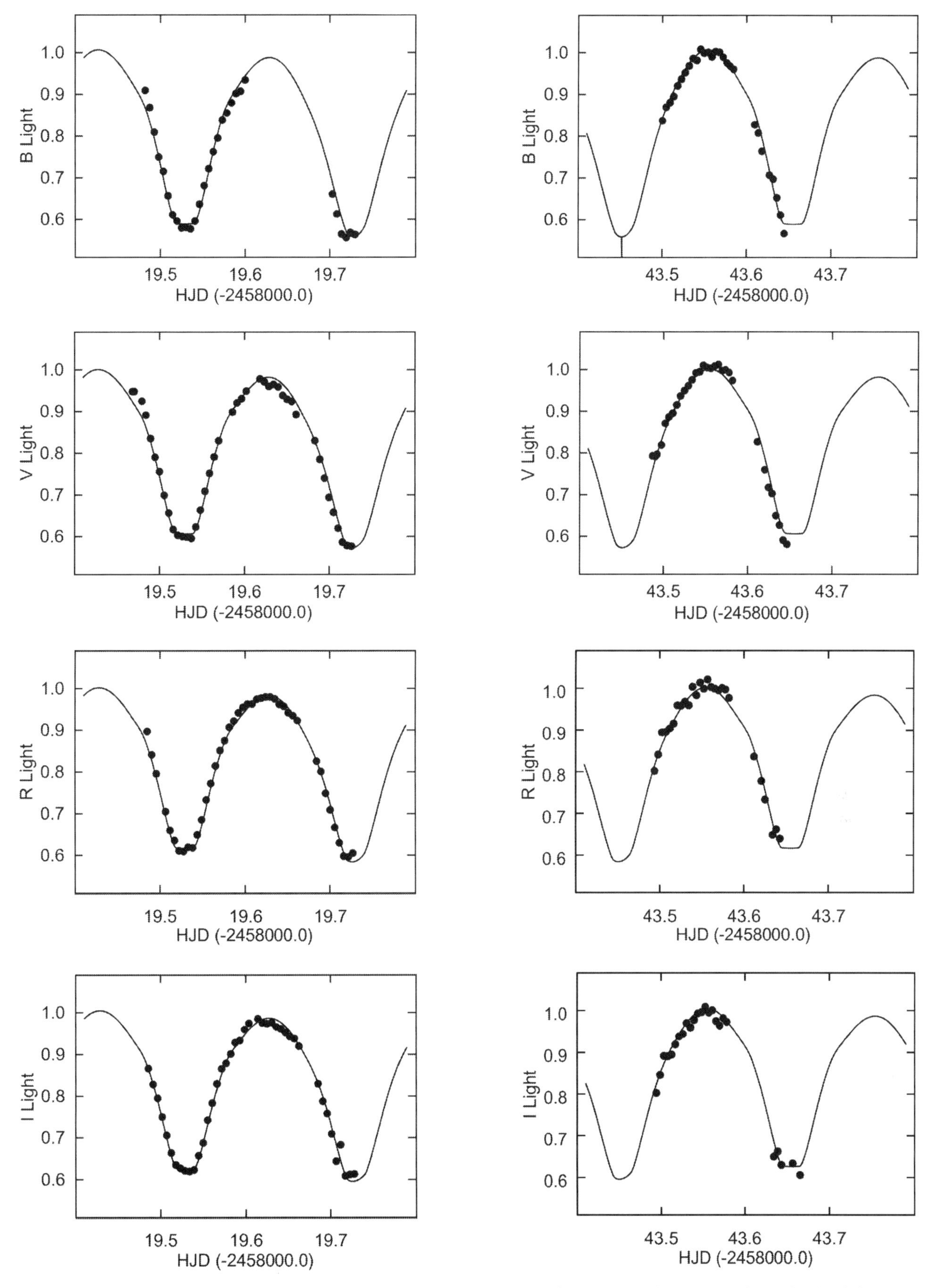

Figure 6. The PY Aqr September 23, 2017 observed and Solution 1 computed light curves.

Figure 7. The PY Aqr October 17, 2017 observed and Solution 1 computed light curves. Light units are normalized flux.

curve which help determine the value of dP/dt. Unfortunately, the lack of observational data between 2009 and 2017 represents a significant gap in the 16-year time base of Equation 3, and the significance of the quadratic term is solely due to the new 2017 light curve data. Futures light curves or times of mid-eclipse are needed to confirm the dP/dt derived here.

For the purpose of future period studies, we extracted individual eclipse timings from the various data sets in the following manner. For eclipses with full phase coverage (four in the 2003 light curve and two in the 2017 curves), we used the DC program to fit single-night sections of curves that contain the eclipse, selecting an initial zero-epoch value near mid-eclipse, and then adjusting the zero-epoch T_0 and luminosity L_1 (to set the light level) only, keeping all other parameters fixed at global solution values. The final zero-epoch time marks a time of conjunction of the two stars that night, and hence, a time of mid-eclipse. For the ASAS light curve with its sparse phase coverage, full eclipses are not available. However, we can identify light curve sections spanning about 200 to 300 days that have at least a few points in or near an eclipse, select an initial zero-epoch near those points, and apply DC to determine the time of conjunction closest to those points from the entire light curve section. As expected, such eclipse timings will have larger errors, but properly weighted they will be useful in future period analyses.

We obtained a total of 15 eclipse timings. They are listed in Table 6 together with their standard errors and eclipse type (primary or secondary). Least-squares fits (weighted, with relative weights inversely proportional to the standard errors squared) yield ephemerides:

$$\mathrm{HJD(min)} = 2455460.00291 \pm 0.00078 + 0.40209363 \pm 0.000000012 \times E, \quad (5)$$

and

$$\mathrm{HJD(min)} = 2455459.9903 \pm 0.0028 + 0.402093617 \pm 0.000000078 \times E + 3.27 \pm 0.71 \times 10^{-10} \times E^2, \quad (6)$$

which corresponds to a dP/dt of $1.63 \pm 0.35 \times 10^{-9}$, in excellent agreement with the dP/dt in Table 4 determined from the whole light curves. Figure 9 shows timing residuals with respect to Eqn. 4 and a fitted quadratic curve. Clearly, the significance of the quadratic term is solely due to the two 2017 eclipse times. Table 5 minima can be combined with future eclipse timings to monitor the period behavior of the system.

5. Discussion

PY Aqr is an over-contact W UMa in possibly a W-type configuration ($T_2 > T_1$). The system has a mass ratio of ~0.34, and a component temperature difference of only ~40 K. One cool region of spots (Tfact ~0.64, ~33-degree radius) was iterated on the primary component in the WD Synthetic Light Curve computations for the new photometry. This temperature is quite normal for average spot temperatures on the Sun (T ~4660). It appears in the Southern hemisphere (colatitude 138 degrees). The Roche Lobe fill-out of the binary is only ~10% with an inclination of ~82°, high enough for total eclipses. Its spectral type indicates a surface temperature of ~5750 K for the primary component, making it a solar-type binary. Such a main sequence star would have a mass of ~0.92 $M_\odot$ and the secondary (from the mass ratio) would have a mass of 0.32 $M_\odot$, making it very much undersized. The W-type phenomena may to be due to saturation of magnetic phenomena on the primary component, suppressing its temperature. The secondary component, which is probably near that of the actual temperature of the primary, has a temperature of ~5800 K.

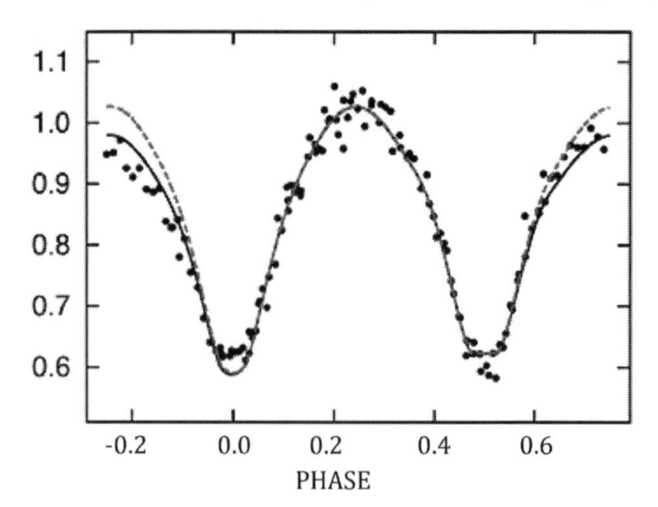

Figure 8. The PY Aqr 2017 V-observed and Solution 1 (see Figure 3) computed light curve phased with the orbit period and the synthetic curve less the dark spot (red dashed line). Light units are normalized flux.

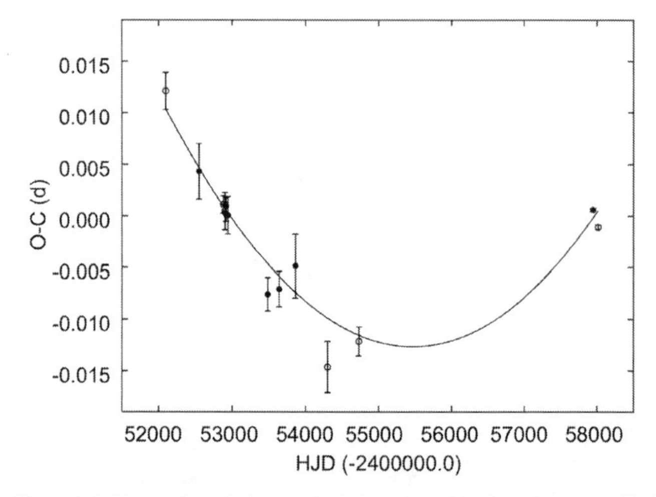

Figure 9. PY Aqr eclipse timing residuals (Equation 4) and quadratic fit. Filled and open circles indicate primary and secondary timings, respectively.

6. Conclusions

The steady period increase does not support the idea of a red nova precursor status for PY Aqr. Such a status would be characterized by a decreasing period at an increasingly rapid rate, shrinking the orbit and leading to a Darwin instability and merger of the two stars (see e.g. Tylenda *et al.* 2011). The phenomenon of long-term increase in the orbital period can be explained by the mass transfer from the less massive component to the more massive component (Qian 2001a, 2001b), which agrees with the TRO theory. The positive quadratic period increase would indicate a mass exchange rate of

$$\frac{dM}{dt} = \frac{\dot{P} M_1 M_2}{3P (M_1 - M_2)} \sim \frac{1.8 \times 10^{-7} M_\odot}{d}. \quad (7)$$

with the primary component being the gainer. However, the period change might point to another possibility. The period increase might be a part of a sinusoidal oscillation, meaning that there is a third body orbiting the system. Alternately, if magnetic braking is also acting (which is likely), the fill-out will be moderated and the components may not separate. A steadily decreasing mass ratio would ultimately lead to an unstable condition and the possible coalescence of the binary. This all points to the need of further efforts to monitor the system for times of mid-eclipse to determine the nature of the orbital evolution. Otherwise, the stars are in fair thermal contact. Obtaining radial velocities will be the next step towards determining astrophysically relevant parameters of the PY Aqr system.

7. Acknowledgements

We thank the Southeastern Association for Research in Astronomy for time on the CTIO SARA telescope. We made use of the SIMBAD database, operated at the CDS, Strasbourg, France. This work has made use of data from the European Space Agency (ESA) mission Gaia (https://www.cosmos.esa.int/gaia), processed by the Gaia Data Processing and Analysis Consortium (DPAC; https://www.cosmos.esa.int/web/gaia/dpac/consortium). Funding for the DPAC has been provided by national institutions, in particular the institutions participating in the Gaia Multilateral Agreement.

References

Bond, H. E., *et al.* 2003, *Nature*, **422**, 405.

Boschi, F., and Munari. U. 2004, *Astron. Astrophys.*, **418**, 869.

Bradstreet, D. H., and Steelman, D. P. 2002, *Bull. Amer. Astron. Soc.*, **34**, 1224.

Chamberlain, H., Samec, R. G., Caton, D. B., and Van Hamme, W. 2018, Amer. Astron. Soc. Meeting #231, 434.03.

Chambliss, Carlson R. 1992, *Publ. Astron. Soc. Pacific*, **104**, 663.

Cox, A. N. 2000, *Allen's Astrophysical Quantities*, 4th ed. AIP Press/Springer, New York.

Demeautis, C., Matter, D., Cotrez, V., Behrend, R., and Rinner, C. 2005, *Inf. Bull. Var. Stars*, No. 5600, 15.

Flannery, B. P. 1976, *Astrophys. J.*, **205**, 217.

Guinan, E. F., and Bradstreet, D. H. 1988, in *Formation and Evolution of Low Mass Stars*, eds. A. K. Dupree, and M. T. V. T. Lago, NATO Adv. Sci. Inst. Ser. C, Vol. 241, Kluwer, Dordrecht, 345.

Han, Q.-w., Li, L.-f., Kong, X.-y., Li, J.-s., and Jiang, D.-k. 2019, *New Astron.*, **66**, 14.

Houdashelt, M. L., Bell, R. A., and Sweigart, A. V. 2000, *Astron. J.*, **119**, 1448.

Kazarovets, E. V., Samus, N. N., Durlevich, O. V., Kireeva, N. N., and Pastukhova, E. N. 2013, *Inf. Bull. Var. Stars*, No. 6052, 1.

Lucy L. B., 1976, *Astrophys. J.*, **205**, 208.

Olah, K., and Jurcsik, J., eds. 2005, *Inf. Bull. Var. Stars*, No. 5600, 1.

Pojmański, G. 2002, *Acta Astron.*, **52**, 397.

Qian, S. 2001a, *Mon. Not. Roy. Astron. Soc.*, **328**, 635.

Qian, S. 2001b, *Mon. Not. Roy. Astron. Soc.*, **328**, 914.

Robertson J. A., and Eggleton P. P., 1977, *Mon. Not. Roy. Astron. Soc.*, 179, 359.

Riello, M., *et al.* 2018, Gaia Data Release 2: Processing of the photometric data, *Astron. Astrophys.*, **616**, id.A4 (arXiv:1804.09368).

Terrell, D., and Wilson, R. E. 2005, *Astrophys. Space Sci.*, **296**, 221.

Tylenda, R., and Kamiński, T. 2016, *Astron. Astrophys.*, **592A**, 134.

Tylenda, R., *et al.* 2011, *Astron. Astrophys.*, **528A**, 114.

Van Hamme, W. 1993, *Astron. J.*, **106**, 2096.

Van Hamme, W., and Wilson, R. E. 2007, *Astrophys. J.*, **661**, 1129.

Wilson, R. E. 1979, *Astrophys. J.*, **234**, 1054.

Wilson, R. E. 1990, *Astrophys. J.*, **356**, 613.

Wilson, R. E. 2012, *Astron. J.*, **144**, 73.

Wilson, R. E., and Devinney, E. J. 1971, *Astrophys. J.*, **166**, 605.

Wilson, R. E., and Van Hamme, W. 2014, *Astrophys. J.*, **780**, 151.

Table 2. PY Aqr: new *BVRI* photometry (variable minus TYC 5191-971-1, the comparison star).

HJD 2457000+	ΔB	*HJD* 2457000+	ΔB	*HJD* 2457000+	ΔB	*HJD* 2457000+	ΔB	*HJD* 2457000+	ΔB
51.5957	1.836	51.7605	1.979	51.9280	1.599	119.5626	1.735	143.5400	1.460
51.6093	1.668	51.7669	2.031	51.9344	1.645	119.5679	1.689	143.5444	1.431
51.6162	1.612	51.7733	2.055	82.8047	1.488	119.5733	1.632	143.5488	1.441
51.6232	1.610	51.7808	2.049	82.8096	1.469	119.5786	1.610	143.5533	1.439
51.6301	1.567	51.7872	2.063	82.8340	1.421	119.5840	1.579	143.5577	1.451
51.6370	1.532	51.7936	1.980	82.8389	1.451	119.5893	1.552	143.5621	1.437
51.6439	1.507	51.8000	1.893	82.8438	1.437	119.5947	1.546	143.5665	1.439
51.6508	1.485	51.8064	1.807	82.8487	1.447	119.6000	1.514	143.5710	1.452
51.6578	1.471	51.8192	1.647	119.4821	1.543	119.7031	1.890	143.5754	1.467
51.6647	1.451	51.8256	1.595	119.4874	1.593	119.7084	1.972	143.5798	1.476
51.6716	1.450	51.8320	1.570	119.4928	1.670	119.7138	2.061	143.5842	1.484
51.6785	1.440	51.8384	1.529	119.4981	1.753	119.7192	2.078	143.6097	1.646
51.6854	1.458	51.8576	1.458	119.5036	1.804	119.7245	2.054	143.6141	1.672
51.6924	1.463	51.8640	1.460	119.5090	1.897	119.7299	2.063	143.6185	1.733
51.6993	1.479	51.8704	1.447	119.5144	1.975	143.5002	1.633	143.6273	1.817
51.7062	1.505	51.8768	1.449	119.5197	2.002	143.5046	1.592	143.6318	1.832
51.7131	1.524	51.8832	1.441	119.5251	2.033	143.5091	1.578	143.6362	1.904
51.7201	1.552	51.8896	1.453	119.5304	2.030	143.5135	1.560	143.6406	1.975
51.7270	1.599	51.8960	1.451	119.5358	2.037	143.5179	1.530	143.6450	2.058
51.7339	1.648	51.9024	1.482	119.5411	2.002	143.5223	1.511		
51.7408	1.727	51.9088	1.490	119.5465	1.930	143.5267	1.493		
51.7477	1.815	51.9152	1.519	119.5518	1.857	143.5311	1.475		
51.7541	1.907	51.9216	1.536	119.5572	1.794	143.5356	1.455		

HJD 2457000+	ΔV	*HJD* 2457000+	ΔV	*HJD* 2457000+	ΔV	*HJD* 2457000+	ΔV	*HJD* 2457000+	ΔV
51.5853	1.898	51.7692	1.924	82.8160	1.325	119.5912	1.410	143.5153	1.416
51.5899	1.833	51.7757	1.904	82.8208	1.315	119.5966	1.398	143.5197	1.391
51.5979	1.748	51.7831	1.909	82.8257	1.310	119.6020	1.377	143.5241	1.376
51.6023	1.681	51.7895	1.891	82.8306	1.300	119.6180	1.344	143.5285	1.363
51.6048	1.651	51.7959	1.812	82.8355	1.319	119.6234	1.352	143.5329	1.347
51.6118	1.593	51.8023	1.715	82.8404	1.335	119.6287	1.364	143.5374	1.329
51.6187	1.518	51.8087	1.625	82.8453	1.341	119.6341	1.359	143.5418	1.326
51.6257	1.475	51.8151	1.561	119.4677	1.378	119.6395	1.366	143.5462	1.309
51.6326	1.445	51.8215	1.498	119.4700	1.378	119.6449	1.389	143.5506	1.315
51.6395	1.409	51.8279	1.439	119.4791	1.405	119.6502	1.400	143.5551	1.317
51.6464	1.393	51.8343	1.431	119.4839	1.445	119.6556	1.406	143.5595	1.311
51.6534	1.363	51.8407	1.404	119.4893	1.516	119.6609	1.443	143.5639	1.307
51.6603	1.349	51.8471	1.301	119.4947	1.576	119.6830	1.523	143.5683	1.323
51.6672	1.348	51.8599	1.334	119.5000	1.624	119.6889	1.583	143.5728	1.321
51.6741	1.340	51.8727	1.328	119.5055	1.709	119.6943	1.647	143.5772	1.328
51.6810	1.353	51.8791	1.326	119.5109	1.777	119.6996	1.717	143.5816	1.349
51.6880	1.353	51.8855	1.327	119.5162	1.845	119.7050	1.774	143.6115	1.527
51.6949	1.365	51.8919	1.328	119.5216	1.869	119.7104	1.840	143.6203	1.619
51.7018	1.383	51.8983	1.345	119.5270	1.875	119.7157	1.900	143.6247	1.681
51.7087	1.410	51.9047	1.359	119.5323	1.877	119.7211	1.915	143.6291	1.703
51.7156	1.429	51.9111	1.377	119.5377	1.883	119.7264	1.918	143.6336	1.788
51.7226	1.463	51.9175	1.398	119.5430	1.834	143.4868	1.572	143.6380	1.827
51.7295	1.500	51.9239	1.436	119.5484	1.765	143.4904	1.574	143.6424	1.892
51.7364	1.562	51.9303	1.486	119.5538	1.694	143.4924	1.567	143.6468	1.910
51.7433	1.630	51.9367	1.536	119.5591	1.630	143.4976	1.537		
51.7500	1.718	82.7997	1.377	119.5645	1.575	143.5020	1.470		
51.7564	1.813	82.8062	1.362	119.5698	1.523	143.5064	1.451		
51.7628	1.873	82.8111	1.334	119.5859	1.436	143.5108	1.440		

Table continued on next page

Table 2. PY Aqr: new *BVRI* photometry (ariable minus TYC 5191-971-1, the comparison star), cont.

HJD 2457000+	ΔR_c	*HJD* 2457000+	ΔR_c	*HJD* 2457000+	ΔR_c	*HJD* 2457000+	ΔR_c	*HJD* 2457000+	ΔR_c
51.5929	1.744	51.7780	1.834	82.8131	1.283	119.5922	1.325	143.5117	1.370
51.5993	1.650	51.7844	1.849	82.8180	1.276	119.5976	1.310	143.5161	1.356
51.6063	1.569	51.7908	1.820	82.8229	1.273	119.6030	1.301	143.5205	1.305
51.6132	1.511	51.7972	1.719	82.8278	1.259	119.6083	1.301	143.5249	1.306
51.6201	1.457	51.8036	1.623	82.8326	1.242	119.6137	1.288	143.5293	1.295
51.6271	1.410	51.8100	1.551	82.8375	1.250	119.6190	1.285	143.5338	1.305
51.6340	1.367	51.8164	1.483	82.8424	1.263	119.6244	1.282	143.5382	1.256
51.6409	1.347	51.8228	1.432	82.8473	1.264	119.6297	1.282	143.5426	1.278
51.6478	1.324	51.8292	1.386	119.4849	1.378	119.6352	1.287	143.5470	1.245
51.6547	1.304	51.8356	1.353	119.4903	1.448	119.6405	1.302	143.5515	1.261
51.6617	1.291	51.8420	1.317	119.4957	1.508	119.6459	1.307	143.5559	1.237
51.6686	1.289	51.8484	1.304	119.5066	1.639	119.6512	1.325	143.5603	1.257
51.6755	1.293	51.8548	1.289	119.5119	1.712	119.6566	1.333	143.5647	1.261
51.6824	1.297	51.8612	1.279	119.5173	1.752	119.6619	1.347	143.5692	1.265
51.6893	1.301	51.8676	1.282	119.5226	1.795	119.6846	1.468	143.5736	1.259
51.6963	1.314	51.8740	1.263	119.5280	1.798	119.6899	1.501	143.5780	1.263
51.7032	1.327	51.8804	1.257	119.5333	1.780	119.6953	1.574	143.5824	1.285
51.7101	1.354	51.8868	1.276	119.5387	1.783	119.7006	1.633	143.6123	1.454
51.7170	1.381	51.8932	1.278	119.5441	1.729	119.7060	1.700	143.6211	1.533
51.7239	1.422	51.8996	1.300	119.5494	1.670	119.7114	1.762	143.6255	1.597
51.7309	1.455	51.9060	1.316	119.5548	1.597	119.7167	1.819	143.6344	1.730
51.7378	1.514	51.9124	1.326	119.5601	1.540	119.7221	1.822	143.6388	1.708
51.7449	1.590	51.9188	1.343	119.5655	1.483	119.7274	1.805	143.6432	1.746
51.7514	1.672	51.9252	1.387	119.5708	1.435	143.4940	1.500		
51.7577	1.754	51.9316	1.437	119.5762	1.405	143.4984	1.448		
51.7641	1.804	82.8033	1.317	119.5815	1.366	143.5028	1.381		
51.7706	1.842	82.8082	1.312	119.5869	1.348	143.5072	1.380		

HJD 2457000+	ΔI_c	*HJD* 2457000+	ΔI_c	*HJD* 2457000+	ΔI_c	*HJD* 2457000+	ΔI_c	*HJD* 2457000+	ΔI_c
51.5938	1.666	51.7789	1.755	82.8073	1.261	119.5769	1.360	143.4988	1.402
51.6002	1.566	51.7853	1.776	82.8122	1.248	119.5823	1.333	143.5032	1.345
51.6072	1.497	51.7917	1.715	82.8171	1.240	119.5876	1.301	143.5076	1.346
51.6141	1.429	51.7981	1.646	82.8220	1.219	119.5930	1.295	143.5121	1.341
51.6211	1.401	51.8045	1.535	82.8268	1.213	119.5984	1.265	143.5165	1.312
51.6280	1.335	51.8109	1.459	82.8317	1.209	119.6037	1.248	143.5209	1.289
51.6349	1.302	51.8173	1.416	82.8366	1.220	119.6144	1.236	143.5253	1.283
51.6418	1.283	51.8237	1.366	82.8415	1.216	119.6198	1.247	143.5297	1.253
51.6487	1.274	51.8301	1.335	82.8464	1.217	119.6251	1.249	143.5342	1.265
51.6557	1.239	51.8365	1.298	119.4856	1.376	119.6305	1.246	143.5386	1.245
51.6626	1.231	51.8429	1.273	119.4911	1.425	119.6359	1.258	143.5430	1.227
51.6695	1.237	51.8493	1.256	119.4964	1.469	119.6413	1.263	143.5475	1.224
51.6764	1.233	51.8557	1.237	119.5019	1.532	119.6466	1.273	143.5519	1.210
51.6834	1.236	51.8621	1.226	119.5073	1.598	119.6520	1.284	143.5563	1.225
51.6903	1.248	51.8685	1.225	119.5127	1.665	119.6573	1.290	143.5607	1.219
51.6972	1.264	51.8749	1.211	119.5180	1.713	119.6627	1.311	143.5651	1.248
51.7041	1.281	51.8813	1.229	119.5234	1.727	119.6853	1.423	143.5696	1.260
51.7110	1.316	51.8877	1.221	119.5287	1.737	119.6907	1.479	143.5740	1.240
51.7180	1.322	51.8941	1.225	119.5341	1.740	119.6960	1.520	143.5784	1.250
51.7249	1.364	51.9005	1.249	119.5395	1.734	119.7014	1.593	143.6348	1.688
51.7318	1.406	51.9069	1.251	119.5448	1.676	119.7067	1.698	143.6392	1.668
51.7387	1.459	51.9133	1.279	119.5502	1.626	119.7121	1.634	143.6436	1.722
51.7458	1.542	51.9197	1.298	119.5555	1.543	119.7175	1.759	143.6569	1.716
51.7522	1.623	51.9261	1.345	119.5609	1.485	119.7228	1.753	143.6657	1.765
51.7586	1.685	51.9325	1.394	119.5662	1.423	119.7282	1.751		
51.7714	1.763	82.8024	1.267	119.5716	1.377	143.4944	1.459		

Table 3. Averaged light curve characteristics of PY Aqr.

Filter	Phase	Magnitude Max. I	Phase	Magnitude Max. II
	0.25		0.75	
ΔB		1.441 ± 0.009		1.477 ± 0.025
ΔV		1.319 ± 0.011		1.355 ± 0.009
ΔR		1.253 ± 0.010		1.286 ± 0.005
ΔI		1.220 ± 0.007		1.241 ± 0.008
Filter	*Phase*	*Magnitude Min. II*	*Phase*	*Magnitude Max. I*
	0.50		0.00	
ΔB		2.033 ± 0.004		2.050 ± 0.012
ΔV		1.878 ± 0.004		1.914 ± 0.008
ΔR		1.789 ± 0.009		1.830 ± 0.017
ΔI		1.730 ± 0.011		1.748 ± 0.009
Filter		*Min. I – Max. I*		*Min. I – Min. II*
ΔB		0.609 ± 0.021		0.017 ± 0.015
ΔV		0.595 ± 0.019		0.036 ± 0.012
ΔR		0.577 ± 0.027		0.041 ± 0.026
ΔI		0.528 ± 0.016		0.018 ± 0.020
Filter		*Max. II – Max. I*		*Min. II– Max. I*
ΔB		0.036 ± 0.034		0.592 ± 0.013
ΔV		0.036 ± 0.020		0.559 ± 0.015
ΔR		0.033 ± 0.014		0.536 ± 0.018
ΔI		0.021 ± 0.015		0.510 ± 0.018

Table 4. Curve-dependent σs and data ranges.

Curve	Band	σ^a	Range (HJD)
2003	V	0.0225	2452877.3–2452913.5
2017	B	0.0144	2457951.5–2458043.7
—	V	0.0160	—
—	R	0.0115	—
—	I	0.0105	—
ASAS	V	0.0683	2452025.8–2455144.6

Note a: In units of light at phase $0_p.25$.

Table 5. PY Aqr light curve solutions.

Parameter	Solution 1	Solution 2
a^a ($R_\odot$)	2.52	2.52
i (deg)	83.57 ± 0.40	83.36 ± 0.44
T_1^b (K)	5750	5750
T_2 (K)	5883 ± 16	5873 ± 17
Ω_1	2.483 ± 0.011	2.481 ± 0.011
Ω_2	2.48296	2.48138
Fill-out[c]	0.1870	0.1693
M_2/M_1	0.3249 ± 0.0045	0.3224 ± 0.0051
T_0 (HJD − 2455460.0)	−0.0091 ± 0.0013	0.00294 ± 0.00034
P0 (d)	0.402093472 ± 0.000000048	0.402093519 ± 0.000000051
dP/dt	+1.54 ± 0.16 × 10−9	—
$L_1/(L_1+L_2)V$	0.7107 ± 0.0037	0.7134 ± 0.0036
$L_1/(L_1+L_2)B$	0.7028 ± 0.0033	0.7058 ± 0.0035
$L_1/(L_1+L_2)V$	0.7107 ± 0.0028	0.7134 ± 0.0029
$L_1/(L_1+L_2)R$	0.7143 ± 0.0026	0.7171 ± 0.0027
$L_1/(L_1+L_2)I$	0.7170 ± 0.0026	0.7196 ± 0.0027
$L_1/(L_1+L_2)V$	0.7107 ± 0.0046	0.7134 ± 0.0048
χ^2	1.24	1.36
2003 Spot		
Co-latitude (deg)	109 ± 56	108 ± 114
Longitude (deg)	70.6 ± 8.5	73 ± 11
Radius (deg)	16 ± 24	16 ± 36
$T_{spot}/T_{surface}$	0.64 ± 2.4	0.70 ± 1.97
Time of Onset (HJD)	2451000	
Start of Maximum (HJD)	2452050	
End of Maximum (HJD)	2452950	
Time of Disappearance (HJD)	2457000	
2017 Spot		
Co-latitude (deg)	155.7 ± 5.9	154.7 ± 4.9
Longitude (deg)	21.5 ± 2.2	22.0 ± 2.3
Radius (deg)	36.0 ± 3.7	35.9 ± 3.4
$T_{spot}/T_{surface}$	0.693 ± 0.074	0.707 ± 0.065
Time of Onset (HJD)	2457000	
Start of Maximum (HJD)	2457950	
End of Maximum (HJD)	2458050	
Time of Disappearance (HJD)	2458500	
Auxiliary Parameters		
r_1(pole)	0.4549 ± 0.0015	0.4570 ± 0.0014
r_1(side)	0.4897 ± 0.0020	0.4922 ± 0.0018
r_1(back)	0.5185 ± 0.0022	0.5205 ± 0.0019
$\langle r_1\rangle$d	0.4914 ± 0.0020	0.4911 ± 0.0017
r_2(pole)	0.2831 ± 0.0053	0.2763 ± 0.0056
r_2(side)	0.2969 ± 0.0066	0.2890 ± 0.0068
r_2(back)	0.3411 ± 0.0130	0.3290 ± 0.0129
$\langle r_2\rangle$d	0.2982 ± 0.0017	0.2965 ± 0.0020

Notes: Band-specific parameters are listed in the order of Table3. a: Adopted to produce a primary star of mass ≈ $1M_\odot$. bBased on the color of the system. c: Defined as $(\Omega_{1,c}-\Omega_1)/(\Omega_{1,c}-\Omega_{2,c})$, with $\Omega_{1,c}$ and $\Omega_{2,c}$ the critical potentials at the L_1 and L_2 Lagrangian points, respectively. d: Radius of an equal-volume sphere.

Table 6. PY Aqr eclipse timings.

Timing (HJD)	*Error (d)*	*Type*	*Weight*[a]	*Source*[b]
2452094.6924	0.0018	2	0.309	ASAS
2452545.2305	0.0027	1	0.137	ASAS
2452877.55768	0.00083	2	1.45	IBVS
2452898.8680	0.0018	2	0.309	ASAS
2452908.31772	0.00083	1	1.45	IBVS
2452912.33873	0.00074	1	1.83	IBVS
2452913.34309	0.00073	2	1.88	IBVS
2452940.0822	0.0018	1	0.309	ASAS
2453478.8800	0.0016	1	0.391	ASAS
2453636.0991	0.0017	1	0.346	ASAS
2453860.0675	0.0031	1	0.104	ASAS
2454300.1492	0.0025	2	0.160	ASAS
2454729.5877	0.0014	2	0.510	ASAS
2457951.77773	0.00015	1	44.4	This paper
2458019.52884	0.00022	2	20.7	This paper

Notes: a. Relative weights, inversely proportional to the standard errors. b. Origin of light curves from which timings were extracted. IBVS refers to the 2003 light curve available from IBVS 5600 (Olah and Jurcsik 2005); ASAS (Pojmański 2002).

Photometric Analysis of Two Contact Binary Systems: USNO-A2.0 1200-16843637 and V1094 Cassiopeiae

Surjit S. Wadhwa
Astronomical Society of New South Wales, Sydney, NSW, Australia; surjitwadhwa@gmail.com

Received January 20, 2019; revised February 1, 2019; accepted February 4, 2019

Abstract Ground-based photometry of two contact binary systems—USNO-A2.0 1200-16843637 and V1094 Cas—was analyzed using the Wilson-Devinney method. Both systems were found to be A-Type, with the smaller star being significantly cooler. Both systems show complete eclipses with good physical contact and almost identical mass ratio of approximately 0.23.

1. Introduction

The W Ursae Majoris (W UMa) group of short-period contact eclipsing binaries are important test beds for theories of stellar evolution. Numerous new contact systems have been discovered recently through automated sky survey programs and dedicated observing efforts using small telescopes. Quite a large percentage of the new discoveries remain largely un-analyzed even though data are of sufficient quality to yield at least basic physical information. In previous papers published in this journal I have demonstrated how analysis of survey or small telescope observations of contact binary stars for which little other information is available can yield a satisfactory photometric solution (Wadhwa 2004, 2017). In this paper I present photometric solutions for two such neglected contact systems.

USNO-A2.0 1200-16843637 (R.A. 21^h 01^m 53.0^s., Dec. +34° 25' 02" (2000)) was recognized as a contact binary system by Kryachko *et al.* (2010) with a magnitude range of 0.46 and period of 0.316 day. Approximately 130 observations (available from the website of the cited journal) extending over most of the cycle were available for analysis. Data available in the SIMBAD database would suggest a B–R of 0.7 magnitude corresponding to an effective temperature of 6500 K (Popper 1980).

V1094 Cas (R.A. 01^h 20^m 23.0^s., Dec. +59° 17' 15.7" (2000)) was reported as a contact binary variable by Hambálek (2008) and detailed photometry obtained by Virnina *et al.* (2012; all photometric data are available on the website of the cited journal). They reported a magnitude range of 0.47 and period of 0.514 day. The photometric data are extensive, extending over several years with several different telescopes. To limit the extent of possible errors only the R-band photometry carried out in 2009 was used in the current analysis as this was carried out over a relatively short period and included data over most of the phase cycle. Even so, nearly 1,800 data points were available, and these were binned along the entire phase cycle to yield a more manageable 180-point normalized observed light curve. The normalized curve was used in the analysis. Data available in the SIMBAD database would suggest a J–H of 0.21 magnitude, corresponding to an effective temperature of 6250 K (Popper 1980; Yoshida 2010).

2. Light curve analysis

Light curve analysis was carried out using the Wilson-Devinney code as included in the Windows-based software supplied by Bob Nelson through the Variable Star South website (Nelson 2009). In each case the available data indicated a probable convective envelope, therefore gravity brightening was set at 0.32 and bolometric albedos were set at 0.5. Black body approximation was used for the stars' emergent flux and simple reflection treatment was applied. VanHamme (1993) limb darkening coefficients were used as included in the Bob Nelson software package. The maximum magnitude of the stars is not well known, therefore the photometric data were normalized to the mean magnitude between phases 0.24 and 0.26 in each case. This methodology has previously been applied to the analysis of All Sky Automated Survey and ground-based amateur observations (Wadhwa 2004, 2005).

The mass ratio of a contact binary system is usually determined by radial velocity studies. The mass ratio is then used to determine other features of the system such as the inclination, degree of contact, and temperature variations. However, where radial velocity data are not available, under certain circumstances, such as when the system exhibits at least one total eclipse, the Wilson-Devinney (Terrell and Wilson 2005) method can be sucessfully employed, as the parameter space is well constrained by the presence of the total eclipse and the best fit solution is quickly obtained. As both the systems have a clear well-defined total eclipse, rather than using the tedious grid method to find a starting point for the final iterations, the light curve part of the software, along with direction from Anderson and Shu's (1979) theoretical atlas of contact binary light curves, a very good visual fit was quickly obtained through simple trial and error. For the final iterations, it is well known that the mass ratio as a free parameter when combined with inclination and the potential of the star can lead to strong correlation between the parameters. However, the problem can be solved using multiple subsets in sequence (Wilson and Biermann 1976). The free parameters were divided into two subsets as follows: $\{q, L_1\}$ and $\{i, T_2, \Omega\}$, and iterations were carried out until the error of a parameter was greater than the estimated correction.

3. Individual systems

3.1. USNO-A2.0 1200-16843637

As noted above, this system has an effective temperature of 6500 K (based on SIMBAD database). The visual estimation of the approximate mass ratio was 0.25, cooler smaller star with high fillout of 0.5 and high inclination exceeding 80°.

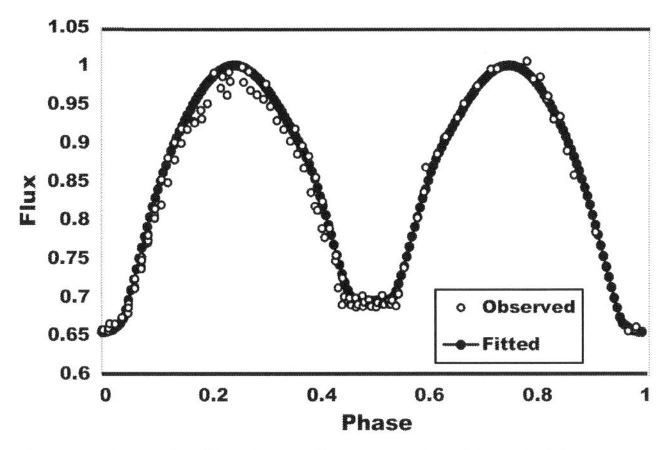

Figure 1. Observed and fitted curve for USNO-A2.0 1200-16843637.

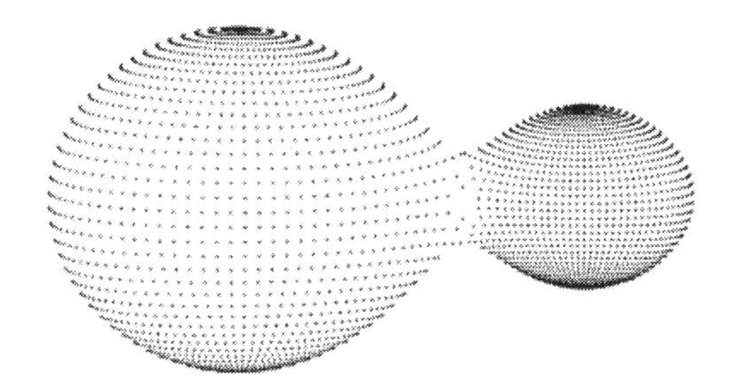

Figure 2. 3D representation of USNO-A2.0 1200-16843637.

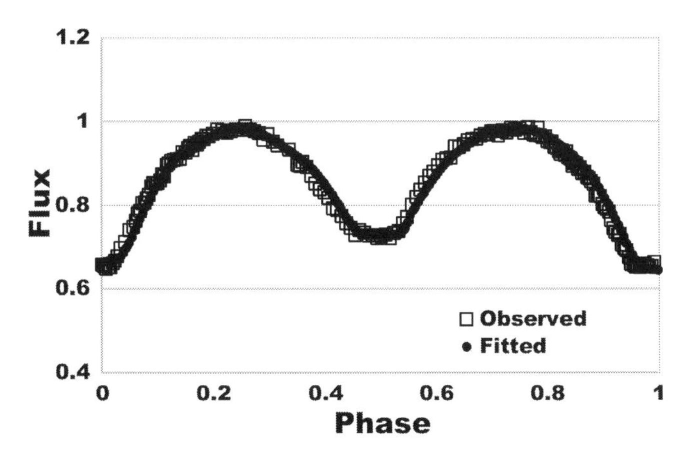

Figure 3. Observed (open squares) and Fitted (solid line) for V1094 Cas.

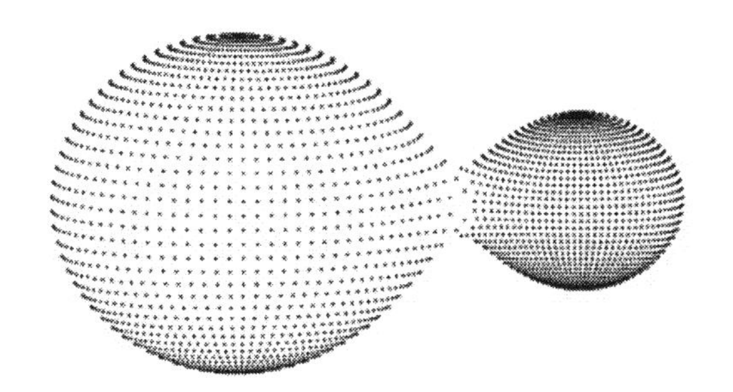

Figure 4. 3D representation of V1094 Cas.

Table 1. Basic photometric elements for USNO-A2.0 1200-1684363.

Parameters	*Value*
T_2	6243 K ± 55 K
Inclination (i)	83.7° ± 1.3°
Potential (Ω)	2.25 ± 0.02
Mass Ratio (q)	0.236 ± 0.003
Fillout	48%

Table 2. Basic photometric elements for V1094 Cas.

Parameters	*Value*
T_2	5568 K ± 51 K
Inclination (i)	80.6° ± 1.7°
Potential (Ω)	2.275 ± 0.02
Mass Ratio (q)	0.235 ± 0.007
Fillout	30%

With these starting parameters the differential correction part of the Wilson-Devinney code was carried out using multiple subsets alternatively, as described above. The results of the best fit are summarized in Table 1 and the curves and three-dimensional representation (Bradstreet 1993) are shown in Figures 1 and 2, respectively. It is clear the system is an A-Type W UMa star with a cooler smaller star in poor thermal contact. There is, however, good physical contact between the stars with a fillout ratio of 0.48.

3.2. V1094 Cas

As noted above, this system has an effective temperature of 6250 K (based on SIMBAD database). The visual estimation of the approximate mass ratio was 0.20, cooler smaller star with mid-range fillout of 0.25 and high inclination exceeding 75°. With these starting parameters the differential correction part of the Wilson-Devenny code was carried using multiple subsets alternatively, as described above. Even after the differential corrections had achieved the best fit the visual inspection suggested the entire curve was slightly out of phase. The differential correction was again run using the previously obtained best solution, but with the phase being the only correctable parameter. This resulted in a significant improvement in the fit. The phase correction required was –0.104. This is likely due to the select normalized curve being folded based on the ephemeris derived from the entire photometry set from multiple instruments over several years. The results of the best fit (after correcting the phase) are summarized in Table 2 and the curves and three-dimensional representation (Bradstreet 1993) are shown in Figures 3 and 4, respectively. It is clear the system is an A-Type W UMa star with a cooler smaller star in quite poor thermal contact. There is, however, reasonable physical contact between the stars with a fillout ratio of 0.30.

4. Conclusion

Photometric analysis using the Wilson Devinney code is presented for two almost identical systems, USNO-A2.0 1200-16843637 and V1094 Cas. Both systems are of A-Type and have a mass ratio of approximately 0.23, high inclination exceeding 80° with poor thermal contact but with good physical contact.

5. Acknowledgement

This research made use of the SIMBAD database, operated by the CDS at Strasbourg, France.

References

Anderson, L., and Shu, F. H. 1979, *Astrophys. J., Suppl. Ser.*, **40**, 667.

Bradstreet, D. H. 1993, BINARY MAKER 2.0 light curve synthesis program, Contact Software, Norristown, PA.

Hambálek, L. 2008, Open Eur. J. Var. Stars, 95, 42.

Kryachko, T., Samokhvalov A, Satovskiy B, and Denisenko, D. 2010, *Perem. Zvezdy Prilozh.*, **10**, No. 2.

Nelson, R. H. 2009, WDWINT56A: Astronomy Software by Bob Nelson (https://www.variablestarssouth.org/bob-nelson).

Popper, D. M. 1980, *Ann. Rev. Astron. Astrophys.*, **18**, 115.

Terrell, D., and Wilson, R. 2005, *Astrophys. Space Sci.*, **296**, 221.

van Hamme, W. 1993, *Astron. J.*, **106**, 2096.

Virnina, N. A., Kocian, R., Hambálek L., Dubovsky, P., Andronov, I. L., and Kudzej, I. 2012, *Open Eur. J. Var. Stars*, **146**, 1.

Wadhwa, S. S. 2004, *J. Amer. Assoc. Var. Star Obs.*, **32**, 95.

Wadhwa, S. S. 2005, *Astrophys. Space Sci.*, **300**, 289.

Wadhwa, S. S. 2017, *J. Amer. Assoc. Var. Star Obs.*, **45**, 11.

Wilson, R. E., and Biermann, P. 1976, *Astron. Astrophys.*, **48**, 349.

Yoshida, S. 2010, Home page (http://www.aerith.net/index.html; cited 28/12/18).

A Photometric Study of the Contact Binary V384 Serpentis

Edward J. Michaels
Stephen F. Austin State University, Department of Physics, Engineering and Astronomy, P.O. Box 13044, Nacogdoches, TX 75962; emichaels@sfasu.edu

Chlöe M. Lanning
Stephen F. Austin State University, Department of Physics, Engineering and Astronomy, P.O. Box 13044, Nacogdoches, TX 75962; chloelanning@gmail.com

Skyler N. Self
Stephen F. Austin State University, Department of Physics, Engineering and Astronomy, P.O. Box 13044, Nacogdoches, TX 75962; skyler.self.23@gmail.com

Received January 20, 2019; revised March 26, 2019; accepted April 3, 2019

Abstract In this paper we present the first photometric light curves in the Sloan g', r', and i' passbands for the contact binary V384 Ser. Photometric solutions were obtained using the Wilson-Devinney program which revealed the star to be a W-type system with a mass ratio of q = 2.65 and a f = 36% degree of contact. The less massive component was found to be about 395 K hotter than the more massive one. A hot spot was modeled on the cooler star to fit the asymmetries of the light curves. By combining our new times of minima with those found in the literature, the (O–C) curve revealed a downward parabolic variation and a small cyclic oscillation with an amplitude of 0.0037 day and a period 2.86 yr. The downward parabolic change corresponds to a long-term decrease in the orbital period at a rate of $dP/dt = -3.6 \times 10^{-8}$ days yr^{-1}. The cyclic change was analyzed for the light-travel time effect that results from the gravitational influence of a close stellar companion.

1. Introduction

V384 Ser (GSC 02035-00175) was identified as an eclipsing binary star by Akerlof *et al.* (2000) using data acquired by The Robotic Optical Transient Search Experiment I (ROTSE-I). An automated variable star classification technique using the Northern Sky Variability Survey (NSVS) classified this star as a W UMa contact binary (Hoffman *et al.* 2009). The machined-learned ASAS Classification Catalog gives the same classification (Richards 2012). Using ROTSE-I sky patrol data, Gettel *et al.* (2006) found an orbital period of 0.268739 day, a maximum visual magnitude of 11.853, and an amplitude of variation of 0.475 magnitude. The parallax measured by the Gaia spacecraft (DR2) gives a distance of d = 211 pc (Bailer-Jones *et al.* 2018). Data Release 4 from the Large Sky Area Multi-Object Fiber Spectroscopic Telescope survey (LAMOST) gives a spectral type of K2 (Luo *et al.* 2015). A ROSAT (Röntgen Satellite) survey of contact binary stars confirmed x-ray emission from V384 Ser (Geske *et al.* 2006). Using the Wide-Angle Search for Planets (SuperWASP) archive, Lohr *et al.* (2015) found evidence for a sinusoidal period change, which suggests a third body may be in the V384 Ser system.

Presented in this paper is the first photometric study of V384 Ser. The photometric observations and data reduction methods are presented in section 2, with new times of minima and a period study in section 3. Light curve analysis using the Wilson-Devinney model is presented in section 4. A discussion of the results is given in section 5 with conclusions in section 6.

2. Observations

Multi-band photometric observations were acquired with a robotic 0.36-m Ritchey-Chrétien telescope located at the Waffelow Creek Observatory (http://obs.ejmj.net/index.php). A SBIG-STXL camera equipped with a cooled (–30°) KAF-6303E CCD was used for image acquisition. Images were obtained in the Sloan g', r', and i' passbands on 4 nights in June 2017. These images comprise the first data set (DS1). A second set of data (DS2) was acquired on 13 nights in April and May 2018 which includes 1,555 images in the g' passband, 1361 in r', and 1935 in i'. For DS2 the exposure times were 40 s for the g' and i' passbands and 25 s for the r' passband. The observation's average S/N for V384 Ser in the g', r', and i' passbands was 264, 327, and 291, respectively. Bias, dark, and flat frames were taken each night. Image calibration and ensemble differential aperture photometry of the light images were performed using MIRA software (Mirametrics 2015). Both data sets were processed using the comparison stars shown on the AAVSO Variable Star Plotter (VSP) finder chart (Figure 1). The standard magnitudes of the comparison and check stars were taken from the AAVSO Photometric All-Sky Survey and are listed in Table 1 (APASS; Henden *et al.* 2015). The instrumental magnitudes of V384 Ser were converted to standard magnitudes using these comparison stars. The Heliocentric Julian Date of each observation was converted to orbital phase (φ) using the following epoch and orbital period: $T_o = 2458251.6910$ and P = 0.26872914 d. The folded light curves for DS2 are shown in Figure 2. All light curves in this paper were plotted from orbital phase –0.6 to 0.6 with negative phase defined as φ – 1. The check star magnitudes were plotted and inspected each night with no significant variability noted. The standard deviation of the check star magnitudes from DS2 (all nights) was 8 mmag for the g' passband, 5 mmag for r', and 6 mmag for i'. The check star magnitudes for each passband are plotted in the

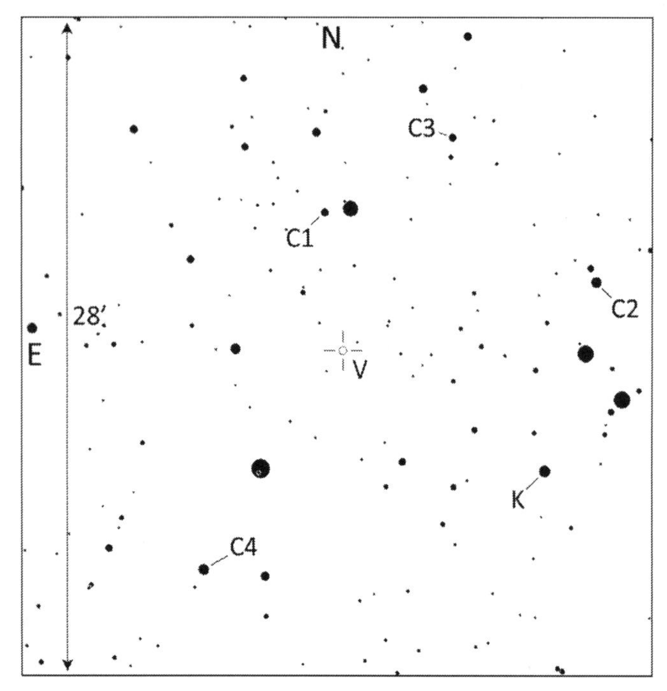

Figure 1. Finder chart for V384 Ser (V) showing the comparison (C1–C4) and check (K) stars.

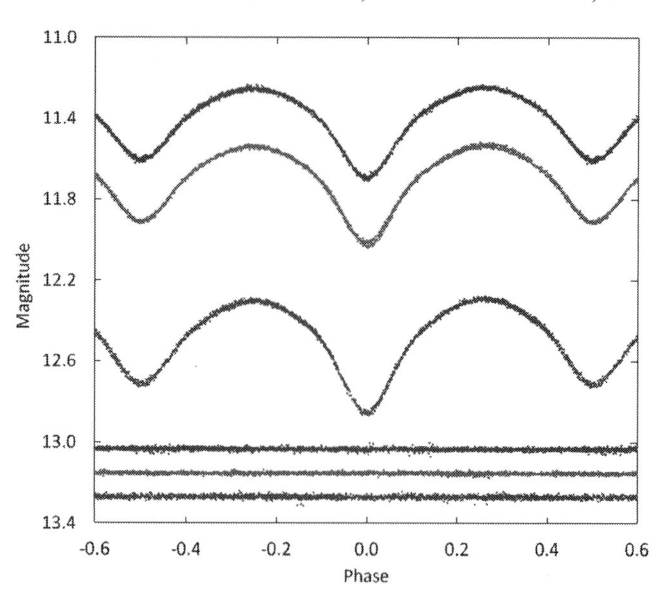

Figure 2. Folded light curves for each observed passband. The differential magnitudes of V384 Ser were converted to standard magnitudes using the calibrated magnitudes of the comparison stars. From top to bottom the light curve passbands are Sloan i', r' and g'. The bottom curves show the offset check star magnitudes in the same order as the light curves (offsets: i' = 1.95, r' = 1.68 and g' = 0.78). Error bars are not shown for clarity.

Table 1. Stars used in this study.

Star	*R.A. (2000)* h	*Dec. (2000)* °	*g'*	*r'*	*i'*
V384 Ser	16.03154	+24.87153			
[1]GSC 02035-00369 (C1)	16.03254	+24.97094	13.400	12.679	12.394
			±0.230	±0.074	±0.087
[1]GSC 02035-00374 (C2)	16.01763	+24.92069	12.452	11.693	11.491
			±0.165	±0.048	±0.085
[1]GSC 02038-00840 (C3)	16.02552	+25.02490	13.085	12.659	12.534
			±0.227	±0.086	±0.105
[1]GSC 02035-00337 (C4)	16.03920	+24.71401	12.067	11.313	10.975
			±0.056	±0.038	±0.056
[2]GSC 02035-00035 (K)	16.02046	+24.78484	12.719	11.469	11.090
			±0.195	±0.101	±0.095
Means of observed K star magnitudes			12.490	11.474	11.083
Standard deviation of observed K star magnitudes			±0.008	±0.005	±0.006

APASS (Henden et al. *2015)* [1]*comparison stars (C1–C4) and* [2]*check star (K) magnitudes.*

bottom panel of Figure 2. New times of minimum light were determined from both the 2017 and 2018 data sets. The 2018 observations can be accessed from the AAVSO International Database (Kafka 2017).

3. Period study

Orbital period changes are an important observational property as well as an important component for understanding contact binaries. The orbital period changes of V384 Ser have not been investigated since its discovery. To study this property, we located 120 CCD eclipse timings in the literature. The minima times are listed in Table 2 along with 22 new eclipse timings from the observations in this study. This data set spans more than 18 years. The first primary minimum in Table 2 and an orbital period taken from The International Variable Star Index (VSX) give the following linear ephemeris:

$$\text{HJD Min I} = 2451247.8121 + 0.268729 \times \text{E}. \quad (1)$$

This ephemeris was used to calculate the $(O–C)_1$ values in Table 2 with the corresponding $(O–C)_1$ diagram shown in the top panel of Figure 3 (black dots). A long-term decrease in the orbital period is apparent in the $(O–C)_1$ diagram (dashed line). In addition, a small amplitude cyclic variation is also clearly visible. We therefore combined a downward parabolic and a sinusoidal variation to describe the general trend of $(O–C)_1$ (solid line in Figure 3). By using the least-squares method, we derived

$$\begin{aligned}\text{HJD Min I} &= 2458251.6910(2) + 0.26872913(7) \times E \\ &-1.31(23) \times 10^{-11} \times E^2 \\ &+ 0.0037(2) \sin(0.00162(1) \times E + 5.55\ (11)). \quad (2)\end{aligned}$$

The bottom panel of Figure 3 shows the residuals from Equation 2.

The quadratic term in Equation 2 gives the rate for the secular decrease in the orbital period, dP/dt = –3.6(8) x 10^{-8} days yr^{-1}, or 0.31 second per century. Subtraction of this continuous downward decrease gives the $(O–C)_2$ values shown in the middle panel of Figure 3. It displays the small amplitude periodic oscillation that overlaid the secular period decrease. The results of this period study will be discussed further in section 5.

4. Analysis

4.1. Temperature, spectral type

The temperature and spectral type of V384 Ser can be

Table 2. Times of minima and (O–C) residuals from Equation 1.

Epoch HJD 2400000+	*Error*	*Cycle*	*$(O–C)_1$*	*References*
51247.8121	0.0001	0.0	0.00000	Blättler and Diethelm 2002
51287.7189	0.0007	148.5	0.00054	Blättler and Diethelm 2002
52019.8715	0.0001	2873.0	0.00098	Nelson 2002
52038.8169	0.0001	2943.5	0.00099	Nelson 2002
52359.4103	0.0011	4136.5	0.00069	Blättler and Diethelm 2002
52360.4871	0.0007	4140.5	0.00258	Blättler and Diethelm 2002
52360.6191	0.0011	4141.0	0.00021	Blättler and Diethelm 2002
52365.4569	0.0008	4159.0	0.00089	Blättler and Diethelm 2002
52365.5911	0.0005	4159.5	0.00072	Blättler and Diethelm 2002
52368.4142	0.0018	4170.0	0.00217	Blättler and Diethelm 2002
52368.5471	0.0003	4170.5	0.00071	Blättler and Diethelm 2002
52395.4223	0.0017	4270.5	0.00301	Blättler and Diethelm 2002
52395.5540	0.0003	4271.0	0.00034	Blättler and Diethelm 2002
52409.3972	0.0007	4322.5	0.00400	Blättler and Diethelm 2002
52409.5282	0.0006	4323.0	0.00063	Blättler and Diethelm 2002
52415.5762	0.0002	4345.5	0.00223	Blättler and Diethelm 2002
52763.4509	0.0002	5640.0	0.00724	Diethelm 2003
53216.3884	0.0004	7325.5	0.00201	Diethelm 2005
53541.4207	0.0008	8535.0	0.00659	Diethelm 2005
53917.5096	0.0009	9934.5	0.00925	Diethelm 2006
54197.3869	0.0009	10976.0	0.00530	Diethelm 2007
54516.6359	0.0005	12164.0	0.00424	Hübscher, *et al.* 2009a
54570.3803	0.0003	12364.0	0.00284	Hübscher, *et al.* 2009b
54583.4154	0.0003	12412.5	0.00459	Hübscher, *et al.*2009b
54583.5492	0.0003	12413.0	0.00402	Hübscher, *et al.* 2009b
54594.4335	0.0002	12453.5	0.00480	Hübscher, *et al.* 2009a
54594.5664	0.0002	12454.0	0.00333	Hübscher, *et al.* 2009a
54596.4472	0.0002	12461.0	0.00303	Hübscher, *et al.* 2009a
54596.5811	0.0005	12461.5	0.00257	Hübscher, *et al.* 2009a
54597.3894	0.0002	12464.5	0.00468	Hübscher, *et al.* 2009a
54597.5232	0.0002	12465.0	0.00412	Hübscher, *et al.* 2009a
54604.1058	—	12489.5	0.00285	Kazuo 2009
54610.4225	0.0002	12513.0	0.00442	Hübscher, *et al.* 2009a
54610.5568	0.0004	12513.5	0.00436	Hübscher, *et al.* 2009a
54636.4897	0.0002	12610.0	0.00491	Hübscher, *et al.* 2009a
54684.4597	0.0006	12788.5	0.00678	Diethelm 2009a
54703.4042	0.0002	12859.0	0.00589	Hübscher, *et al.* 2009a
54934.3748	0.0003	13718.5	0.00391	Hübscher, *et al.* 2010
54934.5081	0.0001	13719.0	0.00285	Hübscher, *et al.* 2010
54943.3768	0.0008	13752.0	0.00349	Hübscher, *et al.* 2010
54943.5111	0.0006	13752.5	0.00343	Hübscher, *et al.* 2010
54959.4998	0.0003	13812.0	0.00275	Hübscher, *et al.* 2010
54961.6506	0.0005	13820.0	0.00372	Diethelm 2009b
54961.7836	0.0001	13820.5	0.00236	Diethelm 2009b
54961.9198	0.0010	13821.0	0.00419	Diethelm 2009b
54996.4497	0.0003	13949.5	0.00241	Hübscher, *et al.* 2010
55029.3681	0.0003	14072.0	0.00151	Hübscher, *et al.* 2010
55029.3688	0.0004	14072.0	0.00221	Diethelm 2010a
55029.5003	0.0003	14072.5	–0.00065	Hübscher, *et al.* 2010
55038.3694	0.0006	14105.5	0.00039	Diethelm 2010a
55038.5057	0.0004	14106.0	0.00233	Diethelm 2010a
55049.3857	0.0005	14146.5	–0.00120	Hübscher, *et al.* 2011
55269.8770	0.0001	14967.0	–0.00204	Diethelm 2010b
55293.3921	0.0081	15054.5	–0.00073	Hübscher, *et al.* 2011
55293.5257	0.0002	15055.0	–0.00150	Hübscher, *et al.* 2011
55304.4085	0.0002	15095.5	–0.00222	Hübscher, *et al.* 2011
55304.5437	0.0003	15096.0	–0.00138	Hübscher, *et al.* 2011
55309.5149	0.0002	15114.5	–0.00167	Hübscher, *et al.* 2011
55376.4290	0.0005	15363.5	–0.00109	Hübscher, *et al.* 2011
55397.5233	0.0004	15442.0	–0.00202	Hübscher, *et al.* 2011
55629.5769	0.0016	16305.5	0.00409	Hübscher, *et al.* 2012
55653.8944	0.0001	16396.0	0.00162	Diethelm 2011
55662.4937	0.0003	16428.0	0.00159	Hübscher and Lehmann 2012
55689.5043	0.0004	16528.5	0.00492	Hübscher and Lehmann 2012
55754.4014	0.0002	16770.0	0.00397	Hübscher and Lehmann 2012
55754.5363	0.0005	16770.5	0.00451	Hübscher and Lehmann 2012
55775.3623	0.0009	16848.0	0.00401	Hübscher and Lehmann 2012
56008.4824	0.0003	17715.5	0.00170	Hübscher, *et al.* 2013
56008.6162	0.0001	17716.0	0.00114	Hübscher, *et al.* 2013
56035.8890	0.0030	17817.5	–0.00206	Diethelm 2012
56045.4316	0.0002	17853.0	0.00066	Hübscher, *et al.* 2013
56045.5651	0.0001	17853.5	–0.00020	Hübscher, *et al.* 2013
56065.4508	0.0002	17927.5	–0.00045	Hübscher, *et al.* 2013
56080.3651	0.0001	17983.0	–0.00061	Gürsoytrak *et al.* 2013
56080.4991	0.0006	17983.5	–0.00097	Gürsoytrak *et al.* 2013
56087.3527	0.0003	18009.0	0.00004	Terzioğlu, *et al.* 2017
56087.4848	0.0008	18009.5	–0.00223	Terzioğlu, *et al.* 2017
56094.4726	0.0003	18035.5	–0.00138	Hübscher, *et al.* 2013
56132.3628	0.0011	18176.5	–0.00197	Hübscher, *et al.* 2013
56132.4991	0.0004	18177.0	–0.00003	Hübscher, *et al.* 2013
56407.4080	0.0006	19200.0	–0.00090	Hübscher 2013
56407.5396	0.0002	19200.5	–0.00366	Hübscher 2013
56475.3965	0.0004	19453.0	–0.00084	Hübscher 2013
56475.5292	0.0003	19453.5	–0.00250	Hübscher 2013
56505.3579	0.0015	19564.5	–0.00272	Hübscher 2014
56505.4949	0.0005	19565.0	–0.00009	Hübscher 2014
56834.4254	0.0008	20789.0	0.00612	Hübscher and Lehmann 2014
56856.4598	0.0004	20871.0	0.00474	Hübscher and Lehmann 2014
56864.3876	0.0003	20900.5	0.00504	Hoňková, *et al.* 2015
56924.3124	0.0030	21123.5	0.00327	Hübscher 2015
57066.6013	0.0001	21653.0	0.00016	Jurysek, *et al.* 2017
57122.3618	0.0003	21860.5	–0.00060	Hübscher 2016
57122.4959	0.0002	21861.0	–0.00087	Hübscher 2016
57132.4396	0.0022	21898.0	–0.00014	Hübscher 2017
57132.5732	0.0026	21898.5	–0.00091	Hübscher 2017
57133.5137	0.0001	21902.0	–0.00096	Hübscher 2016
57134.4542	0.0002	21905.5	–0.00101	Hübscher 2016
57134.5884	0.0001	21906.0	–0.00117	Hübscher 2016
57153.3994	0.0003	21976.0	–0.00120	Hübscher 2016
57153.5338	0.0004	21976.5	–0.00117	Hübscher 2016
57158.3709	0.0034	21994.5	–0.00119	Hübscher 2016
57158.5038	0.0036	21995.0	–0.00266	Hübscher 2016
57225.6858	0.0002	22245.0	–0.00291	Samolyk 2016
57238.4509	0.0002	22292.5	–0.00243	Hübscher 2016
57241.4065	0.0002	22303.5	–0.00285	Hübscher 2016
57266.3980	0.0004	22396.5	–0.00315	Hübscher 2017
57499.3842	0.0002	23263.5	–0.00499	Hübscher 2017
57499.5191	0.0002	23264.0	–0.00446	Hübscher 2017
57508.3868	0.0001	23297.0	–0.00481	Hübscher 2017
57508.5205	0.0001	23297.5	–0.00548	Hübscher 2017
57513.7615	0.0001	23317.0	–0.00469	Nelson 2017
57514.4331	0.0001	23319.5	–0.00492	Hübscher 2017
57514.4351	0.0038	23319.5	–0.00292	Hübscher 2017
57514.5660	0.0018	23320.0	–0.00638	Hübscher 2017
57514.5677	0.0001	23320.0	–0.00468	Hübscher 2017
57515.3725	0.0010	23323.0	–0.00607	Hübscher 2017
57515.3740	0.0001	23323.0	–0.00457	Hübscher 2017
57515.5092	0.0017	23323.5	–0.00373	Hübscher 2017
57516.4489	0.0001	23327.0	–0.00458	Hübscher 2017
57516.5825	0.0005	23327.5	–0.00535	Hübscher 2017
57517.3889	0.0002	23330.5	–0.00513	Hübscher 2017
57517.5243	0.0003	23331.0	–0.00410	Hübscher 2017
57921.6990	0.0002	24835.0	0.00222	this paper
57921.8317	0.0002	24835.5	0.00049	this paper
57924.7878	0.0001	24846.5	0.00062	this paper
57932.7168	0.0001	24876.0	0.00211	this paper
57933.6566	0.0002	24879.5	0.00129	this paper
58224.8184	0.0001	25963.0	–0.00476	this paper
58225.7596	0.0001	25966.5	–0.00407	this paper
58225.8931	0.0001	25967.0	–0.00494	this paper
58231.8050	0.0001	25989.0	–0.00512	this paper
58244.7038	0.0001	26037.0	–0.00527	this paper
58245.6448	0.0001	26040.5	–0.00482	this paper
58245.7788	0.0001	26041.0	–0.00519	this paper
58246.7199	0.0001	26044.5	–0.00464	this paper
58247.7942	0.0001	26048.5	–0.00526	this paper
58248.7347	0.0001	26052.0	–0.00531	this paper
58248.8698	0.0001	26052.5	–0.00462	this paper
58249.6764	0.0001	26055.5	–0.00415	this paper
58249.8096	0.0001	26056.0	–0.00532	this paper
58250.7509	0.0001	26059.5	–0.00457	this paper
58250.8845	0.0001	26060.0	–0.00537	this paper
58251.6908	0.0001	26063.0	–0.00523	this paper
58257.7379	0.0001	26085.5	–0.00453	this paper

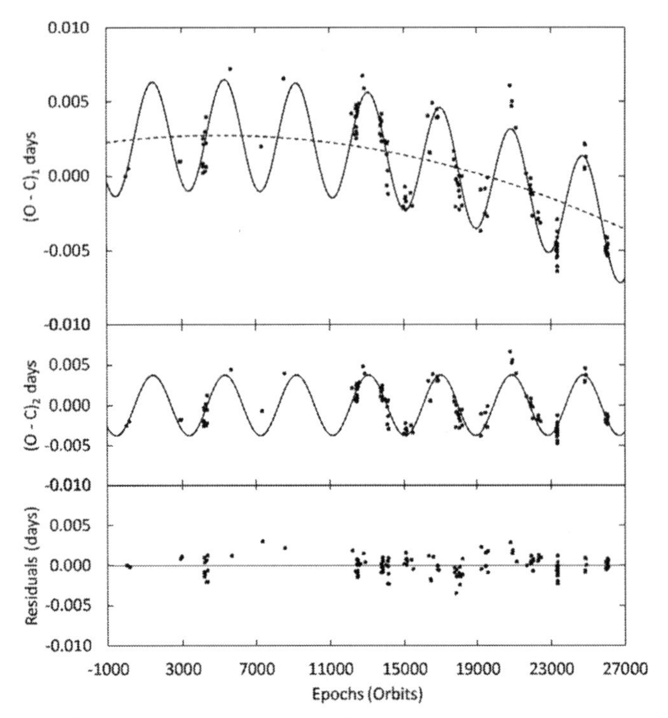

Figure 3. The top panel shows the $(O\text{–}C)_1$ diagram for all minimum times for V384 Ser. Black dots are residuals calculated from the linear ephemeris of Equation 1. The solid line corresponds to Equation 2 which is the combination of a long-term decrease and a small-amplitude cyclic variation. The dashed line refers to the quadratic term in this equation. In the middle panel the quadratic term of Equation 2 is subtracted to show the periodic variation more clearly. The bottom panel shows the residuals after removing the downward parabolic change and the cyclic variation.

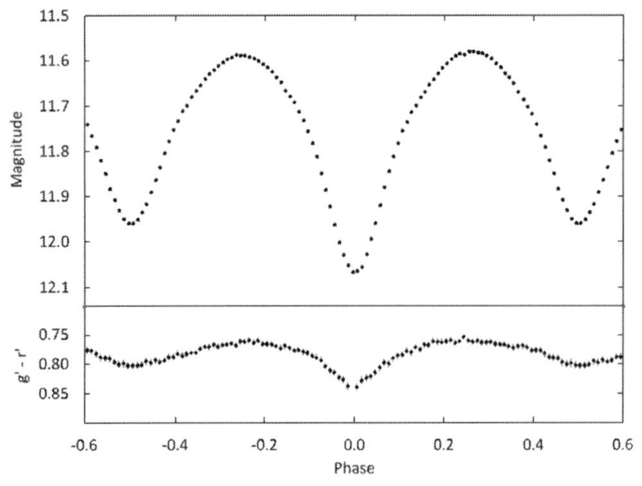

Figure 4. Light curve of all Sloan r'-band observations in standard magnitudes (top panel). The observations were binned with a phase width of 0.01. The errors for each binned point are about the size of the plotted points. The g'–r' colors were calculated by subtracting the linearly interpolated binned g' magnitudes from the linearly interpolated binned r' magnitudes.

measured from the star's color or its spectrum. The average (B–V) color index was determined from the DS2 observations. The phase and magnitude of the g' and r' observations were binned with a phase width of 0.01. The phases and magnitudes in each bin were averaged. The binned r' magnitudes were then subtracted from the linearly interpolated g' magnitudes. Figure 4 displays the binned r' magnitude light curve, with the bottom panel showing the (g'–r') color index. The average of the (g'–r')

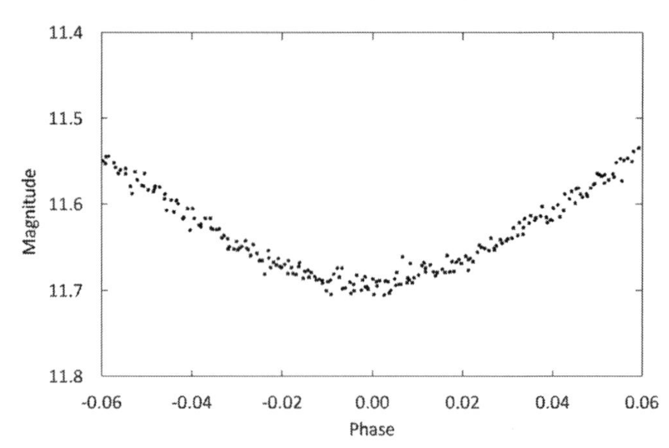

Figure 5. Observations of the primary eclipse portion of the Sloan r' light curve. Error bars are not shown for clarity.

values over the entire phase range gives a color index of (g'–r') = 0.782 ±0.004. The (B–V) color was found using the Bilir *et al.* (2005) transformation equation,

$$(B\text{–}V) = \frac{(g'\text{–}r') + 0.25187}{1.12431} \quad . \tag{3}$$

The average observed color of V384 Ser is (B–V) = 0.920 ± 0.003. This star's spectrum was acquired by the LAMOST telescope on April 19, 2014. The LAMOST DR4 catalog gives an effective temperature of $T_{eff} = 4976 \pm 18$ K and a spectral class of K2. Using this temperature, the color was interpolated from the tables of Pecaut and Mamajek (2013), $(B\text{–}V)_o = 0.924 \pm 0.009$. This value agrees well with the observed photometric color, indicating the color excess for this star is very small. This result is not surprising, given the proximity of V384 Ser and its location well above the galactic equator (galactic latitude +47.4°).

4.2. Synthetic light curve modeling

The DS2 observations were used in the light curve analysis. The light curves showed only slight asymmetries and a small O'Connell effect with Max I ($\varphi = 0.25$) brighter than Max II ($\varphi = 0.75$) by only 0.009 magnitude in the g′ passband. Figure 5 shows a closeup of primary minimum, clearly showing the eclipse is not total. To decrease the total number of points used in modeling and to improve precision in the light curve solution, the observations were binned in both phase and magnitude with a phase interval of 0.01. On average, each binned data point was formed by 16 observations in the g' band, 14 in the r' band, and 19 in the i' band. For light curve modeling the binned magnitudes were converted to relative flux.

The BINARY MAKER 3.0 (BM3; Bradstreet and Steelman 2002) program was used to make the initial fit to each observed light curve using standard convective parameters and limb darkening coefficients from Van Hamm's (1993) tabular values. An initial mass ratio of q = 2.81 was computed using the period-mass relation for contact binaries,

$$\log M_1 = (0.352 \pm 0.166) \log P - (0.262 \pm 0.067), \tag{4}$$
$$\log M_2 = (0.755 \pm 0.059) \log P + (0.416 \pm 0.024), \tag{5}$$

where M_1 is the mass of the less massive star and P is the orbital period in days (Gazeas and Stępień 2008). In the BM3 analysis it was necessary to add a third light to fit the minima of the synthetic light curves to the observed light curves. The parameters resulting from the initial fits to each light curve were averaged. These averages were used as the initial input parameters for the computation of simultaneous three-color light curve solutions using the 2015 version of the Wilson-Devinney program (WD; Wilson and Devinney 1971; Van Hamme and Wilson 1998). The contact configuration (Mode 3) was set in the program since the observed light curves are typical of a short-period contact binary (W-type). Each binned input data point was assigned a weight equal to the number of observations forming that point. The temperature for the star eclipsed at primary minima was fixed at $T_1 = 4976$ K. The other fixed inputs include standard convective parameters: gravity darkening coefficients $g_1 = g_2 = 0.32$ (Lucy 1968) and bolometric albedos $A_1 = A_2 = 0.5$ (Ruciński 1969). Linear limb darkening coefficients were calculated by the program. The adjustable parameters include the orbital inclination (i), mass ratio ($q = M_2/M_1$), dimensionless surface potential (Ω, $\Omega_1 = \Omega_2$), temperature of star 2 (T_2), the normalized flux for each wavelength (L), and third light (ℓ).

The mass ratio (q) for V384 Ser is not known since there are no photometric or spectroscopic solutions available. Symmetrical light curves and total eclipses are very useful in determining reliable photometric solutions (Wilson 1978; Terrell and Wilson 2005). Since total eclipses are not seen in the light curves, we decided a mass ratio search (q-search) should be the first step in the solution process. A series of WD solutions were completed, each using a fixed mass ratio that ranged from 2.3 to 3.0 by steps of 0.02. The plot of the relation between the ΣResiduals^2 and the q values is shown in Figure 6. The minimum residual value was located at q = 2.65. This value was used as the starting mass ratio for the final solution iterations where the mass ratio was an adjustable parameter. The final best-fit solution is shown in column 2 of Table 3. The adjusted parameters are shown with errors, with the subscripts 1 and 2 referring to the primary and secondary stars eclipsed at Min I and Min II, respectively. The filling-factor in Table 3 was computed using the method of Lucy and Wilson (1979) given by

$$f = \frac{\Omega_{inner} - \Omega}{\Omega_{inner} - \Omega_{outer}}, \quad (6)$$

where Ω_{inner} and Ω_{outer} are the inner and outer critical equipotential surfaces and Ω is the equipotential that describes the stellar surface. Figure 7 shows the normalized light curves for each passband overlaid by the synthetic solution curves (solid lines) with the residuals shown in the bottom panel.

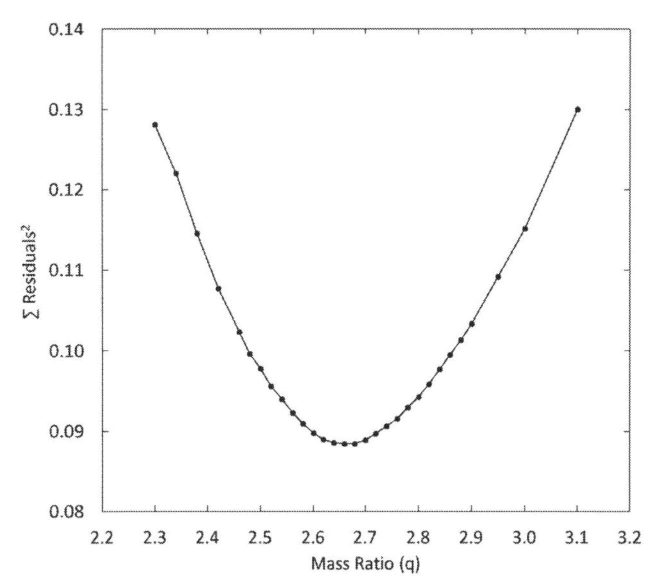

Figure 6. Results of the q-search showing the relation between the sum of the residuals squared and the mass ratio (q).

Table 3. Results derived from light curve modeling with spots.

Parameter	Solution 1 (no spot)	Solution 1 (spot)	Solution 2 (spot)
phase shift	−0.0009 ± 0.0002	0.0000 ± 0.0001	0.0000 ± 0.0001
filling factor	36%	36%	15%
i (°)	78.7 ± 1.1	78.6 ± 0.6	70.4 ± 0.6
T_1 (K)	[1]4976	[1]4976	[1]4976
T_2 (K)	4566 ± 7	4580 ± 5	4595 ± 4
$\Omega_1 = \Omega_2$	5.94 ± 0.09	5.94 ± 0.04	5.99 ± 0.02
$q(M_2/M_1)$	2.66 ± 0.06	2.65 ± 0.03	2.60 ± 0.01
$L_1/(L_1+L_2)$(g')	0.422 ± 0.007	0.417 ± 0.004	0.411 ± 0.010
$L_1/(L_1+L_2)$(r')	0.393 ± 0.007	0.390 ± 0.004	0.384 ± 0.009
$L_1/(L_1+L_2)$(i')	0.376 ± 0.007	0.373 ± 0.004	0.368 ± 0.009
ℓ_3 (g')	[2]0.24 ± 0.02	[2]0.24 ± 0.01	[2]0.02 ± 0.03
ℓ_3 (r')	[2]0.28 ± 0.02	[2]0.27 ± 0.01	[2]0.04 ± 0.03
ℓ_3 (i')	[2]0.29 ± 0.02	[2]0.29 ± 0.02	[2]0.07 ± 0.02
r_1 side	0.303 ± 0.003	0.305 ± 0.001	0.300 ± 0.001
r_2 side	0.510 ± 0.011	0.487 ± 0.005	0.470 ± 0.002
Σres^2	0.088	0.044	0.042
Spot Parameters		*Star 2—hot spot*	*Star 2—hot spot*
colatitude (°)		88 ± 7	92 ± 2
longitude (°)		12 ± 9	8 ± 5
spot radius (°)		10 ± 5	10 ± 4
temp.-factor		1.15 ± 0.05	1.15 ± 0.04

[1]*Assumed.*
[2]*Third lights are the percent of light contributed at orbital phase 0.25.*
The subscripts 1 and 2 refer to the star being eclipsed at primary and secondary minimum, respectively.
Note: The errors in the stellar parameters result from the least-squares fit to the model. The actual uncertainties of the parameters are considerably larger.

4.3. Spot model

The cool stars of contact binaries have a deep common convective envelope. Stars with this property produce a strong dynamo and display solar type magnetic activity. This activity manifests itself as cool regions (dark spots) or hot regions such as faculae in the star's photosphere. The O'Connell effect, where the light curves display unequal maxima, is usually attributed to spots on one or both stars. For V384 Ser, the DS2 light curves (Figure 2) show only a very weak O'Connell effect, but 11 months earlier the DS1 light curves had a pronounced O'Connell effect. This change can be seen in Figure 8, which shows the r' passband light curve for the 2017 observations (open circles) overlaid by the 2018 observations. Not only are season-to-season changes occurring in this star, but night-to-night changes were also observed in the 2018 data. These observations confirm V384 Ser is magnetically active with changing spot configurations. It should also be noted that V384 Ser is an x-ray source, which is another key indication of magnetic activity (Geske *et al.* 2006).

The fit between the synthetic and observed light curves shows excess light between orbital phase 0.2 and 0.4 and a small light loss between 0.6 and 0.8 (see Figure 7). To fit these asymmetries, an over-luminous spot was modeled with BM3 in the neck region of the larger cooler star. The spot parameters, latitude, longitude, spot size, and temperature were adjusted until asymmetries were minimized. The resulting spot parameters were then incorporated into a new WD model. The spot model resulted in an improved fit between the observed and synthetic light curves, with a 50% reduction in the residuals compared to the spotless model. The final solution parameters for the spot model are shown in column 3 of Table 3. Figure 9 displays the model fit (solid lines) to the observed light curves and the residuals. Figure 10 shows a graphical representation of the spotted model that was created using BM3 (Bradstreet and Steelman 2002).

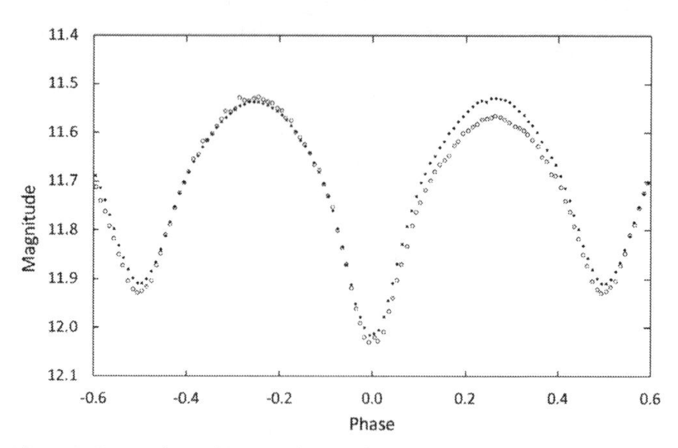

Figure 8. Comparison of 2017 and 2018 Sloan r' band light curves in standard magnitudes. The observations were binned with a phase width of 0.01. The 2017 observations (open circles) were acquired about 11 months before the 2018 observations (black dots).

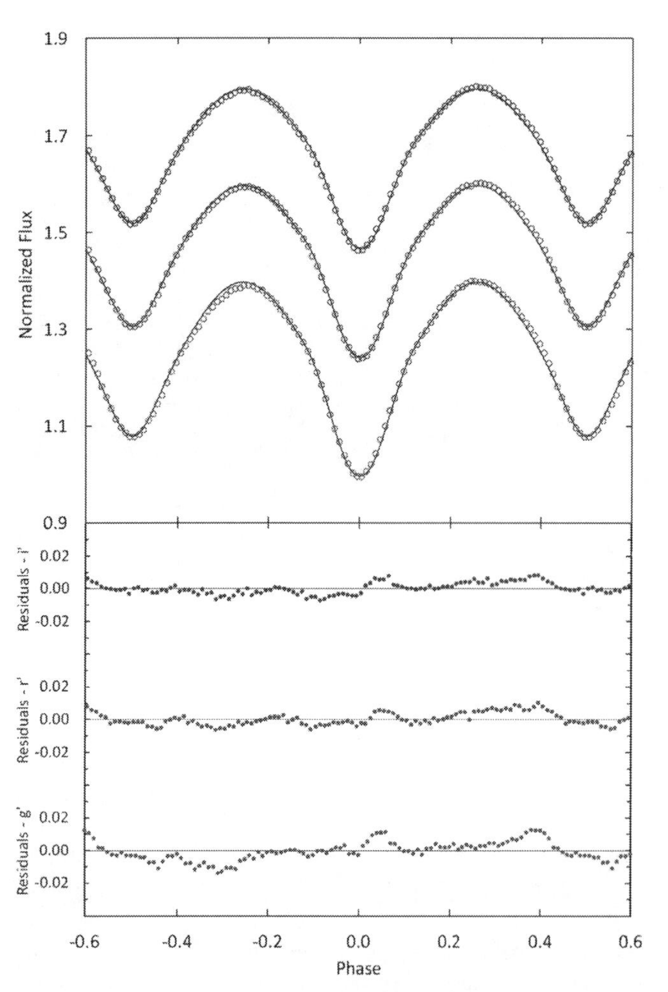

Figure 7. The observational light curves (open circles) and the fitted light curves (solid lines) for the spotless WD Solution 1 model (top panel). From top to bottom the passbands are Sloan i', r', and g' (each curve offset by 0.2). The residuals for the best-fit spotless model are shown in the bottom panel. Error bars are omitted from the points for clarity.

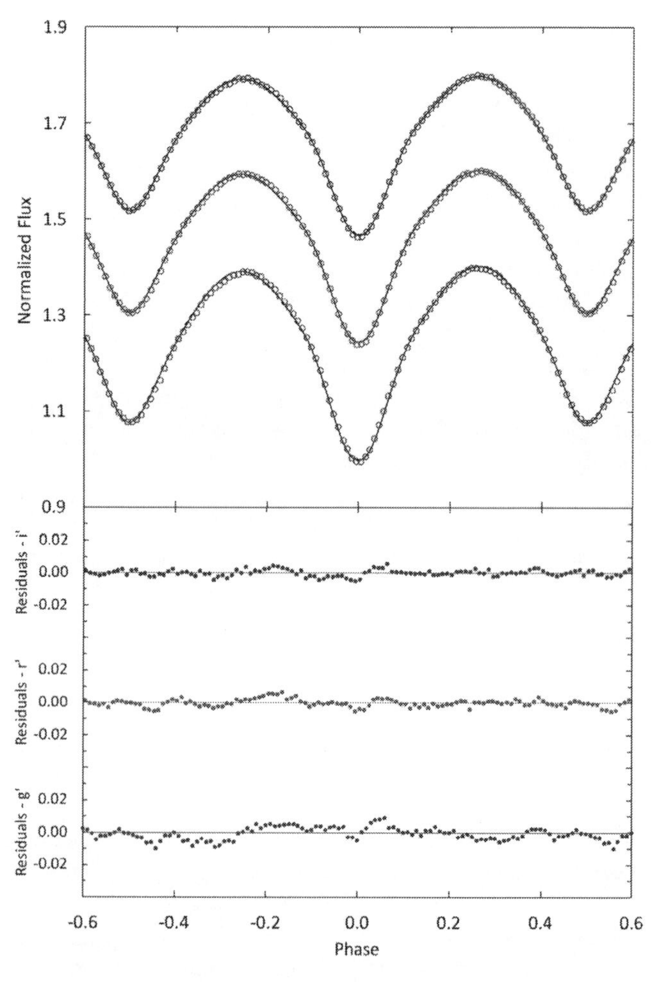

Figure 9. The observational light curves (open circles) and the fitted light curves (solid lines) for the spotted WD Solution 1 model (top panel). From top to bottom the passbands are Sloan i', r', and g' (each curve offset by 0.2). The residuals for the best-fit spot model are shown in the bottom panel. Error bars are omitted from the points for clarity.

5. Discussion

The absolute parameters of the component stars can be determined if their masses are known. Using the mass-period relation for contact binaries (Equation 5), the estimated mass of the larger cooler secondary star is $M_2 = 0.97 \pm 0.09\,M_\odot$ and a derived a primary mass gives $M_1 = 0.36 \pm 0.04\,M_\odot$. The distance between the mass centers, $1.93 \pm 0.05\,R_\odot$, was calculated using Kepler's Third Law. With this orbital separation, the WD light curve program (LC) calculated the stellar radii, luminosities, bolometric magnitudes, and surface gravities. The estimated absolute stellar parameters are collected in Table 4.

The luminosity of V384 Ser was calculated from the measured distance, the observed apparent V magnitude, and the bolometric correction (BCv). The Gaia parallax (DR2) gives a distance of d = 211 ± 2 pc (Bailer-Jones *et al.* 2018). The observed visual magnitude was determined from the DS2 observations using the average g' and r' passband values and the transformation equation of Jester *et al.* (2005),

$$V = g' - 0.59\,(g' - r') - 0.1. \qquad (7)$$

The resulting magnitude, V = 12.01 ± 0.04, agrees well with the APASS (DR9) value of V = 12.01 ± 0.19. As shown in section 4.1, the color excess for this star was very small, therefore, extinction was not applied to the V magnitude. The bolometric correction, $BC_v = -0.328$, was interpolated from the tables of Pecaut and Mamajek (2013) using the color from the LAMOST spectrum. The calculated absolute visual magnitude, visual luminosity, bolometric magnitude, and luminosity are given by $M_v = 5.38 \pm 0.8$, $L_v = 0.62 \pm 0.05\,L_\odot$, $M_{bol} = 5.06 \pm 0.08$, and $L = 0.75\,L_\odot \pm 0.05$, respectively. This luminosity is in good agreement with the value from Gaia DR2, $L = 0.71 \pm 0.01\,L_\odot$ (Gaia 2016, 2018).

The period study of section 3 found a short-term cyclic period change superimposed on a long-term secular decrease in the orbital period. A secular decreasing period could be explained by magnetic braking or by conservative mass exchange. For conservative mass exchange, transfer of matter from the larger more massive star to the smaller hotter companion would be required. For this case, the rate of mass transfer calculated from the well-known equation,

$$\frac{dM}{dt} = \frac{\dot{P} M_1 M_2}{3P(M_1 - M_2)}, \qquad (8)$$

gives a value of $7.09\,(0.01) \times 10^{-11}\,M_\odot$ / day (Reed 2011). The sinusoidally varying component of the ephemeris could be caused by magnetic activity (Applegate 1992) or the result of light-travel time effects caused by the orbital motion of the binary around a third body (Liao and Qian 2010; (Qian *et al.* 2013; Pribulla and Ruciński 2006). The modulation time of the orbital period due to magnetic activity can be estimated from the empirical relationship derived by Lanza and Rodonò (1999),

$$\log P_{mod} = -0.36(\pm 0.10) \log \Omega + 0.018, \qquad (9)$$

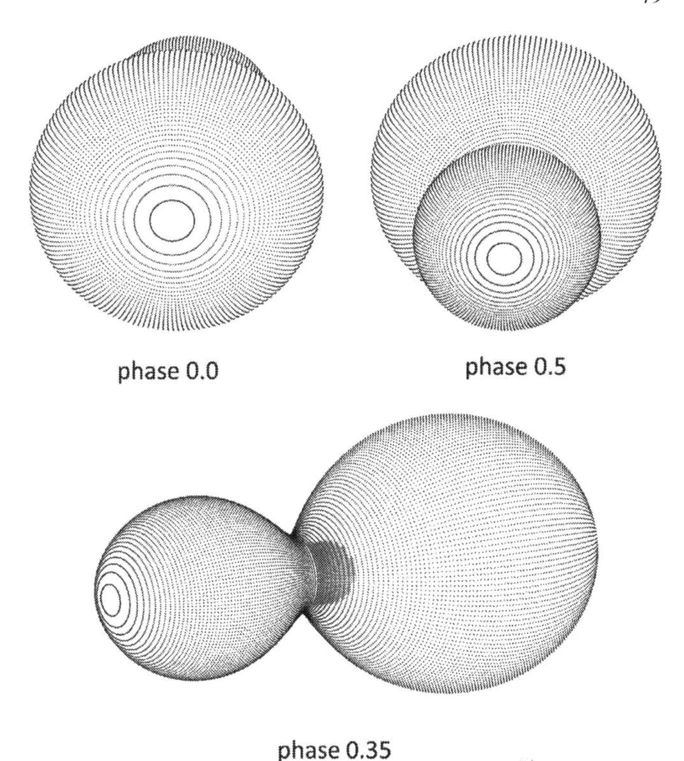

Figure 10. Roche Lobe surfaces of the best-fit WD spot model with orbital phase shown below each diagram.

Table 4. Estimated absolute parameters for V384 Ser.

Parameter	*Symbol*	*Value*
Stellar masses	M_1 ($M_\odot$)	0.36 ± 0.04
	M_2 ($M_\odot$)	0.97 ± 0.09
Semi-major axis	a ($R_\odot$)	1.93 ± 0.05
Mean stellar radii	R_1 ($R_\odot$)	0.62 ± 0.01
	R_2 ($R_\odot$)	0.94 ± 0.03
Stellar luminosity	L_1 ($L_\odot$)	0.21 ± 0.01
	L_2 ($L_\odot$)	0.35 ± 0.02
Bolometric magnitude	$M_{bol,1}$	6.43 ± 0.05
	$M_{bol,2}$	5.88 ± 0.07
Surface gravity	log g_1 (cgs)	4.41 ± 0.04
	log g_2 (cgs)	4.47 ± 0.04

Note: The calculated values in this table are provisional. Radial velocity observations are necessary for direct determination of M_1, M_2, and a.

where $\Omega = 2\pi / P$, P_{mod} is in years and P in seconds. Using the orbital period of V384 Ser gives a modulation period of about 20 years. This is about seven times longer than the observed modulation period, which makes magnetic activity an unlikely cause of the periodic variation. We analyzed the cyclic oscillation in the $(O-C)_2$ diagram (Figure 3, middle panel) for the light-travel time effect caused by a third stellar body orbiting V384 Ser. The sinusoidal term of Equation 2 gives the oscillation amplitude, $A_3 = 0.0037 \pm 0.0002$ days, and the third body's orbital period, $P_3 = 2.86 \pm 0.01$ yr. The orbit is likely circular, given the good sinusoidal fit over the several orbits covered by the observations. Assuming an orbital eccentricity of zero, the projected distance between the barycenter of the triple system and the binary was calculated from the equation:

$$a'_{12} \sin i' = A_3 \times c, \quad (10)$$

where i' is the orbital inclination of the third body and c is the speed of light. The mass function was determined from the following equation:

$$f(m) = \frac{4\pi^2}{GP_3^2} \times (a'_{12} \sin i')^3, \quad (11)$$

where G is the gravitational constant. By using the masses of the primary and secondary stars determined previously, the mass and orbital radius for the third stellar body were calculated from the following equation:

$$f(m) = \frac{(M_3 \sin i')}{(M_1 + M_2 + M_3)}. \quad (12)$$

For coplanar orbits (i' = 78.6°), the computed third body's mass and orbital radius are $M_3 = 0.49 \pm 0.03$ $M_\odot$ and $a_3 = 1.80 \pm 0.05$ AU. The derived parameters are shown in Table 5, and the relation between the orbital inclination and the mass and orbital radius of the third body are shown in Figure 11. The properties of the third stellar body can now be approximated. Subtracting the luminosity for each binary component from the system luminosity gives a third body luminosity of $L_3 = 0.19 \pm 0.06$ $L_\odot$. A main-sequence star of this luminosity has a color of (B–V) = 1.10, a temperature of T_{eff} = 4620 K, and a mass of 0.73 $M_\odot$ (Pecaut and Mamajek 2013). For comparison, the third star's color and temperature can be estimated from the third light values of Solution 1. Interpolating from the tables of Pecaut and Mamajek (2013) gives a color of (B–V) = 1.01 and a temperature of T_{eff} = 4800 K, which are reasonably close to the values found above. The estimated spectral type for the tertiary component is K3 or K4 with a mass between 0.7 – 0.8 $M_\odot$. For the estimated mass, the orbital inclination (i') of the third body would be about 45° (see Table 5 and Figure 11).

Close binaries in triple systems have resulted in spurious photometric solutions and V384 Ser is a good example (Gazeas and Niarchos 2006). The light curve analysis for this star resulted in a second WD solution that is shown in column 4 of Table 3 (Solution 2). The fit between the synthetic and observed light curves for Solution 2 are nearly identical to Solution 1. The residuals for Solution 2 are slightly smaller than Solution 1. The parameter sets differed primarily in orbital inclination, third light, and the filling factor, which are two very different solutions. To determine the best solution, we compared the observed total system luminosity to the luminosity of the binary. For Solution 1, the luminosity of the binary is $L_{12} = 0.56 \pm 0.03$ $L_\odot$. The binary contributes about 74% of the total system light with the remaining 26% coming from a third source. This is a close match to the third lights found in Solution 1 (24%–29%). For Solution 2, the binary contributes 70% to the total system light with 30% coming from a third source. The third lights from Solution 2 are much smaller (2%–7%). The results from this analysis, plus the observed near total primary eclipse, supports Solution 1 with its higher orbital inclination.

Table 5. Parameters of the tertiary component.

Parameter	*Value*	*Units*
P_3	2.86 ± 0.01	years
A_3	0.0037 ± 0.0002	days
e'	0.0	assumed
a'_{12} sin i'	0.65 ± 0.03	AU
f(m)	0.033 ± 0.005	$M_\odot$
M_3 (i' = 90°)	0.47 ± 0.03	$M_\odot$
M_3 (i' = 80°)	0.48 ± 0.03	$M_\odot$
M_3 (i' = 70°)	0.51 ± 0.03	$M_\odot$
M_3 (i' = 60°)	0.57 ± 0.03	$M_\odot$
M_3 (i' = 50°)	0.66 ± 0.04	$M_\odot$
M_3 (i' = 40°)	0.83 ± 0.05	$M_\odot$
a_3 (i' = 90°)	1.80 ± 0.05	AU
a_3 (i' = 80°)	1.80 ± 0.05	AU
a_3 (i' = 70°)	1.78 ± 0.05	AU
a_3 (i' = 60°)	1.75 ± 0.05	AU
a_3 (i' = 50°)	1.69 ± 0.05	AU
a_3 (i' = 40°)	1.60 ± 0.06	AU

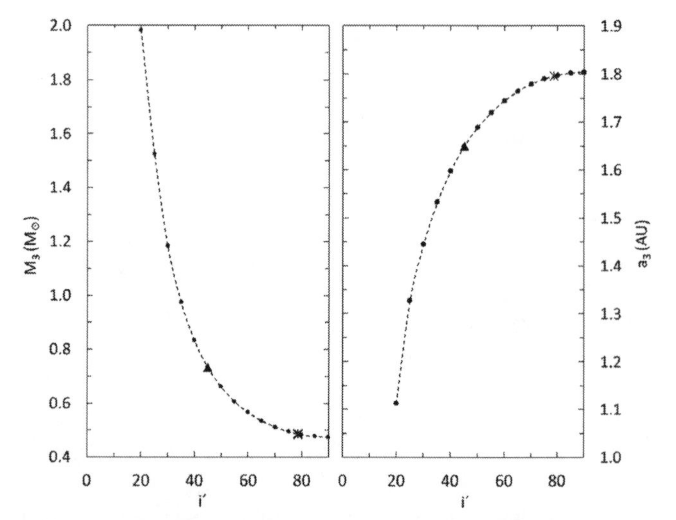

Figure 11. The relation between the third body's mass M_3 and the orbital inclination is shown in the left panel. The right panel shows the relation between the orbital radius and orbital inclination for the third body. The asterisk gives the mass and orbital radius for the tertiary component that is coplanar with V384 Ser and the solid triangle locates the orbital inclination (45°) for the estimated mass of the tertiary component.

6. Conclusions

This paper presents and analyzes the first complete set of photometric CCD observations in the Sloan g', r', and i' passbands for the eclipsing binary V384 Ser. This study confirms it is a W-type contact binary, where the larger more massive star is cooler and has less surface brightness than its companion. The best-fit WD solution gives a mass ratio of q = 2.65, a fill-out of f = 36%, and a temperature difference of 376 K between the component stars. This star was found to be magnetically active, as evidenced by changes in the light curves between observing seasons. The period analysis revealed V384 Ser is a triple system with a cool stellar companion having an orbital radius of about 1.7 AU. Early dynamical interaction between the stars may have had a significant influence in the evolution of this system. A spectroscopic study would be invaluable in confirming the stellar masses and mass ratio found in the photometric solution presented here. In addition, the third stellar body may also have sufficient luminosity to be detected by high resolution spectroscopy.

7. Acknowledgements

This research was made possible through the use of the AAVSO Photometric All-Sky Survey (APASS), funded by the Robert Martin Ayers Sciences Fund. This research has made use of the SIMBAD database and the VizieR catalogue databases (operated at CDS, Strasbourg, France). This work has made use of data from the European Space Agency (ESA) mission Gaia (https://www.cosmos.esa.int/gaia), processed by the Gaia Data Processing and Analysis Consortium (DPAC, https://www.cosmos.esa.int/web/gaia/dpac/consortium). Funding for the DPAC has been provided by national institutions, in particular the institutions participating in the Gaia Multilateral Agreement. Data from the Guo Shou Jing Telescope (the Large Sky Area Multi-Object Fiber Spectroscopic Telescope, LAMOST) was also used in this study. This telescope is a National Major Scientific Project built by the Chinese Academy of Sciences. Funding for the project has been provided by the National Development and Reform Commission. LAMOST is operated and managed by National Astronomical Observatories, Chinese Academy of Sciences.

References

Akerlof, C., *et al.* 2000, *Astron. J.*, **119**, 1901.
Applegate, J. H. 1992, *Astrophys. J.*, **385**, 621.
Bailer-Jones, C., Rybizki, J., Fouesneau, M., Mantelet, G., and Andrae, R. 2018, *Astron. J.*, **156**, 58.
Bilir, S., Karaali, S., and Tunçel, S. 2005, *Astron. Nachr.*, **326**, 321.
Blättler, E., and Diethelm, R. 2002, *Inf. Bull. Var. Stars*, No. 5295, 1.
Bradstreet, D. H., and Steelman, D. P. 2002, *Bull. Amer. Astron. Soc.*, **34**, 1224.
Diethelm, R. 2003, *Inf. Bull. Var. Stars*, No. 5438, 1.
Diethelm, R. 2005, *Inf. Bull. Var. Stars*, No. 5653, 1.
Diethelm, R. 2006, *Inf. Bull. Var. Stars*, No. 5713, 1.
Diethelm, R. 2007, *Inf. Bull. Var. Stars*, No. 5781, 1.
Diethelm, R. 2009a, *Inf. Bull. Var. Stars*, No. 5871, 1.
Diethelm, R. 2009b, *Inf. Bull. Var. Stars*, No. 5894, 1.
Diethelm, R. 2010a, *Inf. Bull. Var. Stars*, No. 5920, 1.
Diethelm, R. 2010b, *Inf. Bull. Var. Stars*, No. 5945, 1.
Diethelm, R. 2011, *Inf. Bull. Var. Stars*, No. 5992, 1.
Diethelm, R. 2012, *Inf. Bull. Var. Stars*, No. 6029, 1.
Gaia Collaboration, *et al.* 2016, *Astron. Astrophys.*, **595A**, 1.
Gaia Collaboration, *et al.* 2018, *Astron. Astrophys.*, **616A**, 1.
Gazeas, K., and Niarchos, P. G., 2006, *Mon. Not. Roy. Astron. Soc.*, **370**, L29.
Gazeas, K., and Stępień, K. 2008, *Mon. Not. Roy. Astron. Soc.*, **390**, 1577.
Geske, M., Gettel, S., and McKay, T. 2006, *Astron. J.*, **131**, 633.
Gettel, S. J., Geske, M. T., and McKay, T. A. 2006, *Astron. J.*, **131**, 621.
Gürsoytrak, H., *et al.* 2013, *Inf. Bull. Var. Stars*, No. 6075, 1.
Henden, A. A., *et al.* 2015, AAVSO Photometric All-Sky Survey, data release 9, (http:www.aavso.org/apass).
Hoffman, D. I., Harrison, T. E., and McNamara, B. J. 2009, *Astron. J.*, **138**, 466.
Hoňková, K., *et al.* 2015, *Open Eur. J. Var. Stars*, **168**, 1.
Hübscher, J. 2013, *Inf. Bull. Var. Stars*, No. 6084, 1.
Hübscher, J. 2014, *Inf. Bull. Var. Stars*, No. 6118, 1.
Hübscher, J. 2015, *Inf. Bull. Var. Stars*, No. 6152, 1.
Hübscher, J. 2016, *Inf. Bull. Var. Stars*, No. 6157, 1.
Hübscher, J. 2017, *Inf. Bull. Var. Stars*, No. 6196, 1.
Hübscher, J., Braune, W., and Lehmann, P. B. 2013, *Inf. Bull. Var. Stars*, No. 6048, 1.
Hübscher, J., and Lehmann, P. B. 2012, *Inf. Bull. Var. Stars*, No. 6026, 1.
Hübscher, J., and Lehmann, P. B. 2014, *Inf. Bull. Var. Stars*, No. 6149, 1.
Hübscher, J., Lehmann, P. B., Monninger, G., Steinbach, H.-M., and Walter, F. 2010, *Inf. Bull. Var. Stars*, No. 5918, 1.
Hübscher, J., Lehmann, P. B., and Walter, F. 2012, *Inf. Bull. Var. Stars*, No. 6010, 1.
Hübscher, J., and Monninger, G. 2011, *Inf. Bull. Var. Stars*, No. 5959, 1.
Hübscher, J., Steinbach, H.-M., and Walter, F. 2009a, *Inf. Bull. Var. Stars*, No. 5889, 1.
Hübscher, J., Steinbach, H.-M., and Walter, F. 2009b, *Inf. Bull. Var. Stars*, No. 5874, 1.
Juryšek, J., *et al.* 2017, *Open Eur. J. Var. Stars*, No. 179, 1.
Kafka, S. 2017, variable star observations from the AAVSO International Database (https://www.aavso.org/aavso-international-database).
Kazuo, N. 2009, *Bull. Var. Star Obs. League Japan*, No. 48, 1.
Lanza, A. F., and Rodonò, M. 1999, *Astron. Astrophys.*, **349**, 887.
Liao, W.-P., and Qian, S.-B. 2010, *Mon. Not. R. Astron. Soc.*, **405**, 1930.
Lohr, M. E., Norton, A. J., Payne, S. G., West, R. G., and Wheatley, P. J. 2015, *Astron. Astrophys.*, **578A**, 136.
Lucy, L. B. 1968, *Astrophys, J.*, **151**, 1123.
Lucy, L. B., and Wilson, R. E. 1979, *Astrophys, J.*, **231**, 502.
Luo, A.-Li, *et al.*, 2015, Res. *Astron. Astrophys.*, **15**, 1095.
Mirametrics. 2015, Image Processing, Visualization, Data Analysis, (http://www.mirametrics.com).
Nelson, R. H. 2002, *Inf. Bull. Var. Stars*, No. 5224, 1.
Nelson, R. H. 2017, *Inf. Bull. Var. Stars*, No. 6195, 1.
Pecaut, M. J., and Mamajek, E. E. 2013, *Astrophys. J., Suppl. Ser.*, 208, 9, (http://www.pas.rochester.edu/~emamajek/EEM_dwarf_UBVIJHK_colors_Teff.txt).
Pribulla, T., and Rucinski, S. M. 2006, *Astron. J.*, **131**, 2986.
Qian, S.-B., Liu, N.-P., Liao, W.-P., He, J.-J., Liu, L., Zhu, L.-Y., Wang, J.-J., and Zhao, E.-G. 2013, *Astron. J.*, **146**, 38.
Reed, P. A. 2011, in *Mass Transfer Between Stars: Photometric Studies, Mass Transfer—Advanced Aspects*, ed. H. Nakajima, InTech DOI: 10.5772/19744 (https://www.intechopen.com/books/mass-transfer-advanced-aspects/mass-transfer-between-stars-photometric-studies), 3.
Richards, J. W., Starr, D. L., Miller, A. A., Bloom, J. S., Butler, N. R., Brink, H., and Crellin-Quick, A. 2012, *Astrophys. J., Suppl. Ser.*, **203**, 32.

Ruciński, S. M. 1969, *Acta Astron.*, **19**, 245.
Samolyk, G. 2016, *J. Amer. Assoc. Var. Star Obs.*, **44**, 164.
Terrell, D., and Wilson, R. E. 2005, *Astrophys. Space Sci.*, **296**, 221.
Terzioğlu, Z., *et al.* 2017, *Inf. Bull. Var. Stars*, No. 6128, 1.
van Hamme, W. 1993, *Astron. J.*, **106**, 2096.
van Hamme, W., and Wilson, R. E. 1998, *Bull. Amer. Astron. Soc.*, **30**, 1402.
Wilson, R. E. 1978, *Astrophys. J.*, **224**, 885.
Wilson, R. E., and Devinney, E. J. 1971, *Astrophys. J.*, **166**, 605.

CCD Photometry, Light Curve Deconvolution, Period Analysis, Kinematics, and Evolutionary Status of the HADS Variable V460 Andromedae

Kevin B. Alton
UnderOak and Desert Bloom Observatories, 70 Summit Avenue, Cedar Knolls, NJ 07927; kbalton@optonline.net

Kazimierz Stępień
Warsaw University Observatory, Al. Ujazdowskie 4, 00-478 Warszawa, Poland

Received March 3, 2019; revised April 8, 2019; accepted April 8, 2019

Abstract Multi-color (BVI_c) CCD-derived photometric data were acquired from V460 And, an intrinsic variable classically defined as a High Amplitude delta Scuti (HADS) type system. Deconvolution of precise time-series light curve data was accomplished using discrete Fourier transformation and revealed a fundamental mode (f_0) of oscillation at ~13.336 d^{-1} along with five other partial harmonics ($2f_0$–$6f_0$). No other statistically significant frequencies were resolved following successive pre-whitening of each residual signal. An assessment of potential period changes over time was performed using six new times-of-maximum light produced from the present study along with other values reported in the literature. These along with sparsely-sampled data collected during the ROTSE-I (1999), Catalina Sky (2005–2013), and SuperWASP (2004–2008) surveys indicate that no substantive change in the primary pulsation period or amplitude (V-mag) has likely occurred over the past 20 years. Recent photometric data from space telescopes have in some cases contradicted traditional classification schemes and clouded the differences between HADS- and SX Phe-like variables. Herein using accurate cosmic distances and proper motions from Gaia DR2, we attempted to exploit potential kinematic differences between established populations of HADS and SX Phe variable stars as an alternate approach for classification. Finally, an investigation with PARSEC models for generating stellar tracks and isochrones provided valuable insight into the evolutionary status and physical character of V460 And.

1. Introduction

The most common A- and F-type stars which exhibit variability are the multi-periodic pulsators known as delta Scuti-like (hereafter δ Sct) stars. As a class these intrinsic variables occupy a narrow area at the intersection of the classical instability strip, pre-main-sequence, and main-sequence (MS) on the Hertzsprung-Russell diagram. Therein they represent a transition from the high-amplitude radial pulsators, such as Cepheid variables, and non-radial multi-periodic pulsators (Breger 2000). Main-sequence δ Sct stars typically range from spectral type F2 to A2 (Rodríguez and Breger 2001), which corresponds to effective temperatures varying between 6300 and 8600 K (Uytterhoeven *et al.* 2011). Hotter δ Sct stars generally have shorter pulsation periods (i.e. higher pulsation mode frequencies) than cooler δ Sct stars.

Similar to Cepheid and RR Lyrae stars (Baker and Kippenhahn 1962, 1965; Zhevakin 1963), pulsations in δ Sct stars are excited by the κ-mechanism operating in the He II partial ionization zone (T~50000 K) which produce low-order pressure (p) modes akin to acoustic waves (Cox 1963; Chevalier 1971). These can produce radial pulsations which evoke symmetrical changes in stellar size and/or non-radial pulsations that give rise to asymmetric changes in shape but not volume. Although shorter periods (< 30 min) have been observed (Holdsworth *et al.* 2014) in some A-type stars, the fundamental radial pulsations of Galactic δ Sct variables with near solar metallicity typically range from 0.05 to 0.25 d. Masses vary from ~1.2 $M_{\odot}$ to ~2.5 $M_{\odot}$ so they are more luminous and larger than our Sun. The luminosity classes for δ Sct variables generally range from III (normal giants) to V (MS stars). δ Sct variables with moderate (40 km s^{-1}) to rapid (250 km s^{-1}) rotational velocities ($v \sin i$) generally have small light amplitudes (ΔV ~0.01–0.03 mag) composed of a multitude of pulsation frequencies, most of them nonradial. Stars with slow rotational velocities (< 30 km s^{-1}) tend to be radial pulsators and have light amplitudes (V-mag) in excess of 0.20–0.30 mag. The latter characteristics define a δ Sct subgroup called High-Amplitude delta Scuti stars (HADS).

HADS represent a very small fraction (< 1%) of all δ Sct variables (Lee *et al.* 2008). They commonly oscillate via low-order single or double radial pulsation modes (Poretti 2003a, 2003b; Niu *et al.* 2013, 2017). A high percentage (~40%) are double pulsators showing simultaneous pulsations in the fundamental and the first overtone mode with amplitudes generally higher in the fundamental mode (McNamara 2000). It should be noted, however, that non-radial pulsations have also been detected with the HADS variable V974 Oph (Poretti 2003a, 2003b). HADS variables have historically been divided according to metallicity relative to our Sun ([Fe/H] = 0 dex). The metal-poor ([Fe/H] << 0) group is called SX Phe stars based on the eponymous prototype SX Phoenicis. Ostensibly they have shorter periods (0.02 < P < 0.125 d) and lower masses (~1.0–1.3 $M_{\odot}$) than their sibling HADS variables possessing near solar metal abundance. SX Phe stars frequently dwell in globular clusters (GC), ancient collections of Population I stars. Therein, the majority of SX Phe variables are classified as blue straggler stars, paradoxically appearing much younger than their GC cohorts. Despite previous claims to the contrary, Balona and Nemec (2012) make a strong case that it is not possible to differentiate between δ Sct and field SX Phe variables based on pulsation amplitude, the number of pulsation modes, period, or even metallicity. Much more sensitive space telescopes like Kepler (Gilliland *et al.* 2010), CoRoT (Baglin 2003), and MOST (Walker *et al.* 2003) have found many examples that violate these basic tenants. They further argue that the evolutionary status of each star is the only way to distinguish between these two classes. One way to get a handle on the age of a star is to exploit potential differences in kinematics. Population II stars often reside away from the Galactic plane in globular clusters, the halo, and thick disc. Arguably, pulsating field stars that reside in these regions of the Milky Way will have high proper motions (μ_α, μ_δ) but more importantly greater tangential velocities (V_{T_α}, V_{T_δ}) than younger MS (Population I) δ Sct variables occupying the thin disk. As will be discussed herein, the availability of very precise proper motion (PM) and parallax values from Gaia DR2 has for the moment muddied the water with regard to differentiation between HADS and SX Phe variables.

The putative variability of V460 And (GSC 02840-01177) was first identified based on an entry (No. 227) in the *Catalogue of the Stars of Suspected Variability* (Kukarkin *et al.* 1951). Its variable nature was further reported (Kinman *et al.* 1982) based on photographic plates taken between 1962 and 1968 but no variable type assignment could be made from the star identified as RRV-26. Following an evaluation of unfiltered photometric data from the ROTSE-I Survey (Akerloft et. al 2000), Khruslov (2005) reported that V460 And (NSVS 4000553) was a HADS star with a period of 0.0749808 d. Photometric (V-mag) data from V460 And (CSS_J023414.3+421427) were also acquired during the Catalina Sky Survey (Drake *et al.* 2009); an assessment herein of the CSS data further confirmed the fundamental period reported by Khruslov (2005). This report marks the first multi-color photometric study on V460 And which also addresses a longstanding, but unproven classification as a HADS variable.

2. Observations and data reduction

2.1. Photometry

Time-series images were acquired at Desert Bloom Observatory (DBO, USA—110.257 W, 31.941 N) with an SBIG STT-1603ME CCD camera mounted at the Cassegrain focus of a 0.4-m Schmidt-Cassegrain telescope. This focal-reduced (f/6.8) instrument produces an image scale of 1.36 arcsec/pixel (bin = 2 × 2) and a field-of-view (FOV) of 11.5' × 17.2'. Image acquisition (75-s) was performed using THESKYX Professional Edition 10.5.0 (Software Bisque 2019). The CCD-camera is equipped with B, V, and I_c filters manufactured to match the Johnson-Cousins Bessell specification. Dark subtraction, flat correction, and registration of all images collected at DBO were performed with AIP4WIN v2.4.0 (Berry and Burnell 2005). Instrumental readings were reduced to catalog-based magnitudes using the APASS star fields (Henden *et al.* 2009, 2010, 2011; Smith *et al.* 2011) built into MPO CANOPUS v10.7.1.3 (Minor Planet Observer 2011). Light curves for V460 And were generated using an ensemble of five non-varying comparison stars. The identity, J2000 coordinates, and APASS color indices (*B–V*) for these stars are provided in Table 1; a corresponding FOV image is rendered in Figure 1. Only data from images taken above 30° altitude (airmass < 2.0) were included; considering the proximity of all program stars, differential atmospheric extinction was ignored. During each imaging session comparison stars typically stayed within ±0.006 mag for V and I_c filters and ±0.009 mag for the B passband.

Table 1. Astrometric coordinates (J2000), V-mag, and color indices (*B–V*) for V460 And and five comparison stars (1–5) used during this photometric study.

FOV ID	*Star Identification*	*R.A. h m s*	*Dec. ° ' "*	*APASS[a] V-mag*	*APASS[a] (B–V)*
T	V460 And	02 34 14.255	+42 14 27.604	13.168	0.238
1	GSC 02840-01355	02 34 05.770	+42 14 05.009	12.334	0.549
2	GSC 02840-01209	02 33 55.975	+42 12 15.444	12.613	0.642
3	GSC 02840-01660	02 33 55.617	+42 08 14.655	12.225	0.578
4	GSC 02840-01826	02 33 04.906	+42 09 43.343	11.767	0.309
5	GSC 02840-00853	02 33 58.663	+42 11 00.420	13.354	0.685

[a] *V-mag and (B–V) for comparison stars derived from APASS database described by Henden* et al. *(2009, 2010, 2011) and Smith* et al. *(2010), as well as on the AAVSO web site (https://www.aavso.org/apass).*

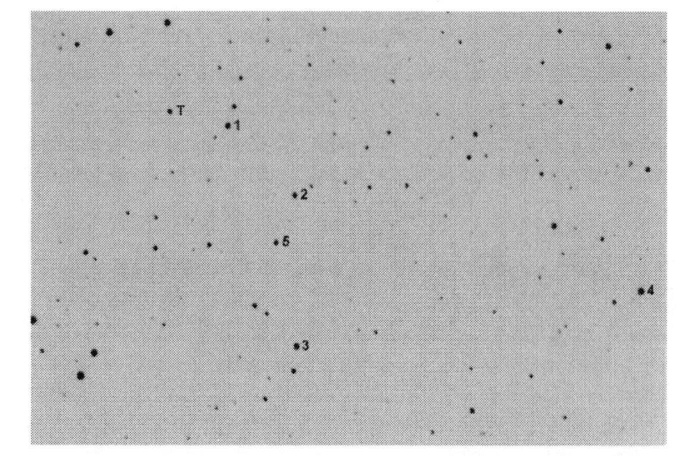

Figure 1. V460 And (T) along with the five comparison stars (1-5) used to reduce time-series images to APASS-catalog based magnitudes.

3. Results and discussion

3.1. Photometry and ephemerides

Photometric values in B (n = 155), V (n = 153), and I_c (n = 150) passbands were separately processed to produce light curves that spanned 4 days between Dec 19 and Dec 23, 2018 (Figure 2). There was no obvious color dependency on the timings such as those reported for other δ Sct variables (Elst 1978); therefore, all BVI_c data were averaged (Table 2) at each time-of-maximum. Period determinations were initially performed using PERANSO v2.5 (Paunzen and Vanmunster 2016) by applying periodic orthogonals (Schwarzenberg-Czerny 1996) to fit observations and analysis of variance (ANOVA) to assess fit quality. In this case a similar period solution for each passband (0.07498 ± 0.00001 d) was obtained. However, folding together (time span = 7,102 d) the sparsely sampled ROTSE-I and Catalina Sky survey data with those (V-mag) acquired at DBO yielded a period at 0.0749808 ± 0.0000010 d (Figure 3). Additionally, the SuperWASP survey (Butters *et al.* 2010) provided a rich source of photometric data taken (30-s exposures) at modest cadence

Table 2. Differences between the times-of-maximum light (HJD) predicted from the updated linear ephemeris (Equation 2) and those observed for V460 And between 2007 and 2018. Cycle No. is determined from the number of pulsations that have occurred since the start time (HJD_0) defined by the reference ephemeris.

HJD 2400000+	*Cycle No.*	*FPPTD*[a]	*Ref.*[b]	*HJD 2400000+*	*Cycle No.*	*FPPTD*[a]	*Ref.*[b]
54135.3080	35745	0.0066	1	55894.2830	59204	0.0065	5
54135.3831	35746	0.0067	1	55894.3580	59205	0.0071	5
54355.5266	38682	0.0067	1	55894.4330	59206	0.0079	5
54355.6013	38683	0.0064	1	55894.5813	59208	0.0079	4
54391.4424	39161	0.0067	1	55894.6564	59209	0.0079	4
54391.5180	39162	0.0073	1	55894.7316	59210	0.0062	4
54391.5923	39163	0.0066	1	55894.8064	59211	0.0064	4
55192.3112	49842	0.0060	2	55894.8812	59212	0.0066	4
55192.3863	49843	0.0061	2	55894.9565	59213	0.0064	4
55452.3457	53310	0.0072	3	55897.2054	59243	0.0062	4
55452.4202	53311	0.0067	3	55897.2808	59244	0.0065	4
55452.4947	53312	0.0062	3	55897.3556	59245	0.0060	4
55452.5701	53313	0.0066	3	56176.5845	62969	0.0064	6
55590.3099	55150	0.0068	4	56176.6597	62970	0.0063	6
55590.3844	55151	0.0063	4	56254.2645	64005	0.0068	6
55590.4591	55152	0.0060	4	56506.5001	67369	0.0070	7
55850.3430	58618	0.0066	5	56566.3358	68167	0.0067	7
55850.5690	58621	0.0076	5	56566.4090	68168	0.0070	7
55856.2666	58697	0.0067	4	56635.3161	69087	0.0081	8
55856.3408	58698	0.0059	4	56742.3147	70514	0.0063	9
55889.6329	59142	0.0066	4	56912.4453	72783	0.0061	9
55889.7076	59143	0.0063	4	56912.5204	72784	0.0071	9
55889.7826	59144	0.0063	4	56962.3075	73448	0.0064	9
55889.8579	59145	0.0066	4	56962.3832	73449	0.0065	9
55889.9325	59146	0.0062	4	57000.4732	73957	0.0064	9
55890.0072	59147	0.0060	4	57006.3208	74035	0.0071	9
55893.6067	59195	0.0064	4	58471.5951	93577	0.0069	10
55893.6820	59196	0.0067	4	58471.6708	93578	0.0060	10
55893.7568	59197	0.0065	4	58473.6204	93604	0.0061	10
55893.8316	59198	0.0063	4	58473.6951	93605	0.0069	10
55893.9067	59199	0.0065	4	58475.6441	93631	0.0070	10
55893.9823	59200	0.0071	4	58475.7196	93632	0.0067	10

[a]*FPPTD = Time difference between observed fundamental pulsation time-of-maximum and that calculated using the reference ephemeris (Equation 2).*
[b]*1. Wils* et al. *2009; 2. Wils* et al. *2010; 3. Wils* et al. *2011; 4. Wils* et al. *2012; 5. Hübscher and Lehmann 2012; 6. Wils* et al. *2013; 7. Wils* et al. *2014; 8. Hübscher 2014; 9. Wils* et al. *2015; 10. This study.*

that repeats every 9–12 min. These data acquired between 2004 and 2008 were period folded with V-mag data collected at DBO and crisply reached superimposition when P = 0.0749807 d (Figure 4). Times-of-maximum light acquired at DBO were estimated using the polynomial extremum fit utility featured in PERANSO (Paunzen and Vanmunster 2016). New maxima from DBO (6) along with published values starting in 2007 (Table 2) were used to analyze fundamental pulse period timings (FPPT). The reference epoch (International Variable Star Index) adopted for initially calculating FPPT differences (FPPTD) was defined by the following linear ephemeris (Equation 1):

$$\mathrm{Max\ (HJD)} = 2451455.114 + 0.0749808\,E. \qquad (1)$$

Secular changes in pulsation period can potentially be uncovered by plotting the difference between the observed FPPT values and those predicted by the reference epoch against cycle number (Figure 5). Thus far, all of the calculated FPPTD values (Table 4) basically describe a straight line relationship

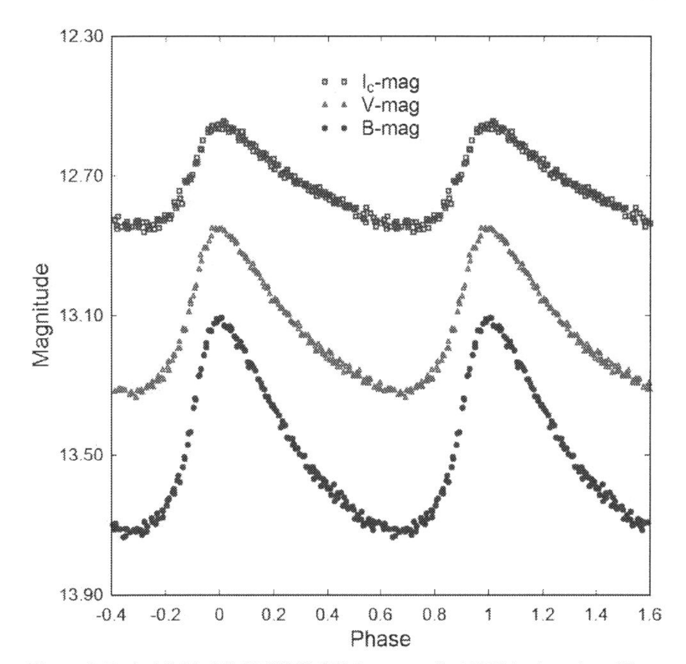

Figure 2. Period-folded (0.0749808 d) light curves for V460 And produced from photometric data obtained between Dec 19 and Dec 23, 2018 at DBO. Light curves shown at the top (I_c), middle (V) and bottom (B) represent catalog-based (APASS) magnitudes determined using MPO CANOPUS.

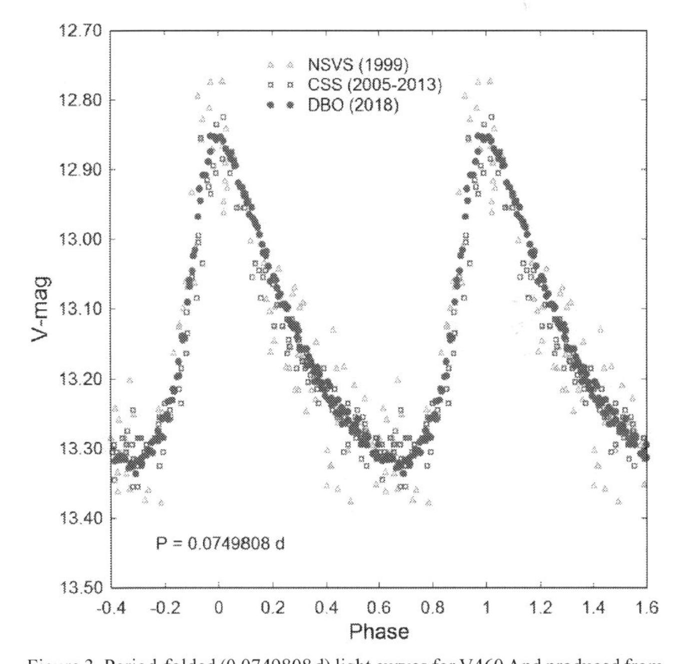

Figure 3. Period-folded (0.0749808 d) light curves for V460 And produced from precise photometric V-mag data obtained at DBO (2018) along with sparsely sampled data from the ROTSE-I (1999) and Catalina Sky (2005-2013) Surveys. Magnitudes were offset to conform with the APASS-derived values from DBO.

(albeit noisy) and suggest that little or no long-term change to the period has occurred since 1999. The updated ephemeris (Equation 2) based on maximum light timing data available through Dec 2018 is as follows:

$$\mathrm{Max\ (HJD)} = 2458475.7195(3) + 0.07498076(5)\,E. \qquad (2)$$

These results along with nearly superimposable period-folded light curves from DBO, ROTSE-I, CSS (Figure 3), and

Table 3. Fundamental frequency (f_0 = d^{-1}) and corresponding partial harmonics detected following DFT analysis of time-series photometric data (BVI_c) from V460 And.

	Freq. (d^{-1})	Freq. Err	Amp. (mag)	Amp. Err	Phase	Phase Err
f_0–B	13.3367	0.0002	0.2682	0.0011	0.2609	0.1497
f_0–V	13.3363	0.0003	0.2045	0.0010	0.9024	0.1627
f_0–I_c	13.3361	0.0004	0.1255	0.0012	0.7262	0.1223
$2f_0$–B	26.6786	0.0007	0.0909	0.0011	0.6918	0.1566
$2f_0$–V	26.6786	0.0008	0.0721	0.0010	0.9546	0.1672
$2f_0$–I_c	26.6793	0.0011	0.0463	0.0012	0.7583	0.1286
$3f_0$–B	40.0132	0.0017	0.0368	0.0011	0.1365	0.1517
$3f_0$–V	40.0093	0.0022	0.0269	0.0009	0.1198	0.1555
$3f_0$–I_c	40.0069	0.0026	0.0184	0.0011	0.8465	0.1209
$4f_0$–B	53.3430	0.0034	0.0198	0.0010	0.8096	0.1588
$4f_0$–V	53.8247	0.0043	0.0131	0.0009	0.4156	0.1482
$4f_0$–I_c	52.8671	0.0056	0.0091	0.0011	0.4305	0.1189
$5f_0$–B	66.6847	0.0058	0.0105	0.0011	0.7239	0.1478
$5f_0$–V	66.6726	0.0083	0.0072	0.0009	0.6484	0.1630
$6f_0$–B	80.0145	0.0083	0.0073	0.0011	0.4521	0.1540
$6f_0$–V	80.0359	0.0170	0.0042	0.0009	0.6981	0.1633

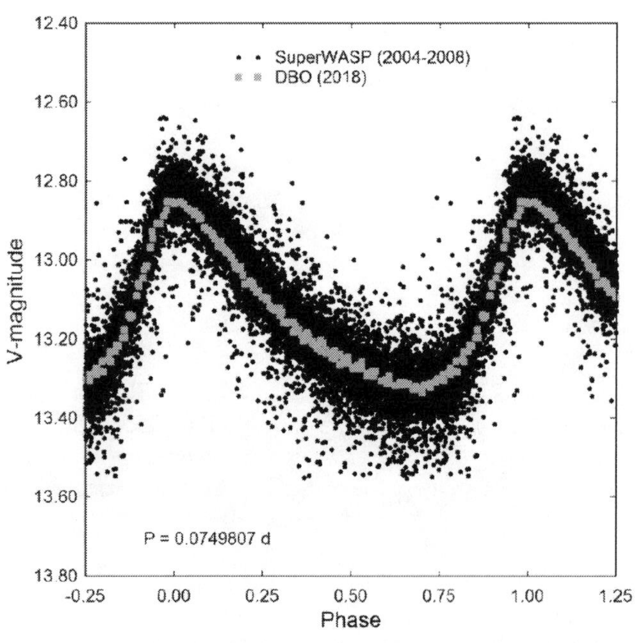

Figure 4. Period-folded (0.0749807 d) light curves for V460 And produced from precise photometric V-mag data obtained at DBO (2018) along with broad-band (400–700 nm) data from the SuperWASP Survey (2004–2008). Magnitudes were offset to conform with the APASS-derived values from DBO.

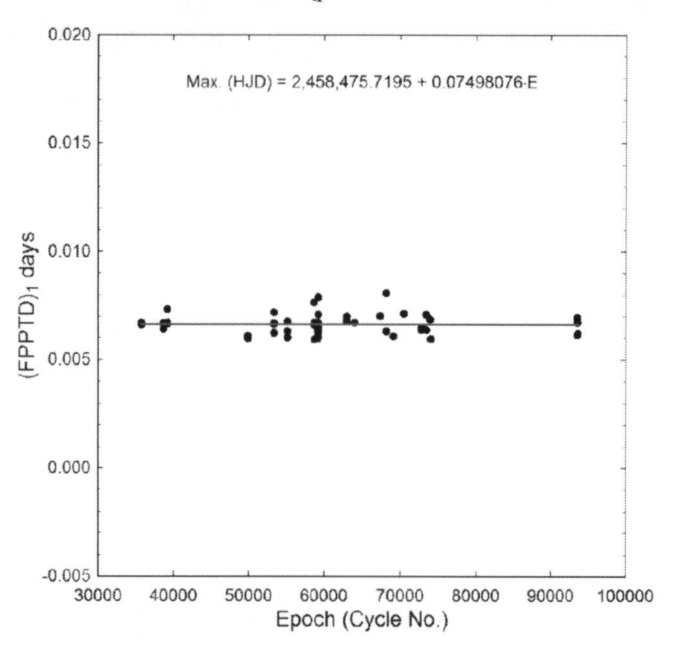

Figure 5. Straight line fit (FPPTD vs. period cycle number) suggesting that little or no change to the fundamental pulsation period of V460 And had occurred between 1999 and 2018.

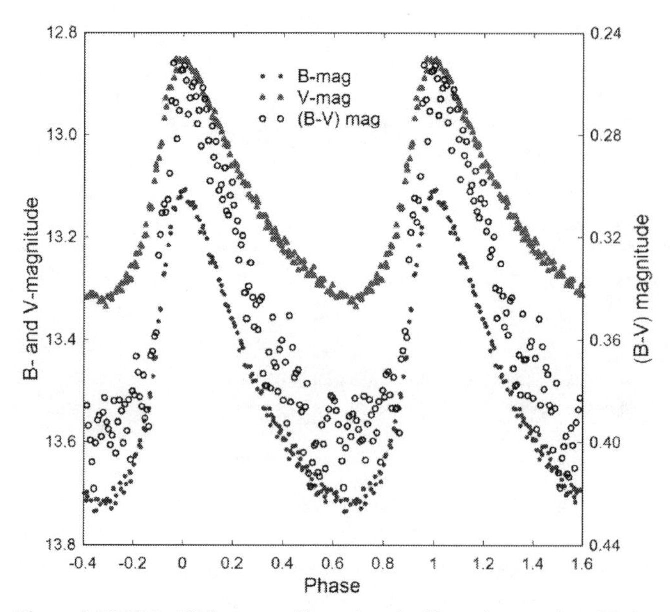

Figure 6. V460 And light curves illustrating significant increase in reddening ($0.261 < (B–V) < 0.393$) as maximum light slowly descends to minimum light. This effect is most closely associated with a decrease in the effective surface temperature during minimum light.

SuperWASP (Figure 4) make a strong case that the fundamental pulsation period has not substantively changed since 1999 nor has the V-mag amplitude changed significantly over the same period of time.

3.2. Light curve behavior

Morphologically, light curves from HADS variables are asymmetrical with a faster rise time from minimum to maximum light than the decline back to minimum brightness. V460 And appears to be a textbook example in this regard (Figure 2). The largest difference between maximum and minimum light is observed in the blue passband (ΔB-mag = 0.61), followed by V (ΔV-mag = 0.47), and finally the smallest difference in infrared (ΔI_c-mag = 0.29). This behavior is typical for pulsating F- to A-type stars. It follows when the B- and V-mag light curves are divided into equal phase intervals and then subtracted from one another, the emerging light curve (*B–V*) exhibits significant reddening during minimum light (Figure 6). In this case color excess (*B–V*) ranges between 0.261 and 0.393 mag. Estimates for interstellar extinction (A_V) vary widely depending on the model selected (Amôres and Lépine 2005, 2007; Burstein and Heiles 1978, 1982; Schlegel *et al.* 1998; Schlafly and Finkbeiner 2011; Drimmel *et al.* 2003). Access to these data is greatly facilitated via the GALextin website at http://www.galextin.org/v1p0/. The median reddening value

(E(B–V) = 0.0803 ± 0.0006 mag), coincidentally from Schlegel *et al.* (1998), corresponds to an intrinsic color index $(B–V)_0$ for V460 And that varies between 0.181 ± 0.006 at maximum light and 0.313 ± 0.010 mag at minimum brightness. Based on the polynomial transformation equations derived by Flower (1996) with the misprints corrected by Torres (2010), the mean effective temperature (T_{eff}) was estimated to be 7385 ± 520 K, with a minimum T_{eff} of ~7150 ± 229 K and a maximum T_{eff} of ~7880 ± 263 K. These results based strictly on (B–V) photometry at DBO are in good agreement with the findings for V460 And (T_{eff} = 7507^{-339}_{+427} K) included in the Gaia DR2 release of stellar parameters (Andrae *et al.* 2018). Furthermore, J- and K-band data from 2MASS (Skrutskie *et al.* 2006) when transformed according to Casagrande *et al.* (2010) predict a T_{eff} range between 7550 and 7100 K, depending on metallicity (–2.0 < [Fe/H] ≤ 0). Although no verifiable classification spectra were found in the literature, the spectral type of this variable would likely range between A9 and F1.

3.3. Light curve analysis by Discrete Fourier Transformation

Light curve deconvolution was performed with PERIOD04 (Lenz and Breger 2005) wherein discrete Fourier transformation (DFT) was used to extract the fundamental pulsating frequency (spectral window = 100 d^{-1}). Pre-whitening steps which successively remove the previous most intense signals were employed to tease out other potential oscillations from the residuals. Only those frequencies with a S/N ≥ 4 in each passband are presented in Table 3. In all cases, uncertainties in frequency, amplitude, and phase were estimated by the Monte Carlo simulation (n = 400) routine built into PERIOD04.

The results strongly indicate that V460 And is a monoperiodic radial pulsator; changes in stellar size during each pulsation cycle are therefore symmetrical. The spectral window and amplitude spectra derived from the B- and V-passband data are illustrated in Figure 7; others are not included since they are essentially redundant with respect to detected frequencies. As would be expected, the fundamental pulsation period (f_0 ≃ 13.336 d–1 ≃ 154.4 μHz) has the greatest amplitude. Successive pre-whitening steps uncovered partial harmonics out as far as $6f_0$; however, they were not statistically significant (S/N < 4) beyond $4f_0$ in the I_c-passband. The amplitude decay appears to be exponential as a function of harmonic order (Figure 8), a behavior that has been observed with other HADS variables such as VX Hya (Templeton *et al.* 2009) and RR Gem (Jurcsik *et al.* 2005). Although no other independent pulsation modes were detected during this short campaign, it is acknowledged that a longer baseline in time from multiple sites would be required to validate this claim (Breger 2000). Representative light curve fits to B-, V-, and I_c-mag time-series data (Dec 21, 2018) following DFT analysis are illustrated in Figure 8.

3.4. Global parameters

Pulsating stars have long served as standard candles for estimating cosmic distances to individual stars, clusters, and galaxies. One of the most important historical events in astronomy occurred when Henrietta Leavitt discovered a period-luminosity (P-L) relationship between 25 Cepheid variables in the Small Magellanic Cloud (Leavitt and Pickering 1912). Since then this P-L relationship has been refined owing to differences between metal-rich (Population I) and metal-poor (Population II) Cepheids (Baade 1956). Like the Cepheids, other variable stars that pulsate via the κ-mechanism were found to obey distinct P-L relationships. Robust P-L relationships in the near infrared (Longmore *et al.* 1986) and mid-infrared (Neeley *et al.* 2015) for the ubiquitous RR~Lyrae-type variables have been established to estimate distances to globular clusters. The earliest descriptions of a P-L relationship for δ Sct variables were published by Frolov (1969) and Dworak and Zieba (1975). A more modern refinement of the P-L relationship for δ Sct variables was reported by McNamara (2011) albeit with Hipparcos parallaxes and not the more accurate values determined by the Gaia Mission (Lindegren *et al.* 2016; Gaia *et al.* 2016, 2018). Nonetheless this empirically-derived expression (Equation 3):

$$M_V = (-2.89 \pm 0.13)\ log(P) - (1.31 \pm 0.10), \quad (3)$$

appears to correspond reasonably well to the main ridge of Gaia DR2-derived P-L data for δ Sct variables determined by Ziaali *et al.* (2018).

Absolute V_{mag} (M_V) was estimated (1.941 ± 0.177) after substituting the fundamental pulsation period P(0.07498076 d) into Equation 3. Using known values for m (V_{avg} = 13.142 ± 0.151), A_V = 0.2489 ± 0.0019), and M_V, the reddening corrected distance modulus (Equation 4):

$$d(pc) = 10^{(m - M_V - A_V + 5)/5)}, \quad (4)$$

produced an estimated distance (1550 ± 166 pc) to V460 And. This value is well within the Gaia DR2 determination of distance (1526^{+128}_{-110} pc) calculated from parallax using the Bailer-Jones bias correction (Bailer-Jones 2015). In the future, investigators using small- to modest-aperture instruments should be able to estimate M_V from parallax (π) data since Gaia DR2 covers a large percentage of stars brighter than G-mag = 15. In this case, since d = 1526 pc, A_V = 0.2489, and V_{avg} = 13.142, the value estimated for M_V is 1.98 ± 0.17, similar to that determined from Equation 3. Gaia DR2 also includes estimates for stellar parameters (Andrae *et al.* 2018) such as radius and luminosity. However, it is worth exploring differences between these values reported for V460 And in Gaia DR2 and those otherwise determined herein. First it should be noted that the Gaia passbands (BP, G, and RP) are unique (Jordi *et al.* 2010) so that transforms are needed should one desire conversion to other conventional photometric systems. For example, when color excess (B–V) is known, Gaia G-magnitudes can be transformed to Johnson-Cousins V_{mag} according to the following expression (Equation 5):

$$G - V = a + b(B - V) + c(B - V)^2 + d(B - V)^3. \quad (5)$$

The appropriate Johnson-Cousins coefficients (a-d) can be found in Table 5.8 of Documentation Release 1.1 from Gaia Data Release 2 (https://gea.esac.esa.int/archive/-documentation/GDR2/). In this case using V460 And values derived for G_{BP} (13.4026±0.0174), G_{RP} (12.92±0.011), and G (13.244±0.0021), V_{mag} was calculated to be 13.296±0.004, a result close to the observed V_{min} (13.288±0.010) for V460 And. Derivation of stellar parameters first released from the Gaia Mission is described in detail by Andrae *et al.* (2018). For the purposes of this paper the steps used to calculate luminosity and radius have been greatly simplified below. Absolute G-band magnitude (M_G) is estimated according to Equation 6:

$$M_G = G - 5 \cdot log_{10}(r) + 5 - A_G, \quad (6)$$

where G is the photometric system magnitude, r = 1/π (arcsec), and A_G is the interstellar extinction. In order to determine stellar luminosity, the calculated value for M_G is adjusted by the bolometric correction (BC_G) using the T_{eff} dependent polynomial coefficients provided in Table 4 and Equation 7 from Andrae *et al.* (2018). In this case BC_G = 0.0608, such that M_{bol} = 2.278 when A_G = 0. The assumed null value for A_G is an important distinction since non-Gaia data such as those determined from independently derived extinction maps are not used to produce absolute magnitude estimates. The luminosity of V460 And in solar units (L_* = 9.65 ± 1.6 $L_\odot$) was calculated according to Equation 7:

$$L_*/L_\odot = 10^{((M_{bol\odot} - M_{bol*})} / 2.5), \quad (7)$$

where $M_{bol\odot}$ = 4.74 and M_{bol*} = 2.278. Finally, the radius of V460 And in solar units (R_* = 1.84 ± 0.20) was estimated using the well-known relationship (Equation 8) where:

$$L_*/L_\odot = (R_*/R_\odot)^2 \, (T_*/T_\odot)^4. \quad (8)$$

It is very challenging to accurately determine the mass of a single isolated field star. Nonetheless, according to a model using MS stars in detached binary systems, Eker *et al.* (2018) developed a mass-luminosity relationship (1.05 < M/$M_\odot$ ≤ 2.40) according to Equation 9:

$$log(L) = 4.329(\pm 0.087) \cdot log(M) - 0.010(\pm 0.019). \quad (9)$$

This expression leads to a mass (M_* = 1.71 ± 0.07) in solar units as derived from the Gaia DR2 stellar parameters where L_* = 9.65 ± 1.55 $L_\odot$. All of these values (M_*, R_*, L_*, and T_{eff}) summarized in Table 4 fall well within expectations for a HADS variable. It bears repeating, however, that these fundamental physical parameters were derived by assuming that A_G = 0 according to Equation 6. As it turns out V460 And is in a region of the Milky Way (Gal. coord. (J2000): *l* = 142.5144; *b* = –16.6884) where interstellar extinction (A_V = 0.2489) should not be ignored. Therefore, the same equations (Equations 6–8) were applied but this time using the data obtained at DBO where V_{avg} = 13.142 ± 0.151, A_V = 0.2489 ± 0.0019, M_V = 1.975, and BC_V = 0.0348. The results summarized in Table 4 indicate that the Gaia DR2 reported values for luminosity and radius appear to be underestimated largely due to different assumptions about interstellar extinction. The greater luminosity (12.36 ± 2.69 $L_\odot$) produced from the DBO data translates into a higher estimate for mass (1.82 ± 0.01) according to Equation 9. Furthermore, stellar radius was independently estimated from an empirically-derived period-radius (P-R) relationship (Equation10) reported by Laney *et al.* (2003) for HADS and classical Cepheids:

$$log(R_*) = a + b \cdot log(P) + c, \quad (10)$$

where a = 1.106 ± 0.012, b = 0.725 ± 0.010, and c = 0.029 ± 0.024. In this case the value for R_* (2.09 ± 0.14 $R_\odot$) was closer to the value obtained from observations at DBO (2.15 ± 0.38 $R_\odot$). Other derived values for density ($\rho_\odot$), surface gravity (log g), and pulsation constant (Q) are also included in Table 4. Stellar density (ρ_*) in solar units (g/cm³) was calculated according to Equation 11:

$$\rho_* = 3 \cdot G \cdot M_* \cdot m_\odot (4\pi (R_* \cdot r_\odot)^3), \quad (11)$$

where G = the gravitational constant (6.67408 · 10^{-8} $cm^3 \cdot g^{-1} \cdot sec^{-2}$), $m_\odot$ = solar mass (g), $r_\odot$ = solar radius (cm), M_* is the mass, and R_* the radius of V460 And in solar units. Using the same algebraic assignments, surface gravity (log g) was determined by the following expression (Equation 12):

$$log g = \log(M_* \cdot m_\odot / (R_* \cdot r_\odot)^2). \quad (12)$$

When attempting to characterize p-mode pulsations (radial) it is helpful to introduce the concept of a pulsation constant (Q). The dynamical time that it takes a p-mode acoustic wave to internally traverse a star is related to its size but more accurately the mean density. This is defined by the period-density relationship (Equation 13):

$$Q = P \sqrt{\bar{\rho}_* / \bar{\rho}_\odot} \quad (13)$$

where P is the pulsation period (d) and $\bar{\rho}_*$ and $\bar{\rho}_\odot$ are the mean densities of the target star and Sun, respectively. The mean density of an isolated field star like V460 And can not be determined without great difficulty. However, it can be expressed in terms (Equation 14) of other measurable stellar parameters where:

$$log(Q) = -6.545 + log(P) + 0.5 \, log(g) + 0.1 \, M_{bol} + log(T_{eff}). \quad (14)$$

The full derivation of this expression is provided in Breger (1990). The resulting Q values (Table 4) derived from observations at DBO are consistent with theory (Q = 0.032 d) and the distribution of Q-values (0.03–0.04 d) from fundamental radial pulsations observed with other δ Sct variables (Breger 1979; Joshi 2015; Antonello and Pastori 1981).

Finally, we attempted to get a relative sense of how the physical size, temperature, and brightness of V460 And changes over the course of a single 1.8-hr pulsation. As shown in Figure 6 there is a significant increase in reddening (B–V) as maximum light descends to minimum light. Intrinsic color reveals that at maximum light, where $(B–V)_0$ = 0.181 ± 0.006, the corresponding effective temperature is 7883 ± 263 K, whereas at minimum light ($(B–V)_0$ = 0.313 ± 0.010) the estimated effective temperature is 7151 ± 229 K. Between these two extremes the putative rise in temperature (+732 K) would correspond to a 1.5-

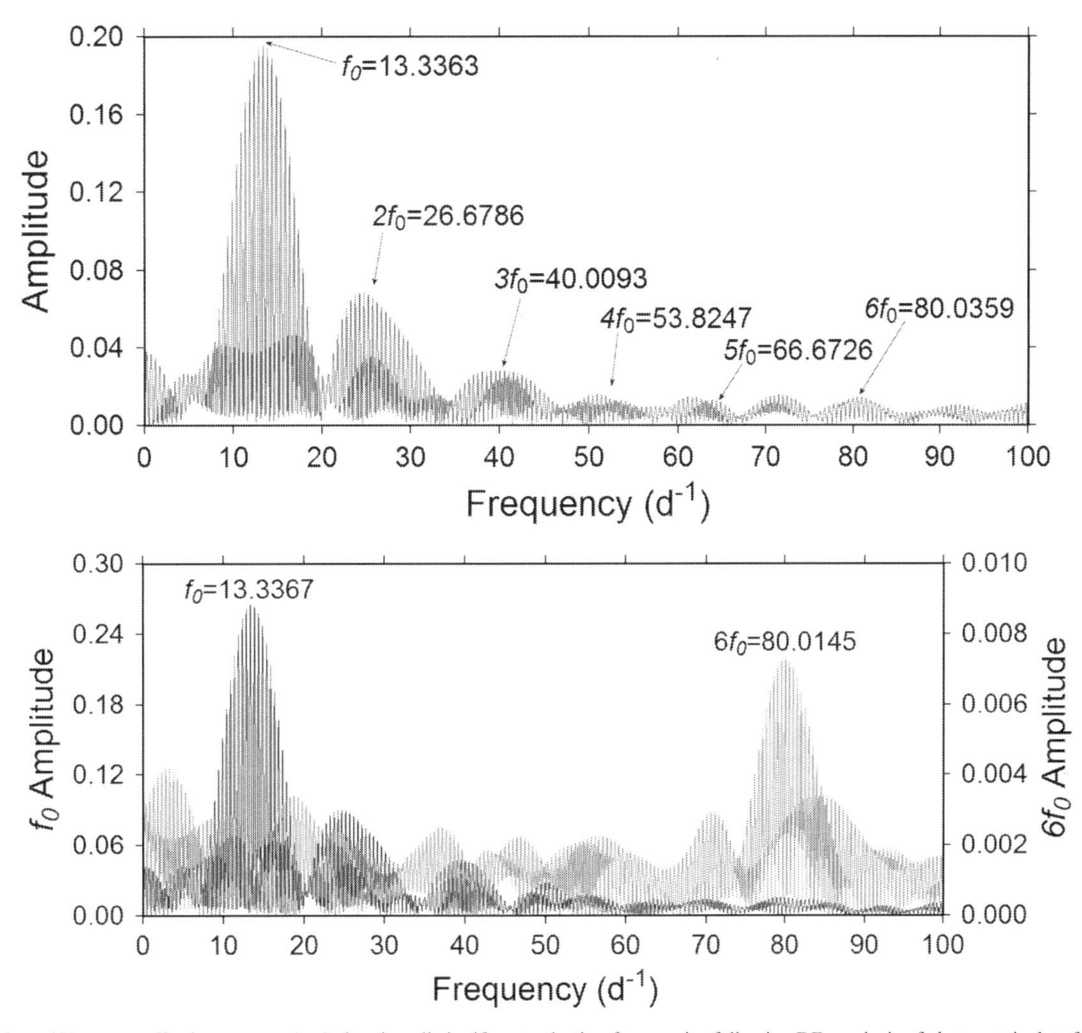

Figure 7. Unwhitened V-mag amplitude spectrum (top) showing all significant pulsation frequencies following DF_T analysis of photometric data from V460 And acquired in 2018 at DBO. The bottom amplitude spectra (B-mag) illustrate the fundamental (f_0) frequency and its highest partial harmonic ($6f_0$), which was clearly detected (S/N ≥ 4) following prewhitening.

fold increase in luminosity but only a relatively small increase (+0.02 $R_\odot$) in radius. This rather crude estimate for changes in stellar radius would be best performed using the Baade-Wesselink method developed by Wesselink (1946) should radial velocity data over an entire oscillation cycle become available for this system.

3.5. Kinematics

Stellar kinematics on field stars have been used (Balona and Nemec 2012, hereafter BN2012) to ostensibly discriminate between Pop. I HADS and its sibling Pop. II SX Phe-type variables found during the Kepler Mission (Gilliland *et al.* 2010). This potentially important observation arrives at a time when new results from space telescopes indicate that SX Phe variables are not necessarily high amplitude, low luminosity, or metal-poor (Nemec *et al.* 2017). It would appear that the canonical definition of SX Phe variables likely suffers from observational bias due to the sensitivity limitations of ground-based telescopes. Gaia DR2 (Gaia *et al.* 2016, 2018) provides highly accurate data for proper motion (PM) and parallax from nearly 80 million sources (G ≤ 15 mag) that fall within the light grasp of small (100 to 400 mm) aperture telescopes. PM (μ_α in R.A. and μ_δ in Dec.) must be understood within the context of where the star resides; it may appear to be large when relatively close to the Sun or diminishingly small at much greater distances. Gaia DR2 (Sartoretti *et al.* 2018) only includes radial velocity (RV) data from stars with an effective temperature between 3500 and 7000 K, thereby eliminating the possibility of calculating space velocity for V460 And along with most other HADS variables. According to BN2012, another discriminating measure of motion relative to the Sun may be tangential velocity (V_T), which factors in distance according to the relationship (Equation 15):

$$V_T = 4.74 \cdot \mu \cdot d \qquad (15)$$

where V_T is in km s^{-1}, PM = μ_α or μ_δ (mas y^{-1}), and d is the distance in kpc. For V460 And, substituting the Gaia DR2 values for PM (μ_α = –0.134 ± 0.061 and μ_δ = –2.976 ± 0.067 mas y^{-1}) and distance (d = 1.526 kpc) lead to V_T values of –0.97 ± 0.45 ($V_{T\alpha}$) and –21.53 – 1.87 ($V_{T\delta}$). By comparison it would appear that the V_T (> 120 km s^{-1} in R.A. or Dec.) of SX Phe variable candidates in the Kepler field (BN2012) far exceed the corresponding velocities observed for V460 And (Table 5). On the strength of

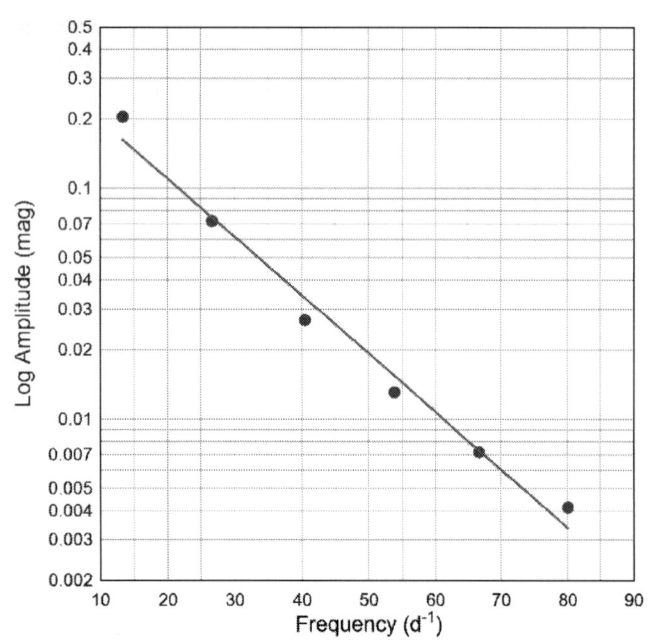

Figure 8. Amplitude decay of the fundamental (f_0) pulsation period and its corresponding partial harmonics ($2f_0$–$6f_0$) observed in the B-passband.

these results, can V460 And be unambiguously classified as a HADS and not an SX Phe variable? Unfortunately, the results generated in BN2012 proved to be much less compelling when comparisons were made using data from Gaia DR2 instead of the UCAC3 catalog (Zacharias *et al.* 2010). This can be seen in Table 5 which compares kinematics from candidate SX Phe stars (Table 1 in BN2012) to those recalculated using PM and distances (Bailer-Jones 2015) from Gaia DR2. This assessment also includes the same kinematics from a group of HADS variables located in Lyra and and Cygnus which are classified as such in the International Variable Star Index (VSX). As illustrated in Figure 10, for the most part the HADS variables identified in VSX co-mingle with the central cluster of the SX Phe candidates and V460 And. Far fewer SX Phe candidates (3 vs. 34) emerge that appear to have V_T values significantly different (circle radius = 1σ) from the mean value. In addition those SX Phe candidates positioned within 1σ of the mean capture the full range of [Fe/H] values provided in BN2012 with no bias towards solar-like metallicity. RR Lyrae variables were included by BN2012 as a positive control since they were known to be very metal-poor (–2.54 > [Fe/H] > –0.42) and estimated to reside at even greater distances (1.5–16.1 kpc). However when V_T was recalculated using Gaia DR2-derived values for PM (Table 6) and distance, less than half (10/22) fell outside the variability (1σ) observed with known HADS

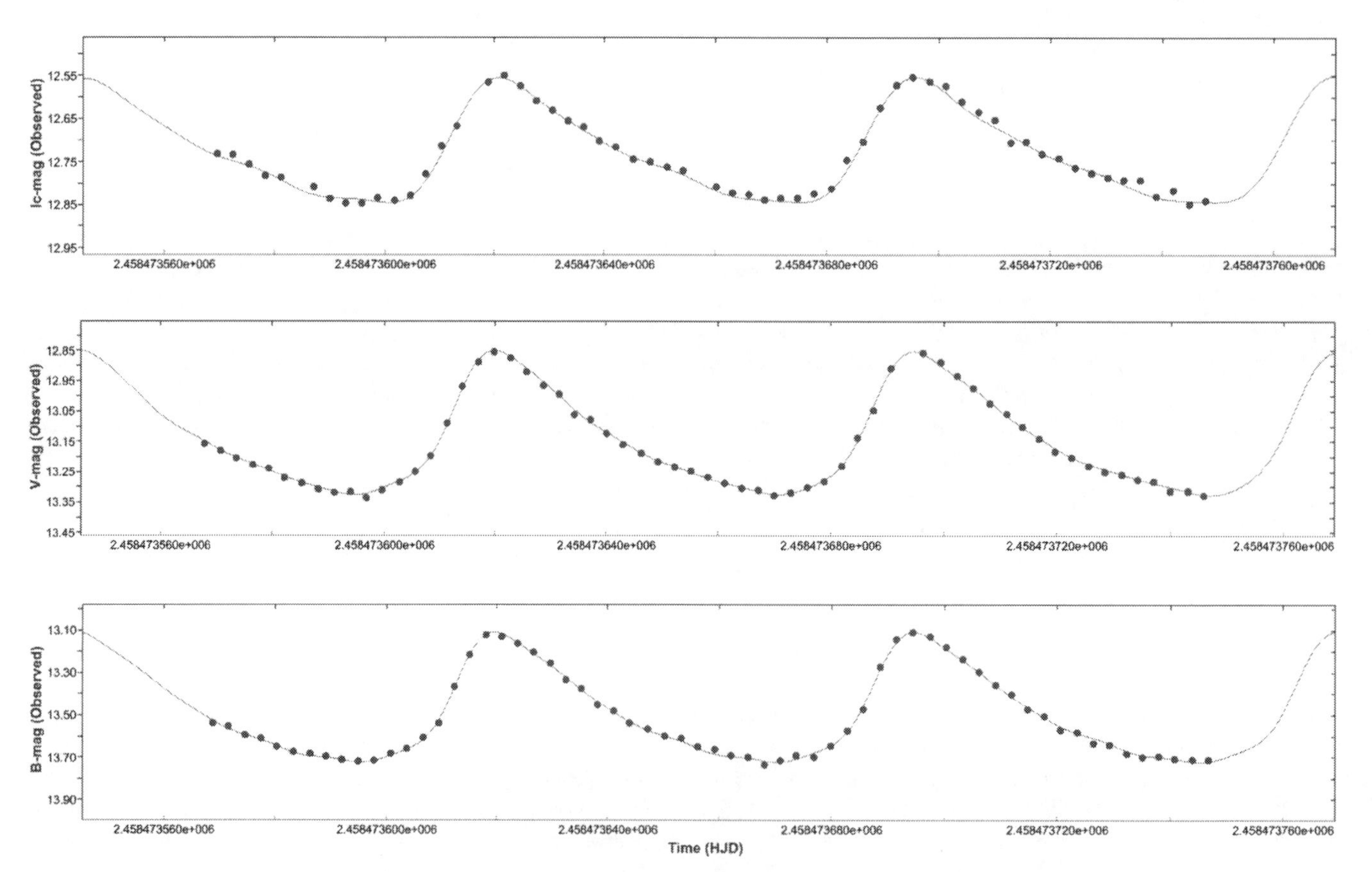

Figure 9. Representative fit of B- (bottom), V- (middle), and I_c-mag (top) time series data based on elements derived from DFT. These data were acquired on Dec. 21, 2018 at DBO.

Table 4. Global stellar parameters for V460 And using Gaia DR2 derived values and those determined directly from observations at DBO.

Parameter	Gaia DR2	DBO
Mean T_{eff} [K]	7507 ± 427	7385 ± 520
Mass [$M_{\odot}$]	1.71 ± 0.07	1.82 ± 0.01
Radius [$R_{\odot}$]	1.84 ± 0.20	2.15 ± 0.38
Luminosity [$L_{\odot}$]	9.65 ± 1.55	12.36 ± 2.69
ρ [g/cm^3]	0.386 ± 0.127	0.258 ± 0.139
log g [cgs]	4.14 ± 0.10	4.03 ± 0.14
Q [d]	0.0393 ± 0.005	0.0321 ± 0.006

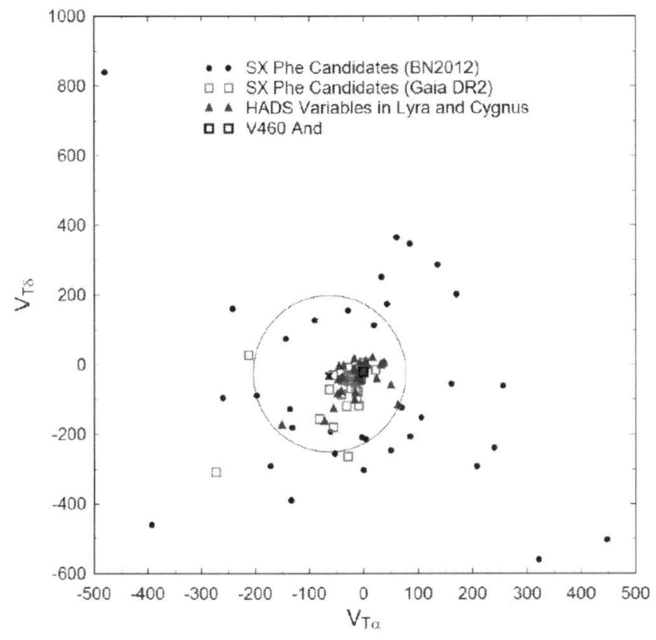

Figure 10. Comparison of tangential velocities ($V_{T\alpha}$ and $V_{T\delta}$) for putative SX Phe stars in the Kepler field using results from BN2012 (●) and those calculated herein with proper motion and distances from Gaia DR2 (□). Known HADS variables in Lyra and Cygnus (VSX) (▲) show very little differentiation from all but three SX Phe candidates.

variables in the same region of the sky (Figure 11). In essence, V_T by itself does not offer a reliable means to determine whether V460 And can be unequivocally classified as a HADS variable. A noteworthy calculation (Equation 16):

$$z = d \cdot sin(b), \tag{16}$$

where d is distance in pc (1526) and b is the Galactic latitude (–16.688°), places V460 And nearly 440 pc below the Galactic plane, territory also occupied by stars in the thick disk. Given these seemingly contradictory results, a more provocative question posits whether there is any unique set of features that allows one to classify a field star as a HADS or SX Phe-type pulsating variable.

3.6. Evolutionary status of V460 And

Knowing the luminosity and effective temperature of V460 And we can attempt to describe the evolutionary status of this variable. These values are plotted in the theoretical Hertzsprung-Russell diagram (HRD) shown in Figure 12. Here, the thick solid line gives the ZAMS position for stars with solar metallicity while two broken lines nearly perpendicular to the ZAMS delimit the blue and red edges of the theoretical instability strip for radial low-p modes (Xiong *et al.* 2016). Asterisks mark the positions of several known HADS, both δ Scuti and SX Phe types (Balona 2018). The open circle indicates the position of V460 And using the DBO-derived parameters and corresponding error estimates provided in Table 4. To determine the mass and age of V460 And from theoretical evolutionary tracks its metallicity, Z, needs to be known. Unfortunately, this star has not been observed spectroscopically so no direct measurement of Z exists, however, we can at least try to estimate its value indirectly. V460 And has low tangential velocity (Figure 10), suggesting its affiliation with the galactic disc. Also its distance from the galactic plane favors a thick disc membership rather than halo. We can therefore assume that V490 And approaches solar metallicity, or at most a few times lower which also corresponds to the metallicity of metal-rich globular clusters classified as Oosterhoff type I.

But what is the true value of the solar metallicity? The numbers obtained in the last few decades range between 0.012 and 0.020, with the recently derived one of $Z_{\odot}=0.0142$ (Asplund *et al.* 2009). However, very recently von Steiger and Zurbuchen (2016) questioned this result and obtained the value of $Z_{\odot}=0.0196\pm0.0014$ based on the analysis of the chemical composition of the solar wind. Yet, Serenelli *et al.* (2016) quickly showed that the derived composition is in serious disagreement with the observables of the basic solar model so it cannot be representative of the solar interior. Obviously, the problem of a precise value for $Z_{\odot}$ still remains open. We plot two series of PARSEC evolutionary models (Bressan *et al.* 2012) in Figure 12 wherein red solid lines show the models with Z=0.020 and blue, dash-dotted lines define the models with Z=0.004. The latter models would correspond to a decrease in metallicity by a factor of 3 to 5, depending on the reference solar metallicity. Assuming Z=0.020, it can be seen (Figure 12) that V460 And has a solar mass of 1.75±0.05, solar radius of 2.15±0.26, and an age of 1.11 Gyr. Alternatively a metal poor (Z=0.004) star would likely be smaller ($R_{\odot}=2.0\pm0.26$), less massive ($M_{\odot}=1.43\pm0.03$), and older (2.0 Gyr). Although V460 And lies closer to ZAMS than most of the plotted variables, it is an MS object which lies well within the instability strip among the other HADS. Uncertainty in the determination of mass will hopefully improve in the future should spectroscopic data become available for the V460 And.

4. Conclusions

This first multi-color (BVI_c) CCD study of V460 And has produced six new times-of-maximum which along with other published values lead to an updated linear ephemeris. Potential changes in the pulsation period assessed using the observed and predicted times-of-maximum suggests that since 1999 no significant change has occurred. Deconvolution of time-series photometric data by discrete Fourier transformation indicates that V460 And is a monoperiodic radial pulsator ($f_0 \simeq 13.336\,d^{-1}$) which also oscillates in at least five other partial harmonics ($2f_0$–$6f_0$). Lacking a definitive classification spectrum, the intrinsic color, $(B–V)_0$, determined from this study was

Table 5. Proper motion (μ) and tangential velocity (V_T) for putative SX Phe stars in the Kepler field reported in BN2012 and those determined herein using data derived from Gaia DR2.

	BN2012					Gaia DR2				
Kepler ID	μ_α[a]	μ_δ[a]	d[b]	$V_{T\alpha}$[c]	$V_{T\delta}$[c]	μ_α	μ_δ	d	$V_{T\alpha}$	$V_{T\delta}$
KIC 1162150	−33.1	37.6	0.7	−90	128	0.0	1.0	1.0	0	5
KIC 3456605	1.3	−48.4	0.9	4	−214	−1.9	−5.5	2.6	−23	−68
KIC 4168579	36.0	−39.2	1.6	208	−292	−2.2	−7.4	1.4	−15	−50
KIC 4243461	−74.4	−21.4	0.9	−260	−96	−2.5	−4.6	2.2	−26	−48
KIC 4662336	0.0	−62.1	1.0	0	−303	0.1	−1.9	1.5	1	−13
KIC 4756040	28.5	−31.6	1.0	106	−153	−0.8	−7.2	1.3	−5	−43
KIC 5036493	−1.1	−45.1	1.0	−3	−209	−5.6	−7.0	1.4	−36	−45
KIC 5390069	−11.9	47.6	0.7	−29	155	−0.9	−7.7	2.7	−12	−100
KIC 5705575	92.4	−78.5	1.3	448	−504	−1.5	−7.4	2.0	−15	−71
KIC 6130500	59.9	−11.0	1.2	256	−62	−4.8	−5.6	1.7	−38	−44
KIC 6227118	21.8	−28.4	4.2	322	−561	−2.3	−7.9	1.1	−12	−42
KIC 6445601	−55.6	−18.7	1.0	−197	−89	−1.0	−7.2	2.2	−11	−77
KIC 6520969	−42.5	21.0	1.6	−242	160	−40.2	5.1	1.1	−212	27
KIC 6780873	−24.5	−56.7	0.7	−61	−193	−1.0	−5.2	1.8	−8	−45
KIC 7020707	−32.6	−22.6	1.2	−135	−127	−0.2	−1.2	1.5	−2	−8
KIC 7174372	−35.7	−76.7	1.1	−133	−390	−5.1	−9.8	3.4	−81	−157
KIC 7300184	−45.7	−46.5	0.8	−131	−182	−2.1	−7.0	5.4	−55	−180
KIC 7301640	44.2	−11.2	1.0	161	−56	−0.4	−0.6	1.3	−3	−4
KIC 7621759	8.9	48.1	1.1	33	251	−0.5	−6.0	1.5	−4	−42
KIC 7765585	27.7	81.2	0.4	43	173	−3.0	−3.6	2.2	−32	−38
KIC 7819024	11.6	−41.1	1.3	50	−246	−1.0	−14.1	1.8	−9	−118
KIC 8004558	−11.1	−38.4	1.4	−53	−256	−4.5	−41.1	1.4	−29	−263
KIC 8110941	−46.8	17.5	0.9	−143	74	−1.6	−6.3	4.0	−31	−120
KIC 8196006	24.4	−31.2	0.8	69	−124	2.7	−1.9	1.7	22	−15
KIC 8330910	39.0	59.2	1.0	135	286	−1.2	−5.7	1.4	−8	−39
KIC 9244992	40.2	−28.2	1.8	240	−240	−7.0	−3.9	1.7	−56	−31
KIC 9267042	−79.0	96.5	1.8	−480	840	−2.8	−0.9	2.2	−29	−9
KIC 9535881	6.9	28.2	0.8	19	113	−3.7	−4.8	1.4	−25	−33
KIC 9966976	19.1	53.8	1.4	84	346	−1.3	−5.6	1.5	−9	−39
KIC 10989032	−68.0	−52.8	1.8	−393	−460	−1.9	−3.1	2.4	−21	−35
KIC 11649497	21.4	−33.6	1.3	85	−206	−4.6	−9.6	1.9	−41	−86
KIC 11754974	−53.2	−57.7	1.1	−172	−291	−51.7	−58.3	1.1	−273	−308
KIC 12643589	98.6	72.4	0.6	170	201	−0.4	−1.9	1.7	−3	−16
KIC 12688835	9.5	35.2	2.2	60	364	−4.0	−4.6	3.3	−63	−72

[a] *μ_α = proper motion (mas y^{-1}) in R.A. and μ_δ = proper motion (mas y^{-1}) in Dec.*
[b] *d = distance in kpc.*
[c] *$V_{T\alpha}$ = tangential velocity (km s^{-1}) in R.A. and $V_{T\delta}$ = tangential velocity (km s^{-1}) in Dec.*

used to estimate a mean effective temperature for V460 And (7385 ± 520 K); this corresponds to spectral type A9-F1. These results along with the distance estimate (1550 ± 166 pc) agreed quite well with the same findings (1526^{+128}_{-110} pc) provided in Gaia DR2. The pulsation period (0.07498076 d), oscillation mode (radial), V_{mag} amplitude (0.47 mag), and light curve morphology are all consistent with the defining characteristics of a HADS variable. Even if a metallicity ([Fe/H]) determination was available, these criteria do not necessarily exclude the possibility that V460 And is an example of a field SX Phe-type pulsator. In this case, the estimated mass of V460 And (1.70–1.80 $M_\odot$) according to Eker *et al.* (2018) exceeds the generally accepted threshold ($M < 1.3\,M_\odot$) for SX Phe stars (McNamara 2011). Furthermore, evolutionary tracks from the PARSEC model which assume near solar abundance ($Z = 0.020$) for V460 And are best matched by a MS star with a mass of $1.75 \pm 0.05\,M_\odot$ and radius of $2.15\,R_\odot$. Given these results, the sum total of evidence points to a HADS rather than an SX Phe variable. Unlike previously published findings (BN2012), a kinematic assessment using data from Gaia DR2 failed to prove that tangential velocity alone could be used to differentiate HADS from SX Phe stars. New results arriving from various space telescopes appear to contradict the traditional definition for each type, belying the notion that field stars like V460 And can be neatly classified as HADS or SX Phe variables.

5. Acknowledgements

This research has made use of the SIMBAD database operated at Centre de Données astronomiques de Strasbourg, France. Time-of-maximum light data from the *Information Bulletin on Varariable Stars* website proved invaluable to the assessment of potential period changes experienced by this variable star. In addition, the Northern Sky Variability Survey hosted by the Los Alamos National Laboratory, the International Variable Star Index maintained by the AAVSO, and the Catalina Sky Survey (CSDR1) maintained at CalTech were mined for critical information. This work also presents results from the European Space Agency (ESA) space mission Gaia. Gaia data are being processed by the Gaia Data Processing

Table 6. Proper motion (μ) and tangential velocity (V_T) of RR Lyae stars in the Kepler field reported in BN2012 and those determined herein using data derived from Gaia DR2.

	BN2012					*Gaia DR2*				
Kepler ID	$\mu_\alpha^{\,a}$	$\mu_\delta^{\,a}$	d^b	$V_{T\alpha}^{\,c}$	$V_{T\delta}^{\,c}$	μ_α	μ_δ	d	$V_{T\alpha}$	$V_{T\delta}$
KIC3733346	−17.1	−3.5	2.7	−221	−45	−17.05	−6.38	2.890	−233.49	−87.42
KIC3864443	−13.3	9.8	10.4	−658	484	−3.23	3.77	5.792	−88.60	103.48
KIC4484128	−22.5	−27.9	9.4	−1000	−1241	−0.60	−1.3	8.357	−23.65	−51.58
KIC5299596	−7.1	9.1	9.5	−322	410	−3.33	−6.01	4.161	−65.66	−118.43
KIC5559631	5.9	10.6	6.7	187	338	5.94	2.00	4.617	130.01	43.85
KIC6070714	−7.8	−1	9.4	−346	−44	−2.18	−5.54	5.226	−54.10	−137.12
KIC6100702	3.9	10.3	3.9	73	190	−0.61	0.77	3.106	−9.01	11.26
KIC6183128	−14.3	18.6	14.2	−959	1251	−0.58	0.26	6.285	−17.22	7.69
KIC6763132	−0.9	−1.9	3.3	−14	−29	−0.09	−1.05	3.054	−1.25	−15.23
KIC6936115	3.4	10.2	3	48	144	7.49	9.67	3.026	107.43	138.70
KIC7505345	4.4	−1.1	5.2	109	−27	1.67	5.51	4.406	34.94	114.96
KIC7742534	−6.4	−6.1	12.6	−380	−364	−0.52	−1.29	5.960	−14.80	−36.50
KIC7988343	−9.6	6.2	6.3	−287	184	−10.22	4.61	4.781	−231.56	104.38
KIC8344381	−2.3	2.2	15.3	−170	159	−1.54	−0.44	4.261	−31.08	−8.97
KIC9578833	0.3	16.5	16.1	21	1261	0.89	−1.13	7.379	31.23	−39.38
KIC9591503	−3.6	9	3.6	−62	154	−5.06	11.90	3.707	−88.98	209.04
KIC9697825	−3.9	10.1	14.2	−261	681	−2.30	0.36	8.210	−89.66	13.82
KIC9947026	1.4	8	3.6	23	137	−5.63	−6.63	3.475	−92.70	−109.23
KIC10136240	6.3	19.3	10.7	317	979	−0.67	0.25	5.194	−16.59	6.11
KIC10789273	5.5	−1.1	4.5	117	−23	5.83	9.05	8.378	231.39	359.22
KIC11125706	−6	−12.8	1.5	−42	−90	−8.07	−14.61	1.546	−59.13	−107.05
KIC11802860	−7.9	−3.9	3.2	−120	−59	−9.89	−6.23	3.031	−142.05	−89.57
KIC12155928	12.9	14.1	8.1	492	539	0.73	0.15	6.516	22.42	4.73

[a]μ_α = *proper motion (mas y*$^{-1}$*) in R.A. and* μ_δ = *proper motion (mas y*$^{-1}$*) in Dec.*
[b]*d = distance in kpc.*
[c]$V_{T\alpha}$ = *tangential velocity (km s*$^{-1}$*) in R.A. and* $V_{T\delta}$ = *tangential velocity (km s*$^{-1}$*) in Dec.*

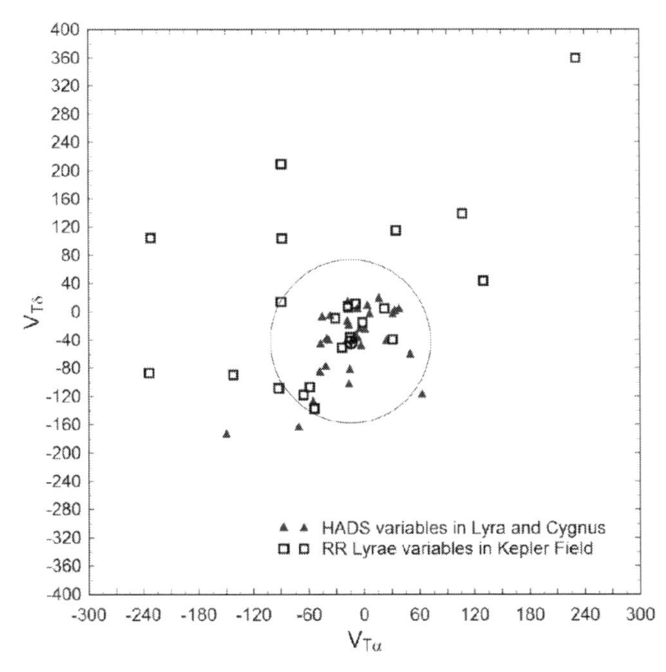

Figure 11. Comparison of tangential velocities ($V_{T\alpha}$ and $V_{T\delta}$) for RR Lyrae stars in Kepler field (□) and known HADS variables in Lyra and Cygnus (▲) using proper motion and distances from Gaia DR2. The significant overlap challenges the notion that V_T alone can be used to classify pulsating variables.

and Analysis Consortium (DPAC). Funding for the DPAC is provided by national institutions, in particular the institutions participating in the Gaia MultiLateral Agreement (MLA). The Gaia mission website is https://www.cosmos.esa.int/gaia. The Gaia archive website is https://archives.esac.esa.int/gaia. This paper makes use of data from the first public release of the WASP data (Butters *et al.* 2010) as provided by the WASP consortium and services at the NASA Exoplanet Archive, which is operated by the California Institute of Technology, under contract with the National Aeronautics and Space Administration under the Exoplanet Exploration Program. The diligence and dedication shown by all associated with these organizations is very much appreciated. We gratefully acknowledge the careful review and helpful commentary provided by an anonymous referee.

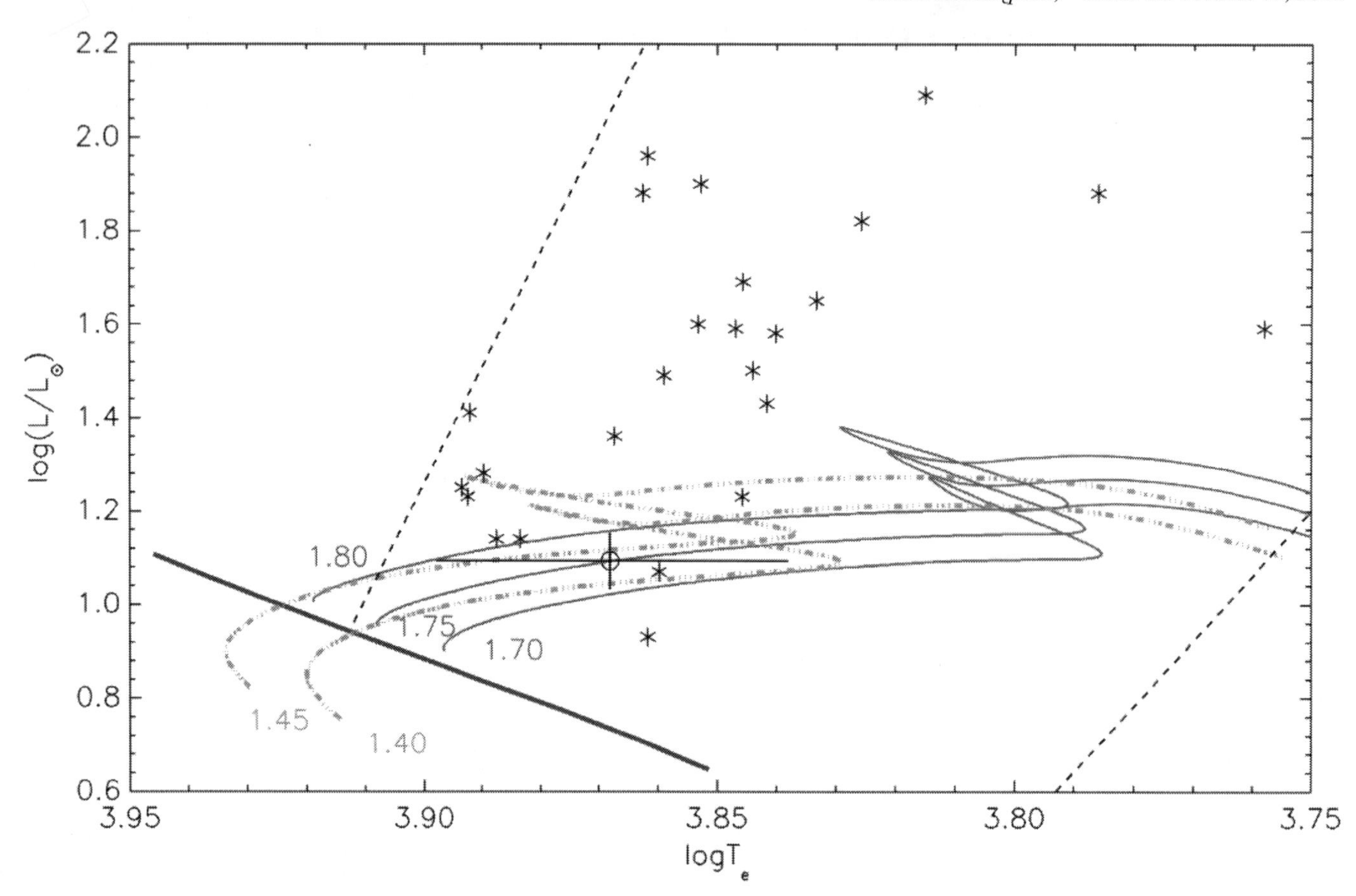

Figure 12. Evolutionary tracks (red solid line: Z =0.020 and blue dashed line: Z = 0.004) derived from PARSEC models (Bressan *et al.* 2012) showing position of V460 And (circle) relative to ZAMS (thick maroon line) within the theoretical instability strip (dashed lines) for radial low-p mode pulsators. Asterisks denote the positions of known HADs (Balona 2018).

References

Akerlof, C., *et al.* 2000, *Astron. J.*, **119**, 1901.

Amôres, E. B., and Lépine, J. R. D. 2005, *Astron. J.*, **130**, 650.

Amôres, E. B., and Lépine, J. R. D. 2007, *Astron. J.*, **133**, 1519.

Andrae, R., *et al.* 2018, *Astron. Astrophys.*, **616A**, 8.

Antonello, E., and Pastori, L. 1981, *Publ. Astron. Soc. Pacific*, **93**, 237.

Asplund, M., Grevesse, N., Sauval, A. J., and Scott, P. 2009, *Ann. Rev. Astron. Astrophys.*, **47**, 481.

Baade, W. 1956, *Publ. Astron. Soc. Pacific*, **68**, 5.

Baglin, A. 2003, *Adv. Space Res.*, **31**, 345.

Bailer-Jones, C. A. L. 2015, *Publ. Astron. Soc. Pacific*, **127**, 994.

Baker, N., and Kippenhahn, R. 1962, Z. Astrophys., **54**, 114.

Baker, N., and Kippenhahn, R. 1965, *Astrophys. J.*, **142**, 868.

Balona, L. A. 2018, *Mon. Not. Roy. Astron. Soc.*, **479**, 183.

Balona, L. A. and Nemec, J. M. 2012, *Mon. Not. Roy. Astron. Soc.*, **426**, 2413.

Berry, R., and Burnell, J. 2005, *The Handbook of Astronomical Image Processing*, 2nd ed., Willmann-Bell, Richmond, VA.

Breger, M. 1979, *Publ. Astron. Soc. Pacific*, **91**, 5.

Breger, M. 1990, *delta Scuti Star Newsl.*, **2**, 13.

Breger, M. 2000, *Baltic Astron.*, **9**, 149.

Bressan, A., Marigo, P., Girardi, L., Salasnich, B., Dal Cero, C., Rubele, S., and Nanni, A. 2012, *Mon. Not. Roy. Astron. Soc.*, **427**, 127.

Burstein, D., and Heiles, C. 1978, *Astrophys. J.*, **225**, 40.

Burstein, D., and Heiles, C. 1982, *Astron. J.*, **87**, 1165.

Butters, O. W., *et al.* 2010, *Astron. Astrophys.*, **520**, L10.

Casagrande, L., Ramírez, I., Meléndez, J., Bessell, M., and Asplund, M. 2010, *Astron. Astrophys.*, **512A**, 54.

Chevalier, C. 1971, *Astron. Astrophys.*, **14**, 24.

Cox, J. P. 1963, *Astrophys. J.*, **138**, 487.

Drake, A. J., *et al.* 2009, *Astrophys. J.*, **696**, 870.

Drimmel, R., Cabrera-Lavers, A., and López-Corredoira, M. 2003, *Astron. Astrophys.*, **409**, 205.

Dworak, T. Z., and Zieba, S. 1975, *Inf. Bull. Var. Stars*, No. 1005, 1.

Eker, Z., *et al.* 2018, *Mon. Not. Roy. Astron. Soc.*, **479**, 5491.

Elst, E. W. 1978, *Inf. Bull. Var. Stars*, No. 1442, 1.

Flower, P. J. 1996, *Astrophys. J.*, **469**, 355.

Frolov, M. S. 1969, *Astron. Tsirk.*, **505**, 1.

Gaia Collaboration, *et al.* 2016, *Astron. Astrophys.*, **595A**, 1.

Gaia Collaboration, *et al.* 2018, *Astron. Astrophys.*, **616A**, 1.

Gilliland, R. L., *et al.* 2010, *Publ. Astron. Soc. Pacific*, **122**, 131.

Henden, A. A., Levine, S. E., Terrell, D., Smith, T. C., and Welch, D. L. 2011, *Bull. Amer. Astron. Soc.*, **43**, 2011.

Henden, A. A., Terrell, D., Welch, D., and Smith, T. C. 2010, *Bull. Amer. Astron. Soc.*, **42**, 515.
Henden, A. A., Welch, D. L., Terrell, D., and Levine, S. E. 2009, *Bull. Amer. Astron. Soc.*, **41**, 669.
Hübscher, J. 2014, *Inf. Bull. Var. Stars*, No. 6118, 1.
Hübscher, J., and Lehmann, P. B. 2012, *Inf. Bull. Var. Stars*, No. 6026, 1.
Holdsworth D. L., *et al.* 2014, *Mon. Not. Roy. Astron. Soc.*, **439**, 2078.
Jordi, C., *et al.* 2010, *Astron. Astrophys.*, **523A**, 48.
Joshi, S., and Joshi, Y. C. 2015, *J. Astrophys. Astron.*, **36**, 33.
Jurcsik, J., *et al.* 2005, *Astron. Astrophys.*, **430**, 1049.
Khruslov, A. V. 2005, *Perem. Zvezdy Prilozh.*, **5**, 5.
Kinman, T. D., Mahaffey, C. T., and Wirtanen, C. A. 1982, *Astron. J.*, **87**, 314.
Kukarkin, B. V., Parenago, P. P., Efremov, Yu. N., and Kholopov, P. N. 1951, *Catalogue of Suspected Variable Stars*, Academy of Sciences USSR Sternberg, Moscow.
Laney, C. D., Joner, M., and Rodriguez, E. 2003, in *Interplay of Periodic, Cyclic and Stochastic Variability in Selected Areas of the H-R Diagram*, ed. C. Sterken, ASP Conf. Ser. 292, Astronomical Society of the Pacific, San Francisco, 203.
Leavitt, H. S., and Pickering, E. C. 1912, *Circ. Harvard Coll Obs.*, No. 173, 1.
Lee, Y.-H., Kim, S. S., Shin, J., Lee, J., and Jin, H. 2008, *Publ. Astron. Soc. Japan*, **60**, 551.
Lenz, P., and Breger, M. 2005, *Commun. Asteroseismology*, **146**, 53.
Lindegren, L., *et al.* 2016, *Astron. Astrophys.*, **595A**, 4.
Longmore, A. J., Fernly, J. A., and Jameson, R. F. 1986, *Mon. Not. Roy. Astron. Soc.*, **220**, 279.
McNamara, D. H. 2000, in *Delta Scuti and Related Stars*, ASP Conf. Ser., 210, ed. M. Breger, M. Montgomery, Astronomical Society of the Pacific, San Francisco, 373
McNamara, D. H. 2011, *Astron. J.*, **142**, 110.
Minor Planet Observer. 2011, MPO Software (http://www.minorplanetobserver.com), BDW Publishing, Colorado Springs.
Neeley, J. R., *et al.* 2015, *Astrophys. J.*, **808**, 11.
Nemec, J. M., Balona, L. A., Murphy, S. J., Kinemuchi, K., and Jeon, Y.-B. 2017, *Mon. Not. Roy. Astron. Soc.*, **466**, 1290.
Niu, J.-S., Fu, J.-N., and Zong, W.-K. 2013, *Res. Astron. Astrophys.*, **13**, 1181.
Niu, J.-S., *et al.* 2017, *Mon. Not. Roy. Astron. Soc.*, **467**, 3122.
Paunzen, E., and Vanmunster, T. 2016, *Astron. Nachr.*, **337**, 239.
Poretti, E. 2003a, *Astron. Astrophys.*, **409**, 1031.
Poretti, E. 2003b, in *Interplay of Periodic, Cyclic and Stochastic Variability in Selected Areas of the H-R Diagram*, ed. C. Sterken, ASP Conf. Ser. 292. Astronomical Society of the Pacific, San Francisco, 145.
Rodríguez E., and Breger, M. 2001, *Astron. Astrophys.*, **366**, 178.
Sartoretti, P., *et al.* 2018, *Astron. Astrophys.*, **616A**, 6.
Schlafly, E. F., and Finkbeiner, D. P. 2011, *Astrophys. J.*, **737**, 103.
Schlegel, D. J., Finkbeiner, D. P., and Davis, M. 1998, *Astrophys. J.*, **500**, 525.
Schwarzenberg-Czerny, A. 1996, *Astrophys. J., Lett.*, **460**, L107.
Serenelli, A., Scott, P., Villante, F. L., Vincent, A. C., Asplund, M., Basu, S., Grevesse, N., and Peña-Garay, C. 2016, *Mon. Not. Roy. Astron. Soc.*, **463**, 2.
Skrutskie, M. F., *et al.* 2006, *Astron. J.*, **131**, 1163.
Smith, T. C., Henden, A. A., and Starkey, D. R. 2011, in *The Society for Astronomical Sciences 30th Annual Symposium on Telescope Science*, The Society for Astronomical Sciences, Rancho Cucamonga, CA, 121.
Software Bisque. 2013, THESKYX Professional Edition 10.5.0 (http://www.bisque.com).
Templeton, M. R., Samolyk, G., Dvorak, S., Poklar, R., Butterworth, N., and Gerner, H. 2009, *Pub. Astron. Soc. Pacific*, **121**, 1076.
Torres, G. 2010, *Astron. J.*, **140**, 1158.
Uytterhoeven, K., *et al.* 2011, *Astron. Astrophys.*, **534A**, 125.
von Steiger, R., and Zurbuchen, T. H. 2016, *Astrophys. J.*, **816**, 13.
Walker, G., *et al.* 2003, *Publ. Astron. Soc. Pacific*, **115**, 1023.
Wesselink, A. J. 1946, *Bull. Astron. Inst. Netherlands*, **10**, 91.
Wils, P., *et al.* 2009, *Inf. Bull. Var. Stars*, No. 5878, 1.
Wils, P., *et al.* 2010, *Inf. Bull. Var. Stars*, No. 5928, 1.
Wils, P., *et al.* 2011, *Inf. Bull. Var. Stars*, No. 5977, 1.
Wils, P., *et al.* 2012, *Inf. Bull. Var. Stars*, No. 6015, 1.
Wils, P., *et al.* 2013, *Inf. Bull. Var. Stars*, No. 6049, 1.
Wils, P., *et al.* 2014, *Inf. Bull. Var. Stars*, No. 6122, 1.
Wils, P., *et al.* 2015, *Inf. Bull. Var. Stars*, No. 6150, 1.
Xiong, D. R., Deng, L., Zhang, C., and Wang, K. 2016, *Mon. Not. Roy. Astron. Soc.*, **457**, 3163.
Zacharias, N., *et al.* 2010, *Astron. J.*, **139**, 2184.
Zhevakin, S. A. 1963, *Ann. Rev. Astron. Astrophys.*, **1**, 367.
Ziaali, E., *et al.* 2018, Physics of Oscillating Stars, Banyuls-sur-mer, France, DOI 10.5281/zenodo.1494351.

KAO-EGYPT J064512.06+341749.2 is a Low Amplitude and Multi-Periodic δ Scuti Variable Star

Ahmed Essam
Mouhamed Abdel-Sabour
Gamal Bakr Ali
National Research Institute of Astronomy and Geophysics, 11421 Helwan, Cairo, Egypt; essam60@yahoo.com, sabour2000@hotmail.com, mrezk9@yahoo.com

Received March 11, 2019; revised April 23, 30, 2019; accepted April 7, 2019

Abstract CCD photometric observations with *BVRI* filters using the 1.88-m telescope of Kottamia Astronomical Observatory (KAO), Egypt, revealed that the star KAO-EGYPT J064512.06+341749.2 is a low-amplitude ($\Delta m < 0.3$ mag) δ Scuti star. The peak-to-peak amplitude is 0.014 magv. Two modes are present ($f_1 = 23.600 \pm 0.133$ c/d and $f_2 = 18.314 \pm 0.202$ c/d). The frequency ratio, $f_2 / f_1 = 0.776$, suggests that the star is a radial pulsator. By using the empirical relations for KAO-EGYPT J064512.06+341749.2, we determined the global physical parameters.

1. Introduction

δ Scuti stars are pulsating variable stars useful for studies of stellar structure and evolution. The class of δ Scuti stars includes stars situated under the classical Cepheids in the instability strip on the main sequence or moving from the main sequence to the giant branch. δ Scuti stars normally have small amplitude variations, with radial pulsation, non-radial p-mode pulsation, and short periods. Many modes can be excited simultaneously. In general, the period range is limited from 30 minutes to 6 hours and the masses range from 1.0 to 3.0 $M_{\odot}$. Observations for several decades revealed that the low amplitude δ Scuti stars (LADS) show a large variety of non-radial modes, complex light variability, multi-periodicity, and phase and amplitude variations. High amplitude (> 0.3 mag.) δ Scuti stars (HADS) have been thought to be classical radial pulsating stars, mostly mono-periodic, though double mode in some cases, but always pulsating in radial modes (Kjurkchieva *et al.* 2013). However, recently many HADS have been found to be multi-periodic variable stars with non-radial as well as radial pulsations (Zhou 2002; Poretti 2003; Poretti *et al.* 2005). The differentiation between LADS and HADS is that the non-radial modes in HADS have much smaller amplitudes than the radial modes. The period and amplitude variations can be considered as small perturbations of a mode visible in the light curve.

The variability of the star KAO-EGYPT J064512.06+341749.2 was discovered by Essam (2013). It has many alternative identifications, including 2MASS J06451206+3417492, GSC 02444-00241, UCAC4 622-035906, and USNO-B1.0 1242-0138204. Table 1 contains the basic data for the comparison and check stars used as well as for the variable.

2. Observations

Observations were carried out by Essam (2013) on two consecutive nights, 7/8 February and 8/9 February 2013, in addition to the night of 20/21 January 2015 (data available at https://www.aavso.org/apps/webobs/results/?star=KAO-EGYPT+J064512.06%2B341749.2&obscode=EAEA&num_results=200&obs_types=all), at the Newtonian focus (f/4.84) of the 1.88-meter telescope of the Kottamia Astronomical Observatory (KAO), Egypt (for more details about KAO see Azzam *et al.* 2010). The observations were performed using the back-illuminated EEV 42-40 CCD chip with 2048 × 2048 pixels. The pixel size, scale, and total field of view were 13.5μ, 0.305"/pixel, and 10 × 10 arcmin, respectively. The standard *BVRI* Johnson photometric system was used. All raw images are bias-subtracted and flat-fielded corrected. The exposures were 180, 60, 20, and 10 sec. in the *B*, *V*, *R*, and *I* bands, respectively. The light curves were then produced by computing the magnitude differences between the variable KAO-EGYPT J064512.06+341749.2 and the comparison star.

Table 1. Coordinates, magnitudes, and color index of the variable and comparison stars.

Star	*Name*	*R.A. (2000)* h m s	*Dec. (2000)* ° ' "	*V*	*B–V*
V[1]	KAO-EGYPT J064512.06+341749.2	06 45 12.064	+34 17 49.17	13.452 ± 0.08	0.319 ± 0.094
	UCAC4 622-035906				
C1[2]	USNO-A2.01200-05104801	06 44 58.05	+34 23 52.8	14.037 ± 0.061	0.648 ± 0.087
C2[2]	USNO-A2.01200-05103968	06 44 54.65	+34 24 09.6	13.564 ± 0.071	0.713 ± 0.074

Notes: 1. UCAC4 catalogue (Zacharias et al. *2013). 2. USNO-A2.0 catalogue (Monet* et al. *1998).*

3. Light curve analysis

The Δb, Δv, Δr, and Δi light curves of the star KAO-EGYPT J064512.06+341749.2 are presented in Figure 1. KAO-EGYPT J064512.06+341749.2 is a pulsating variable with a total amplitude of 0.083, 0.068, 0.042, and 0.051 mag. in the *B, V, R,* and *I* bands, respectively. Using the standard *B* and *V* magnitudes of the comparison star ($B = 14.681$, $V = 14.052$) to determine the standard magnitude of the variable, we found that the average magnitude and color index for the variable are $V = 13.391 \pm 0.013$ and $B-V = 0.373 \pm 0.005$.

4. Frequency analysis

All light curves of KAO-EGYPT J064512.06+341749.2 were examined in more detail using the Phase-Dispersion-Method within the software PERANSO (Vanmunster 2013). Also, PERIOD04 (Lenz and Breger 2005) was used to make Fourier transformations of the light curves to search for the significant peaks in the amplitude spectra; the results are listed in Table 2. The DFT method can be used to detect a signal, remove the detected frequency and its harmonics from the data, and search for additional frequencies in the residuals. Also, the first step was to construct the "periodogram" by fitting a sinusoid to the highest-amplitude period obtained from an initial fit to the observed magnitudes. The derived sinusoid was then subtracted from the original magnitudes. The analysis was repeated on the pre-whitened data in an iterative fashion, until no more significant periods were found. The analysis of the *B*-band light curve shows that there are two frequencies in the periodogram, at 0.0413 d (24.175 c/d) and 0.0556d (18.681 c/d) (see Figure 2).

The light curve analysis of the present observations, i.e. from 2013 and 2015, indicates that the amplitudes of the two frequencies are changing with time. We checked the amplitudes for KAO-EGYPT J064512.06+341749.2 but unfortunately we do not have sufficient observations in all bands. The amplitude in V-band for KAO-EGYPT J064512.06+341749.2 dropped from 13.12 mmag in 2013 to 10.51 mmag during 2015; other filters in 2015 have more scatter. This phenomenon was also found in other δ Scuti stars, such as BR Cancri (Zhou *et al.* 2001). More observations are needed to confirm this phenomenon in KAO-EGYPT J064512.06+341749.2.

Breger *et al.* (1993) found that the solutions with frequencies whose signal-to-noise ratios (S/N) are larger than 4.0 are accurate for distinguishing between peaks due to pulsation and noise for the average amplitude in the range 15 to 25 c/d. Present results are in the same range for the two frequencies: S/N=07.862 and 10.732 in the first and second frequencies,

Table 2. Fourier parameters of the best-fitting sinusoids for the *B* light curve of the variable star KAO-EGYPT J064512.06+341749.2.

Filter	*Frequency (c/d)*	*Amplitude (magnitude)*	*Phase*	*S/N*
B_1	24.175 ± 0.086	0.022 ± 0.001	0.989 ± 0.007	07.862
B_2	18.681 ± 0.227	0.008 ± 0.001	0.345 ± 0.019	10.732

Note: The subscripts 1 and 2 refer to the first and the second frequencies, respectively.

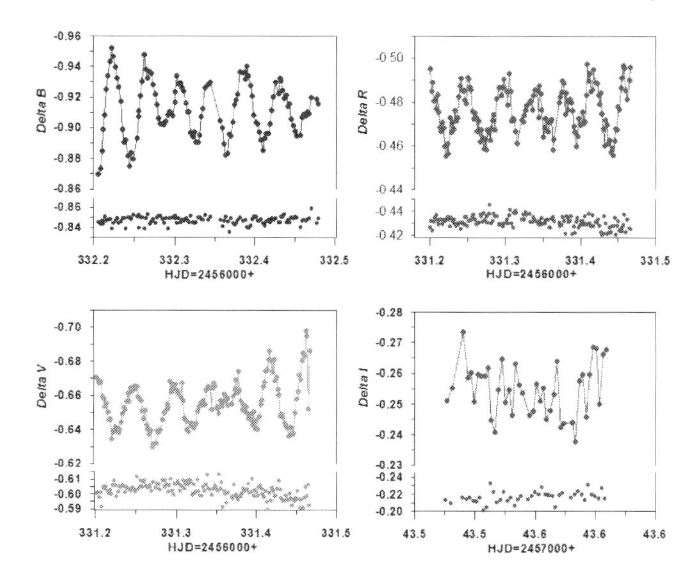

Figure 1. *BVRI* differential magnitude light curves of the pulsating variable KAO-EGYPT J064512.06+341749.2. The dots below each curve represent the magnitude difference between the comparison and check stars.

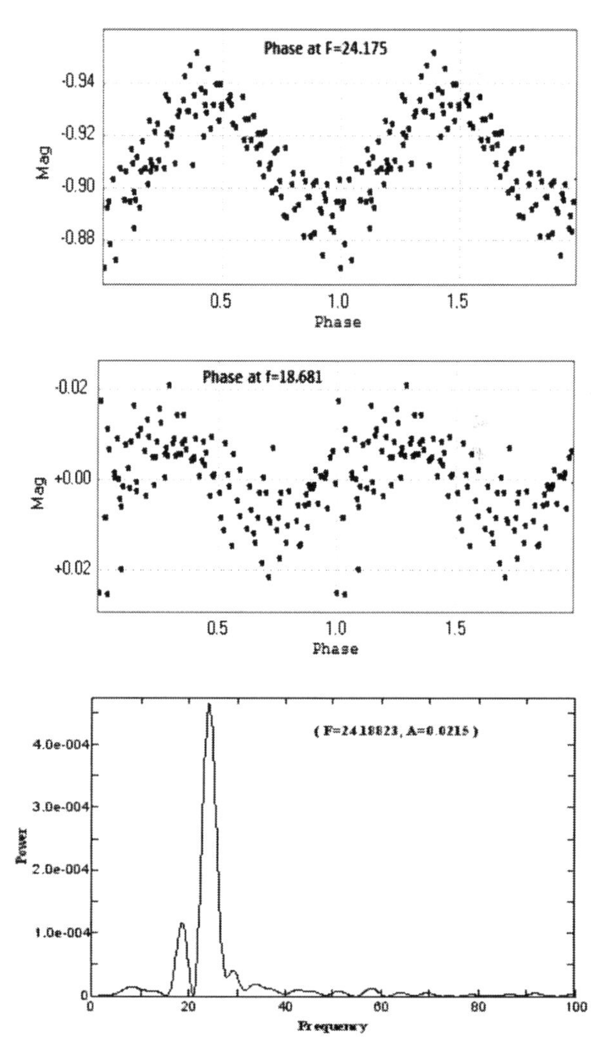

Figure 2. The *B*-band phased light curves at two frequencies' (24.175 c/d) and (18.681 c/d) periods are presented in the upper panels. Amplitude spectra for the same frequencies are presented in the lower panel.

respectively. The uncertainties in frequency, amplitude, and phase are obtained from a Monte Carlo simulation. The spectral window of KAO-EGYPT J064512.06+341749.2 in *B*-band and the phased light curves at two frequencies' (24.175 c/d) and (18.681 c/d) periods are presented in Figure 2.

The light curve was fitted by using two frequencies (24.321 and 18.377) with residuals = 0.009 and 0.007, respectively, as shown (upper and lower plots) in Figure 3. More time series photometric and spectroscopic observations are needed to determine a more accurate spectral type and to study the multi-periodic nature of the pulsation and amplitude variations.

We carried out detailed frequency analysis of the available data, and we obtained improved frequencies f_1 (24.321) and f_2 (18.377) and detected oscillations corresponding to the interaction between f_1 and f_2, (f_1+f_2) and (f_1-f_2). The period ratio, $P_1/P_0=0.756$, is a little higher than the mean canonical value of the range 0.75–0.79, with the minimum value corresponding to metal-strong stars (Z ~ 0.01), while the maximum value corresponds to metal-poor stars (Z ~ 0.001) (Poretti *et al.* 2005).

5. Pulsating mode identification of KAO-EGYPT J064512.06+341749.2

In order to obtain the global physical parameters of KAO-EGYPT J064512.06+341749.2 we used the following relations for the pulsating stars. We attempted to identify the observed frequencies of pulsation with pulsation modes. The basic solar parameters were T_{eff} = 5777 K, log g = 4.44, and M_{bol} = 4.75, which we used in the following equations. We also used the effective temperature of our system from the Gaia web site (http://sci.esa.int/gaia/), T_{eff} = 7776 K.

The absolute magnitude of the star was calculated in *V*-band in a recent paper by McNamara (2011):

$$M_v = (-2.89 \pm 0.13) \log(p) - (1.31 \pm 0.10). \quad (1)$$

The bolometric correction (BC) relation was evaluated by Reed (1998):

$$BC = -8.499\,[\log(T)-4]^4 + 13.421\,[\log(T)-4]^3 - 8.131\,[\log(T)-4]^2 - 3.901\,[\log(T)-4] - 0.438. \quad (2)$$

The bolometric magnitude, M_{bol}, is further given by $M_\lambda = M_{bol} - BC$.

By using the last equations, the absolute magnitude of the star was calculated in *V-filter* as M_V = 2.696 ± 0.078 mag. We found that BC = – 0.127 ± 0.003, M_{bol} = 2.823 ± 0.078.

Using the mass relation by Cox (2000) for δ Scuti stars, $Log\ M = 0.46 - 0.10\ M_{bol}$, the stellar mass of the system equals M = 1.506 ± 0.072$M_\odot$.

The stellar radius was calculated from a polynomial fit to the temperature/radius relation by using Gray (1992, equation 3), or from the formulae by Tsvetkov (1988), $\log R = 8.472 - 2\log T_{eff} - 0.2 M_{bol}$. The results are $R/R_\odot = 1.330 \pm 0.012$ and 1.259 ± 0.053, respectively.

We can use the following equation to calculate stellar luminosity:

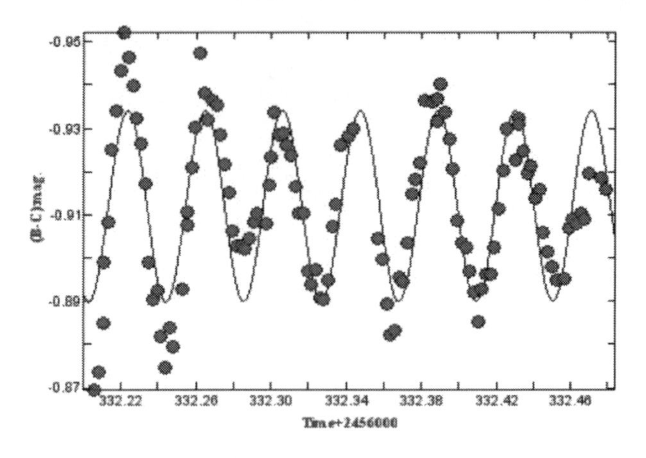

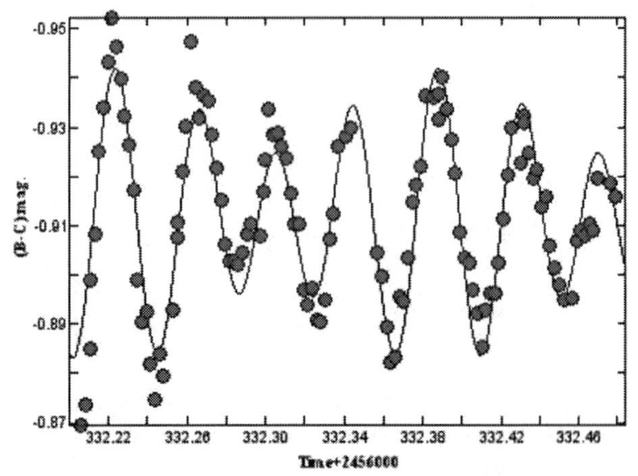

Figure 3. Observed *B*-band light curve (dots) of the present observations together with the fitted two frequencies (solid line). The upper plot is the fit by using the first frequency (residual = 0.0088) and the lower plot is the fit by using the second frequency (residual = 0.0066).

$$M_* = M_\Theta - 2.5\log(L_*/L_\Theta). \quad (3)$$

Surface gravity can be evaluated by using the following equation and its range should be between 3.4 and 4.4 for δ Scuti stars (Alcock *et al.* 2000):

$$\frac{g}{g_\Theta} = \left(\frac{M_*}{M_\Theta}\right) \frac{(T_*/T_\Theta{}^4)}{(L_*/L_\Theta)} \quad (4)$$

The luminosity ratio $L/L_\odot$ is found to equal 5.899 and log g = 4.363 ± 0.111.

The pulsation constant (Q) can be determined using the following equation (Breger and Bregman 1975):

$$\log Q = 0.5 \log g + 0.1\, M_{bol} + \log T_{eff} + \log P - 6.456, \quad (5)$$

where Q depends on the physical parameters of the star. The pulsation frequency of 24.321 c/d has a Q value corresponding to 0.0325, and 0.0277 for frequency 18.377 c/d.

6. Discussion and conclusion

With the *BVRI* photometric observations obtained using the 1.88-m KAO telescope, we discovered the variability of KAO-EGYPT J064512.06+341749.2 (Essam 2013) as a δ Scuti pulsating variable. The present analysis shows that KAO-EGYPT J064512.06+341749.2 is clearly a double-mode variable star with two radial modes. Photometric analysis conducted by the authors yields a period of 0.0425 day and a peak-to-peak amplitude of 0.014 magnitude. There are two clear frequencies: the first overtone mode ($f_1 = 24.3212678 \pm 0.081$ c/d with amplitude $= 0.022 \pm 0.001$ mag., S/N = 07.862) and the fundamental mode ($f_0 = 18.681 \pm 0.227$ c/d with amplitude $= 0.008$ mag., S/N = 10.732), with the ratio $f_0 / f_1 = 0.768$. The frequency ratio of 0.77 seen in the confirmed double mode stars indicates that they are pulsating primarily in the fundamental and first overtone modes, suggesting that this star must lie nearer the red edge of the instability strip. The physical parameters of this newly discovered pulsating star KAO-EGYPT J064512.06+341749.2 are determined as: $M_{bol} = 2.823 \pm 0.078$, $M / M_\odot = 1.506 \pm 0.072$, $R / R_\odot = 1.330 \pm 0.012$, $L / L_\odot = 5.899$, $\log g = 4.363 \pm 0.111$, and Q values of $0.032\,d \pm 0.002$ for the pulsation frequency 24.321 c/d and $0.0277\,d \pm 0.001$ for the pulsation frequency 18.377 c/d. All values are based on solar units and a bolometric correction of -0.127 ± 0.003 mag. Typical values of pulsation constants of the fundamental, first, and second overtone radial p modes in δ Sct stars are $0.022 \leq Q \leq 0.033$ d (Breger and Bregman 1975), thus the Q values indicate overtone radial or non-radial pulsation ($k \geq 0$) (Breger 1990).

Our results show a good agreement with the work thesis by Bowman (2016, figure 4.12) for low-frequency pulsations ($\nu \leq 25\,d-1$), and physical evolution with that expounded by Flower (1996), as seen in Table 3. From the study of the evolution status of KAO-EGYPT J064512.06+341749.2 we found that its age (τ) is equal to $6.6 \pm 4.5 \times 10^8$ years by using the database of stellar evolutionary tracks and isochrones of Mowlavi *et al.* (2012). Also, we predict that this pulsating star is in the red edge of the instability strip as compared with theoretical results by Baglin *et al.* (1973) and Christiansen *et al.* (2007). More photometric and spectroscopic observations are needed to confirm the possible variation in amplitude and period found in the present analysis of this pulsating star.

7. Acknowledgement

The authors are very grateful to Dr. Luis Balona (South African Astronomical Observatory) for his very helpful discussion.

Table 3. Frequency ratio as detected in the *B*-band.

Mode	*Frequency (c/d)*	*Amplitude (magnitude)*	*Phase*	*Epoch 2456000+*
f_1	24.321 ± 0.081	0.0213	0.657	331.963
f_0	18.377 ± 0.206	0.0086	0.402	331.964
$f_1 + f_0$	42.698 ± 0.915	0.0020	0.553	331.981
$f_1 - f_0$	5.944 ± 1.525	0.0016	0.684	331.843

References

Alcock, C. A., *et al.* 2000, *Astrophys. J.*, **536**, 798

Azzam, Y. A., Ali, G. B., Elnagahy, F., Ismail, H. A., Haroon, A., Selim, I. M., and Essam, A. 2010, in *Proceedings of the Third UN/ESA/NASA Workshop on the International Heliophysical Year 2007 and Basic Space Science*, eds. H. J. Haubold, A. M. Mathai, Springer, Heidelberg, Dordrecht, London, New York, 175.

Baglin, A., Breger, M., Chevalier, C., Hauck, B., Le Contel, J. M., Sareyan, J. P., and Valtier, J. C. 1973, *Astron. Astrophys.*, **23**, 221.

Bowman, D. M. 2016, Ph.D. thesis, Institute for Mathematics, Physics and Astronomy, University of Central Lancashire at KU Leuven, Belgium.

Breger, M. 1990, *Astron. Astrophys.*, **240**, 308.

Breger, M., *et al.* 1993, *Astron. Astrophys.*, **271**, 482.

Breger, M., and Bregman, J. N. 1975, *Astrophys. J.*, **200**, 343.

Christiansen, J. L., Derekas A. M., Ashley, C. B., Webb, J. K., Hidas, M. G., Hamacher, D. W., and Kiss, L. L. 2007, *Mon. Not. Roy. Astron. Soc.*, **382**, 239.

Cox, A. N., ed. 2000, *Allen's Astrophysical Quantities*, 4th ed. AIP Press (Springer-Verlag), New York.

Essam, A. 2013, AAVSO VSX entry for KAO-EGYPT J064512.06+341749.2 (https://www.aavso.org/vsx/index.php?view=detail.top&oid=305018).

Flower, P. J. 1996, *Astrophys. J.*, **469**,355.

Gray, D. F. 1992, *The Observation and Analysis of Stellar Photospheres*, 2nd ed., Cambridge University Press, Cambridge.

Kjurkchieva, D. P., Dimitrov, D. P., Ibryamov, S. I., and Srinivasa Rao, M. 2013, *Bull. Astron. Soc. India*, **41**, 173.

Lenz, P., and Breger, M. 2005, *Commun. Asteroseismology*, **146**, 53.

McNamara, D. H. 2011, *Astron. J.*, **142**, 110.

Monet, D., *et al.* 1998, *USNO-A V2.0 Catalog of Astrometric Standards*, U.S. Naval Observatory, Flagstaff, AZ.

Mowlavi, N., Eggenberger, P., Meynet, G., Ekström, S., Georgy, C., Maeder, A., Charbonnel, C., and Eyer, L. 2012, *Astron. Astrophys.*, **541A**, 41.

Poretti, E. 2003, *Astron. Astorphys.*, **409**, 1031.

Poretti, E., *et al.* 2005, *Astron. Astrophys.*, **440**, 1097.

Reed, B. C. 1998, *J. Roy. Astron. Soc. Canada*, **92**, 36.

Tsvetkov, Ts. G. 1988, *Astrophys. Space Sci.*, **150**, 357.

Vanmunster, T. 2013, light curve and period analysis software, PERANSO v.2.50 (http://www.cbabelgium.com/peranso).

Zacharias, N., Finch, C. T., Girard, T. M., Henden, A., Bartlett, J. L., Monet, D. G., and Zacharias, M. I. 2013, *Astron. J.*, **145**, 44.

Zhou, A.-Y., 2002, in *Radial and Nonradial Pulsations as Probes of Stellar Physics*, ASP Conf. Proc. 259, eds. C. Aerts, T. R. Bedding, J. Christensen-Dalsgaard, Astronomical Society of the Pacific, San Francisco, 332.

Zhou, A.-Y., Rodriguez, E., Liu, Z.-L., and Du, B.-T. 2001, *Mon. Not. Roy. Astron. Soc.*, **326**, 317.

Low Resolution Spectroscopy of Miras—X Octantis

Bill Rea
6A Bygrave Place, Bishopdale, Christchurch, 8053, New Zealand; rea.william@gmail.com

Received March 12, 2019; revised May 9, 14, 2019; accepted May 14, 2019

Abstract We present a photometric and spectroscopic study of X Oct and five selected comparison stars. We show that the changes in spectral classification of X Oct over the course of its pulsation cycle can be determined by comparison between its spectra and that of the five comparison stars. In particular, we conclude that the ratio between the counts at the continuum point at 754nm and the TiO absorption line at 719nm is the single most reliable feature for determining spectral type. We suggest that X Oct may cycle between M3 and M7 as it pulsates, rather than the published range of M3 to M6. We also undertook a frequency analysis of CQ Oct and report two significant pulsation periods of 52.9 and 37.4 days, respectively.

1. Introduction

Mira variables have long been a favorite observing target for amateur astronomers because their long periods and large amplitudes meant that they only needed to be observed about once a week and small uncertainties in the estimation of the visual magnitudes did not detract significantly from the quality of the light curves.

Over recent decades CCD photometry with multiple filters has become commonplace among amateur astronomers, with a selection of UBV filters from the Johnson system and RI from the Cousins system being the most common. The collection of multiple color estimates allows the observer to estimate the effective temperature of a variable star, particularly using the B–V index. However, it has long been known that for spectral types K5 through M8, which encompasses most Miras, the B–V index is nearly constant (Smak and Wing 1979) despite the temperature dropping by nearly a factor of two over this range. The presence of large numbers of TiO absorption bands depresses almost the entire spectrum between 4300 Å in the blue and 7500 Å in the infra-red. Kirkpatrick *et al.* (1991) state that in the region 6300 to 9000 Å there are, at best, only six points which could possibly be labelled as continuum points, and they are listed in Table 1.

The problem of trying to determine a spectral class for these stars from purely photometric data has led to the development of some narrow band systems such as the three- and eight-filter systems described in Wing (1992) and White and Wing (1978). Bessel *et al.* (1989) provide an example of the White and Wing (1978) system applied to M giants. Recently, Azizi and Mirtorabi (2015) proposed a modification to some of the filters in the eight-filter system of White and Wing (1978) to overcome known problems with the spectra of the coolest stars.

Table 1. The six possible continuum ponts in spectral class K5 to M9 identifed by Kirkpatrick *et al.* (1991).

Continuum Point	*Wavelength (Å)*
C1	6530
C2	7040
C3	7560
C4	8130
C5	8840
C6	9040

The problems associated with methods of obtaining spectral classification from photometry were discussed in some detail in Wing (2011).

As Wing (1997) notes: "If we think of Miras as variable stars with time-dependent spectra, it is clearly desirable to record both their spectroscopic and photometric behavior." While many Miras pulsate in a stable manner, a number have evolved significantly on time scales considerably less than a human lifespan. Changes can include the shapes of light curves, pulsation period, and pulsation amplitude. Templeton *et al.* (2005) studied the pulsation periods of 547 Miras and reported 57, or slightly more than 10%, of these had changes in period which were significant at the 2σ level, 21 at the 3σ level, and eight at the 6σ level. At least some of these changes are thought to be the result of a helium flash (see Hawkins *et al.* (2001) for example), with potential for the star to change its spectral type (see Uttenthaler et al. (2016) for an example). It is these types of Miras which would benefit most from long-term spectroscopic monitoring, although monitoring of stable Miras is also worthwhile.

With the readily available, low cost, filter-wheel grating spectroscopes it is now possible to routinely observe Mira spectra as part of an CCD photometric observing program. Although the resolution of these types of spectrographs is very low, with typical R ($\lambda/\Delta\lambda$) values in the range 50–200, the recent introduction of the AAVSO spectroscopic database means these observations can now be combined with similar observations by other observers, increasing their scientific value.

The main problem with determining the range of spectral classifications for a Mira over the course of its cycle is that spectroscopy is usually done differentially. That is, the spectrum of a star to be classified is compared with the spectra of known standards and a best fit is obtained. Sometimes the best fit involves statistical testing of the goodness-of-fit (see Kirkpatrick *et al.* (1991) for an example). Gray and Corbally (2009) provide considerable detail on spectra of a wide range of spectral types.

For Miras we would usually require red giant stars of spectral class M for spectral comparison but all such stars are variable to a greater or lesser extent. Hence there are no true standards to work with. Nevertheless, with a filter wheel grating spectroscope an amateur can obtain both photometric and spectroscopic data for target Miras and a range of suitable comparison stars so that the photometric and spectroscopic variability in the comparisons can be quantified.

The present paper presents some results from an on-going pilot study on the feasibility and potential contributions to the study of Miras through the addition of low-resolution spectroscopy to a photometric observing program. Some previous results can be found in Martin *et al.* (2016a) and Martin *et al.* (2016b).

The research question addressed here is: can the spectral type of an M-type Mira be reliably determined for M2-M6 over the course of its pulsation cycle.

The remainder of this paper is structured as follows: section 2 describes the target stars, observing equipment and methods, section 3 presents the results, section 4 contains the discussion, and section 5 explains our conclusions and gives some indication of future directions.

2. Target stars, observing equipment and methods

2.1. Target stars

Table 2 presents a list of stars observed in the current phase of this project. All stars chosen are close to the south celestial pole to enable year-round observation so that no part of the cycle will be lost through being below the horizon and hence unobservable.

2.2. Observing equipment

Three telescopes have participated in this study. They are:

1. I operated an 80-mm f/6 Explore Scientific apochromatic refractor in Christchurch, New Zealand, with an Atik 414E Mono CCD camera using a SONY ICX424AL front-illuminated chip. The plate scale in the imaging plane is 2.77 arcseconds/pixel. I used a Paton Hawksley Star Analyzer 100 grating yielding a first order spectrum with a dispersion of 1.488 nm/pixel.

2. BSM_South of the AAVSOnet's Bright Star Monitors. It is an AstroTech-72ED, a 72-mm, f/6 apochromatic refractor with an SBIG ST8-XME CCD camera located at Ellinbank Observatory in Victoria, Australia. The filter wheel contains Astrodon filters of which the B and V filters were used in this study. (Note that recently BSM_South has recently had an upgrade and these details no longer accurately reflect the camera and filters available.)

3. BSM_Berry is located in Perth, Australia. It also is an AstroTech-72ED with an SBIG ST8-XME CCD camera. Of the filters available we used the B and V for photometry and the grating spectrograph.

The published spectral range of the Mira, X Oct, was M3 to M6. This guided the selection of the five comparison stars listed in Table 2 to cover the same range of spectral types using stars with low variability.

In both the three- and eight-filter systems of Wing (1992) and White and Wing (1978), as well as the modified system of Azizi and Mirtorabi (2015), the deep TiO absorption line at 719 nm was used to establish an estimate of the stellar temperature and spectral classification of M-type Miras. While these systems involved narrow band filters, with spectroscopy, even the low resolution spectroscopy presented here, there is no question where the lowest point of the absorption line and the peak of the nearby continuum point are (754 nm in the Wing system and 704 nm in the Azizi system). Accordingly, in the remainder of this paper we will refer to the ratio of counts at 754 nm/719 nm as the Wing ratio and 704 nm/719 nm as the Azizi ratio.

The spectra obtained with the gratings were measured with SAOImage DS9 and analyzed with custom write R code (R Foundation 2015).

3. Results

A total of 69 spectra of X Oct were collected between 20 October 2015 and 25 September 2018 and cover approximately four pulsation cycles.

Spectra for the five comparison stars were collected between 9 April 2017 and 25 September 2018. Spectra for the comparison stars only began after the early part of this study indicated that something like comparison stars, long used in photometry, were also required for spectroscopy.

The final two columns in Table 2 presents the mean Wing and Azizi ratios (described in section 2) for each of the comparisons together with the standard deviations in brackets.

Figure 1 presents the V-band light curves for the five comparison stars listed in Table 2. The data were obtained largely from the AAVSOnet telescopes described in section 2; some were from other observers who contributed to the AAVSO International Database. The light curve for each comparison was created by subtracting the mean observed magnitude from each observation and then stacking the resulting curves at one-magnitude intervals. Vertical error bars have been added to the points in Figure 1 but in most cases the uncertainty in magnitude

Table 2. The details of the published variability in spectral type and magnitude of the primary target and a selection of five comparison stars.

Star	*Spectral Class*	*Variable Type*	*Brightness Range (V mag)*	*Period (days)*	*Wing Ratio*	*Azizi Ratio*
X Oct	M3/M6IIIe	Mira	6.8-10.9	200		
CV Oct	M3	LB	8.92-9.19	—	2.63 (0.37)	2.27 (0.25)
BQ Oct	M4III	LB	6.8	—	2.21 (0.15)	2.05 (0.15)
CQ Oct	M4/M5III	SRB	8.12-8.59	50.8	3.82 (0.30)	2.76 (0.19)
eps Oct	M5III	SRB	4.58-5.3	55	4.08 (0.29)	2.81 (0.19)
BW Oct	M5-M7III	LB	7.9-9.1	—	5.85 (0.41)	3.26 (0.20)

Note: The comparison stars are ordered in decreasing expected temperature. Magnitude ranges are for the V band. The data in first five columns of this table were obtained from the AAVSO Variable Star Index (VSX). The final two columns are the mean and standard deviation, in brackets, of the Wing and Azizi ratios and are results from this study. They are discussed in more detail in sections 3 and 4 below.

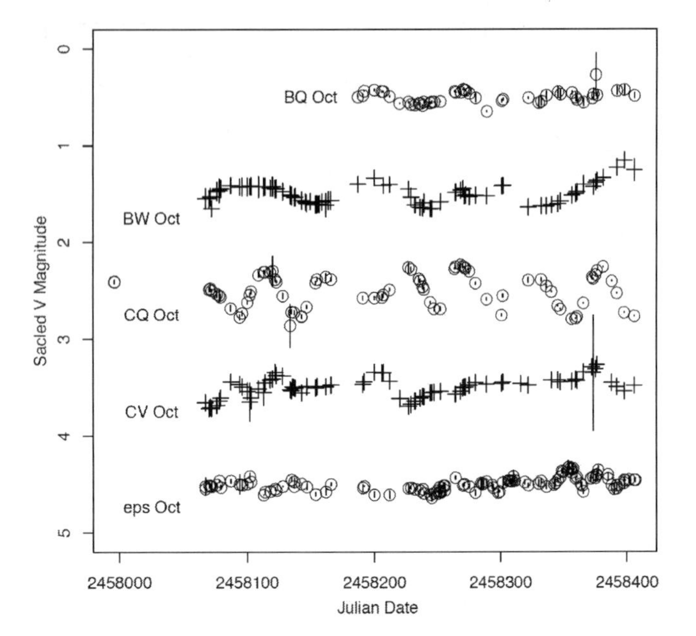

Figure 1. The V-band light curves for the five comparison stars over the course of the study period. Each light curve has had the mean magnitude subtracted from the observed magnitude and the resulting light curves have been stacked at one magnitude intervals. Details of their published magnitude ranges, variable type, and spectral classification can be found in Table 2.

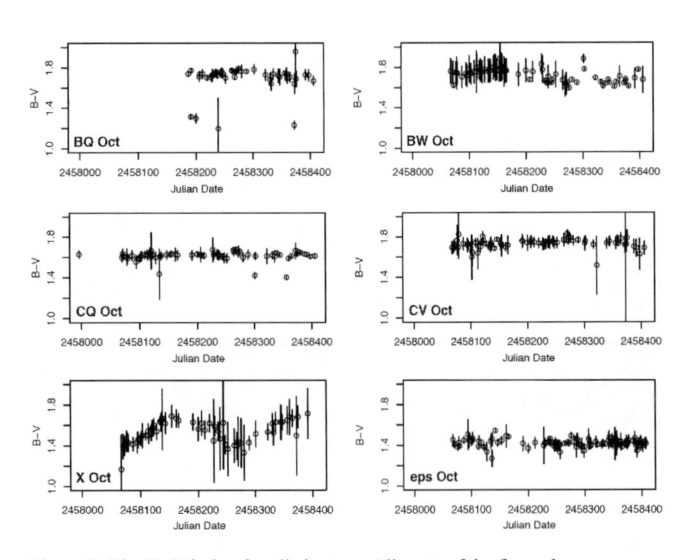

Figure 2. The B–V index for all six stars. All parts of the figure have common scales on their horizontal and vertical axes and so are directly comparable. The uncertainties in the B–V index are plotted as vertical lines on each data point. Details of the stars' spectral classification and variable type are in Table 2.

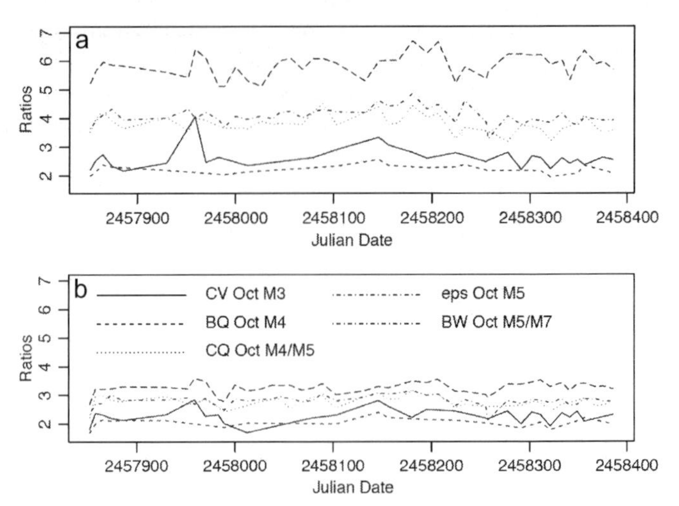

Figure 3. Panel (a) presents the Wing ratios for the five comparison stars over the course of the study period while panel (b) presents the Azizi ratios for the same stars and period. The use of a common scale for the vertical axes allows direct comparison between the two ratios for the same set of stars. The legend in panel (b) is common to both panels. Details of their spectral classification and variable type are in Table 2.

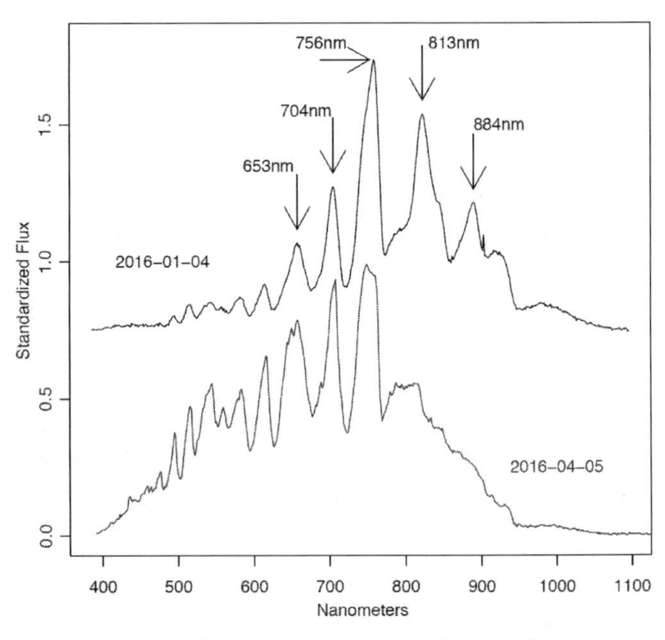

Figure 4. The spectra of X Oct at close to maximum (2016-04-05) and minimum (2016-01-04) light. Each spectrum has been scaled so that it has a range of one and the spectrum from the minimum light has been offset by 0.75 unit so that the two spectra do not overlap. Five of the six possible continuum points discussed in Kirkpatrick *et al.* (1991) and listed in Table 1 are marked on the upper spectrum.

estimates were sufficiently small that they were less than the size of the plotting symbol used and hence appear to be dots inside the circles for BQ Oct, CQ Oct, and ε Oct while for BW Oct and CV Oct the use of the plus symbols usually does not allow the error bars to be seen.

Figure 2 presents the B–V index for X Oct and each of the five comparison stars. As indicated in the caption to the figure, they all have a common scale on both horizontal and vertical axes and error bars have been included on each observation.

Panel (a) of Figure 3 presents the Wing ratios (754/719 nm) for the five comparison stars while panel (b) presents the Azizi ratios (702/719 nm) for the same stars during the study period. The two parts of the figure have common horizontal (date) and vertical axes. The common axes allows a direct comparison between the two ratios for the same set of stars.

Figure 4 presents two representative spectra from near minimum and maximum light for X Oct during the study period. Significant changes in the spectrum can be seen. There is much less output in the 400 to 600 nm region in the 2016-01-04 spectrum, when X Oct is cooler, than in the 2016-04-05 spectrum near maximum light. The relative strengths of the continuum points C1 through C5 have changed. While the C3 point (754nm) is the maximum in both spectra, in the 2016-04-05 spectrum the C1 and C2 points are stronger and the C4 and C5 points weaker than they are in the 2016-01-04 spectrum.

Figure 5 presents the light curve of X Oct over the study period together with both the Wing and Azizi ratios from the spectra. The common vertical axis is in magnitudes when examining the light curve and in ratios of counts at 754/719 nm and 704/719 nm, respectively, for the Wing and Azizi ratios.

Figure 6 presents the Wing ratios and the V band light curve of CQ Oct where the light curve has been shifted up by 5.5 magnitudes to fit on the common vertical axis. Details of CQ Oct can be found in Table 2.

A frequency analysis of the CQ Oct light curves was undertaken with FAMIAS (Zima 2008) because its main pulsation period appeared to be modulated. We found two significant pulsation periods; the stronger of the two was 52.9 days and closely matches the 50.8-day period reported in VSX. A second, lower amplitude pulsation period of 37.4 days was also found. Two relatively closely spaced pulsation periods would account for the observed beating seen in the light curve. There was a third peak in the Fourier periodogram at 67.1 days, but with only 76 observations spread over 409 days it was not possible to either rule in or rule out this periodicity in the pulsations.

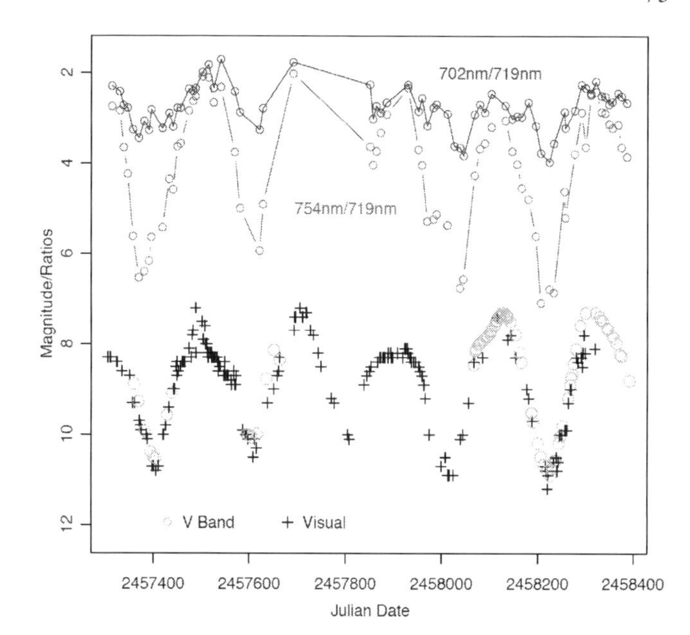

Figure 5. The upper part of the figure presents the Wing and Azizi ratios calculated from the spectra obtained during the study period. On a common set of axes the light curve of X Oct from visual and V band CCD observations over the study period is presented in the lower part of the figure. These data were obtained from the AAVSOnet BSM telescopes described in section 2 and also contain observations by other observers contributed to the AAVSO International Database.

4. Discussion

Figure 4 presents two spectra of X Oct, one taken near maximum light and labelled 2016-04-05, and the other near minimum light and labelled 2016-01-04. The differences in the two spectra are visually obvious. To the extent that the continuum can be determined, it is clear that when X Oct is at its coolest, at or near minimum light, the peak energy output has shifted deeper into the infrared and there is significantly less output in the 400 to 750 nm region than when it is warmer. The cooling between maximum and minimum has resulted in a considerable strengthening of the TiO absorption line at 719 nm. In addition, the strength of the C4 and C5 continuum points have increased significantly as well. In the 2016-04-05 spectrum the C4 continuum point is nothing more than a flattened "bump" in the spectrum, whereas the C5 point gives no discernable peak. By contrast, both points are clearly evident as peaks in the 2016-01-04 spectrum.

When we examine Figure 5 we see that the Wing and Azizi ratios move in a common direction to the light curve. As X Oct dims, it cools, and the TiO absorption line at 719 nm strengthens, giving rise to an increasing ratio between the counts at the continuum point (C2 or C3 as appropriate) and the minimum of the absorption line. Although both the Wing and Azizi ratios show the same pattern the Wing ratio provides a much better picture of the temperature changes in the photosphere for this star.

Comparing the range of the Wing ratio for X Oct (2.02 to 7.09) with the Wing ratios of the five comparison stars in Figure 3 which range from 1.95 for BQ Oct to 6.71 for BW Oct, we see that the comparisons cover the range well. The mean Wing and Azizi ratios for the five comparisons presented in the final two columns of Table 2 are well separated and, apart

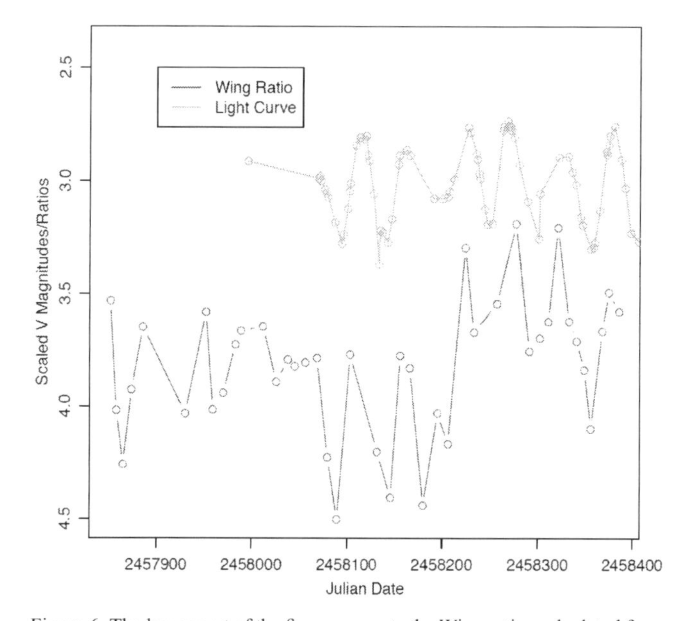

Figure 6. The lower part of the figure presents the Wing ratios calculated from the spectra obtained during the study period. On a common set of axes the light curve of CQ Oct from V band CCD observations from the BSM telescopes is presented in the upper part of the figure where the magnitudes have been scaled up by 5.5 magnitudes to fit on the common vertical axis.

from the obvious discrepancy between BQ Oct and CV Oct, are ordered according to the published spectral types. The minimum value of the Wing ratio for CQ Oct (3.19), which should reflect an M4 spectral type, does not overlap the maximum value for BQ Oct (2.56). However, the range of BQ Oct has a significant overlap with CV Oct. It is reasonable to conclude that both CV Oct and BQ Oct are M3 but with BQ Oct being, on average, the warmer of the two stars.

At maximum light X Oct has a spectral type close to that of BQ Oct. Although the published type is M4 the evidence in Figure 2 suggests that BQ Oct has an earlier spectral type than CV Oct, which has a published estimate of M3. As indicated in the previous paragraph, it would be reasonable to conclude that both CV Oct and BQ Oct have an M3 spectral type and hence so does X Oct at maximum light. This agrees well with the published data for X Oct.

At minimum light the Wing ratio for X Oct of 7.09 is larger than the maximum attained by BW Oct at 6.71, suggesting that at minimum light X Oct is cooler than BW Oct. The published range for BW Oct is M5 to M7. A visual inspection of the spectra of both X Oct and BW Oct at their largest Wing ratios shows that VO molecules are starting to form in the atmosphere and hence beginning to affect the reliability of the 754 nm continuum point (C3). Were either star to cool much more than they do, an alternative method of spectral classification would be needed; see future directions in section 5 below.

Small differences in spectral type can be detected. In Figure 5 the minimum light near Julian date 2457800 is brighter than the other four minima for which we have spectra. The Wing ratio line of the same figure shows that it was the highest of the four minima for which we have data.

For five of the six stars, the B–V index in Figure 2 exhibits low variability, the exception being X Oct, which exhibits a clear color change over the course of its pulsation cycle. It is clearly redder when dimmer and bluer when brighter. The clear pulsation cycle of CQ Oct evident in the light curve in Figure 1 and in the Wing ratio in Figure 6 is not discernable in its B–V index values. When CQ Oct was selected as a comparison star, VSX classified it as an SRB, a semi-regular, late-type giant with poorly defined periodicity. Figures 1 and 6 show that it has quite a regular pulsation and that this is matched by the changes in strength of the TiO absorption line as measured by the Wing ratio. Although this study was aimed at Miras because of their long periods, large pulsation amplitudes, and corresponding large changes in their spectra, an unexpected result is that it appears that the Wing ratio is sufficiently sensitive to changes in the stellar temperatures that lower amplitude variables, such as the semi-regular CQ Oct, can be usefully studied with small, BSM-type, telescopes equipped with filter wheel grating spectrographs.

5. Conclusions and future directions

5.1. Conclusions

For spectral types M2-M6 where there is no contamination of the C3 continuum point by the VO molecule, the Wing ratio of counts at 754/719 nm provides a superior guide to the spectral class than the Azizi ratio of 702/719 nm. Even if there are some uncertainties with the wavelength calibration of the spectrograph, or non-linearities in the first-order spectra, or other factors which affect the conversion of pixels to nanometers, these were sufficiently small that the spectral peaks at the continuum points and the troughs of the absorption lines can easily be determined. If masking is not used (see Martin *et al.* (2016a) and Martin *et al.* (2016b) for details) there is an ever-present problem of potential contamination of the first-order spectra by dim field stars or other luminous sources. Consequently, the spectra should always be visually inspected before measurements are made.

We conclude that over the course of the pulsation cycles we observed that X Oct changes between M3 and somewhere between M6 and M7.

The evidence in Figure 3 suggests that BQ Oct is actually warmer than CV Oct, though the difference is small, and so the assignment of spectral classifications of M4 and M3, respectively, in VSX seems unlikely to be correct; both appear to be M3.

The method of determining spectral type which we present here requires calibration because of differences in the quantum efficiency of among CCD detectors at different wavelengths. However, the use of comparison stars in CCD photometry is routine and should also be used in spectroscopy.

The evidence from the CQ Oct observations indicates that lower amplitude pulsating variables, such as semi-regulars, can be usefully studied with low resolution spectrographs. For CQ Oct, owing to its shorter period, a higher cadence of spectral observation than was used in this study would be beneficial.

5.2. Future directions

We are currently working on more closely identifying the spectral classification of M-type Miras for those which are cooler than M6. For these stars the presence of the VO molecule renders the C3 (754 nm) continuum point unreliable and hence the Wing ratio should not be used to estimate spectral type. This work will primarily feature observations of R Oct and will be presented at a later date.

6. Acknowledgements

The author would like to acknowledge with gratitude the many helpful email discussions with John Martin of the University of Illinois at Springfield on all aspects of this study. I thank Robert Wing for the email discussions about his three- and eight-filter photometry systems and for providing copies of some of his papers I had been unaware of. I thank Arne Henden, Michael Nicholas, George Silvis, and Bill Toomey of the AAVSO for their help in operating the AAVSO's Bright Star Monitor telescopes. I thank Tim Crawford of the AAVSO comparison star team for adding further suitable comparison stars to fields captured by the BSM telescopes used in this project. I thank the AAVSO observers who contributed to the observations AAVSO's International Database used in this study. Without the assistance and encouragement of these people this study would not have been possible.

References

Azizi, F., and Mirtorabi, M. T. 2015, *Astrophys. Space Sci.*, **357**, 96.

Bessel, M. S., Brett, J. M., Scholz, M., and Wood, P. R. 1989, *Astron. Astrophys., Suppl. Ser.*, **77**, 1.

Gray, R. O., and Corbally, C. J. 2009, *Stellar Spectral Classification*, Princeton University Press, Princeton, NJ.

Hawkins, G., Mattei, J. A., and Foster, G. 2001, *Publ. Astron. Soc. Pacific*, **113**, 501.

Kirkpatrick, J. D., Henry, T. J. , and McCarthy, D. W. 1991, *Astrophys. J., Suppl. Ser.*, **77**, 417.

Martin, J. C., Rea, B., McFarland, R., and Templeton, M. 2016a, "Crowd-Source Spectroscopy of Long Period Mira-Type Variables" (http://socastrosci.org/Video2016.html).

Martin, J. C., Barber, H. R., Rea, B., Templeton, M., and McFarland, R. 2016b, in *The Society for Astronomical Sciences 35th Annual Symposium on Telescope Science*, eds. R. K. Buchheim, J. L. Foote, D. Mais, Society for Astronomical Sciences, Rancho Cucamonga, CA, 29.

R Foundation for Statistical Computing. 2015, R: A Language and Environment for Statistical Computing (https://www.R-project.org).

Smak, J., and Wing, R. F. 1979, *Acta Astron.*, **29**, 187.

Templeton, M. R., Mattei, J. A., and Willson, L. A. 2005, *Astron. J.*, **130**, 776.

Uttenthaler, S., Meingast, S., Lebzelter, T., Aringer, B., Joyce, R. R., Hinkle, K., Guzman-Ramirez, L., and Greimel, R. 2016, *Astron. Astrophys.*, **585A**, 145.

White, N. M., and Wing, R. F. 1978, *Astrophys. J.*, **222**, 209.

Wing, R. F. 1992, *J. Amer. Assoc. Var. Star Obs.*, **21**, 42.

Wing, R. F. 1997, *J. Amer. Assoc. Var. Star Obs.*, **25**, 63.

Wing, R. F. 2011, in *Astronomical Photometry: Past, Present and Future*, eds. E. F. Milone, C. Sterken, Springer, Heidelberg, 145.

Zima, W. 2008, *Commun. Asteroseismology*, **155**, 17.

Long-Term Changes in the Variability of Pulsating Red Giants (and One RCB Star)

John R. Percy
Department of Astronomy and Astrophysics, and Dunlap Institute of Astronomy and Astrophysics, University of Toronto, 50 St. George Street, Toronto, ON M5S 3H4, Canada; john.percy@utoronto.ca

Arthur Lei Qiu
Department of Astronomy and Astrophysics, University of Toronto, 50 St. George Street, Toronto, ON M5S 3H4, Canada; arthur.qiu@mail.utoronto.ca

Received December 12, 2018; revised March 15, 2019; accepted March 18, 2019

Abstract We have used many decades of visual observations from the AAVSO International Database, and the AAVSO time-series analysis package VSTAR, to study the long-term changes in period, amplitude, and mean magnitude in about 30 normal pulsating red giants (PRGs), i.e. those without large secular changes in period, as well as a few of the rare PRGs which do have such secular period changes. The periods of the typical PRGs "wander" on time scales of about 40 pulsation periods—significantly longer than the time scales of amplitude variation which are 20–35 pulsation periods, with a mean of 27. We have also studied the range and time scale of the long-term changes in pulsation amplitude and mean magnitude, as well as period, and looked for correlations between these. Very long-term changes in mean magnitude of PRGs have not been extensively studied before, because of the challenges of doing so with visual data. Changes in mean magnitude are larger in stars with larger mean amplitude, but correlate negatively with changes in amplitude. There is a weak positive correlation between the long-term period changes and amplitude changes. The causes of these three kinds of long-term variations are still not clear. We note, from the presence of harmonics in the Fourier spectra, that the longest-period PRGs have distinctly non-sinusoidal phase curves. For studying PRGs, we demonstrate the advantage of studying stars with minimal seasonal gaps in the observations, such as those near the celestial poles. We studied Z UMi, misclassified as a possible Mira star but actually an RCB (R Coronae Borealis) star. We determined times of onset of its fadings, but were not able to determine a coherent pulsation period for this star at maximum, with a visual amplitude greater than 0.05. We did, however, find that the times of onset of fadings were "locked" to a 41.98-day period—a typical pulsation period for an RCB star.

1. Introduction

When low- to medium-mass stars exhaust their nuclear fuel, they expand and cool, and become red giants as they exhaust their core hydrogen, then asymptotic-branch (AGB) stars as they exhaust their core helium. In this paper, we shall lump these together as red giants.

Red giants are unstable to radial pulsation. As they expand, their pulsation period increases from days to hundreds of days. Their visual amplitude increases from hundredths of a magnitude to up to 10 magnitudes.

In the *General Catalogue of Variable Stars* (GCVS; Samus *et al.* 2017), pulsating red giants (PRGs) are classified according to their light curves. Mira (M) stars have reasonably regular light curves, with visual peak-to-peak amplitudes greater than 2.5 magnitudes. Semiregular (SR) stars are classified as SRa if there is appreciable periodicity, and SRb if there is very little periodicity. Irregular (L) stars have no periodicity. Percy and Kojar (2013), Percy and Long (2010), Percy and Tan (2013), and Percy and Terziev (2011) have published detailed analyses of AAVSO observations of SRa, SRb, and L stars. There are several processes which can contribute to non-periodicity or apparent irregularity in PRG light curves, including the following:

- In some stars, both the fundamental and first overtone pulsation mode are excited (Kiss *et al.* 1999). The period ratios can be used to derive potentially useful astrophysical information (Percy and Huang 2015).
- The periods of PRGs "wander" by several percent on time scales of decades (Eddington and Plakidis 1929; Percy and Colivas 1999). This phenomenon can be described or modeled by random, cycle-to-cycle period fluctuations.
- About a third of all PRGs show long secondary periods (LSPs), 5–10 times the pulsation period, depending on the pulsation mode (Wood 2000). The nature and cause of LSPs are unknown, despite two decades of research on the topic.
- The amplitudes of PRGs vary by up to a factor of 10 on time scales of 20–30 pulsation periods (Percy and Abachi 2013; Percy and Laing 2017); the cause is not known.
- In a very few stars, thermal pulses cause large, secular changes in period, amplitude, and mean magnitude (Templeton *et al.* 2008 and references therein).

These processes occur on time scales which are much longer than the pulsation period, which itself can be hundreds of days. Since visual observations of these stars have been made for many decades, these observations—despite their limitations—are the best (and only) tool for studying long-term changes in the variability parameters of these stars. The purpose of this project was to use such visual observations to obtain further information about the long-term changes in period, amplitude, and especially mean magnitude in a sample of PRGs, and any correlations between these.

2. Data and analysis

We analyzed visual observations from the AAVSO International Database (AID; Kafka 2018) using the AAVSO's VSTAR software package (Benn 2013). It includes both a Fourier and wavelet analysis routine. The latter uses the Weighted Wavelet Z-Transform (WWZ) method (Foster 1996). The wavelet scans along the dataset, estimating the most likely value of the period and amplitude at each point in time, resulting in graphs which show the best-fit period and amplitude versus time.

3. Results and discussion

3.1. An alternate way of quantifying the "wandering" periods of pulsating red giants

The wandering periods in PRGs have been known for over a century, and can be modeled as random cycle-to-cycle period fluctuations (Eddington and Plakidis 1929), i.e. a "random walk." This implies a process which takes place on a time scale of approximately one pulsation period. One could also look at the period-versus-time graphs in a more global way, assuming them to represent long-term changes, and then to measure the typical time scale of the variations, in a similar way as was done for the amplitude-versus-time graphs (Percy and Abachi 2013, and especially Percy and Laing 2017). We used the first 20 (O–C) diagrams of Karlsson (2013), rather than period-versus-time plots, to measure the ratio of L, the length of the cycles of period increase and decrease, to the pulsation period P. This same analysis could have been done with wavelet analysis; both it and the (O–C) method can be used to display and measure cycles of period increase and decrease. Conveniently, the Karlsson (O–C) diagrams measure time in units of pulsation periods. See Percy and Abachi (2013) and Percy and Laing (2017) for a discussion of the uncertainties of determining L. The values of L/P were as follows: R And (35), T And (51), V And (38), W And (37), X And (32), Y And (67), RR And (60), RW And (44), SV And (35), SX And (35), SZ And (49), TU And (56), UU And (30), UZ And (40), V Ant (56), T Aps (40), R Aqr (28), S Aqr (133), T Aqr (60), W Aqr (40). The median value of L/P is about 40. This ratio is significantly larger than that for amplitude increases and decreases (20–35, mean 27), and much larger than the ratio of LSP/P (5–10), i.e. the time scales are different. We emphasize, though, that the wandering periods may still be a result of accumulated cycle-to-cycle fluctuations, rather than any long-term process.

3.2. Measuring the changing mean magnitudes of pulsating red giants

We have previously studied the long-term changes in the periods and amplitudes of PRGs, but not the mean magnitudes. Some stars have LSPs, of course, but we wondered whether there were even longer-term variations in mean magnitude, an order of magnitude longer, possibly correlated with longterm variations in period or amplitude. Both the light curves and the Fourier spectra suggest that such variations might be present.

One complication is the possible interaction of the pulsational variations and the seasonal gaps. It can produce apparent long-term variations in mean magnitude. These correspond to alias peaks in the Fourier spectrum which lie close to zero frequency. One strategy would be to analyze stars with minimal seasonal gaps, those near the celestial poles; we have done this in section 3.4.

We used the wavelet routine within VSTAR to determine and graph the long-term changes in mean magnitude. Although mean magnitude is not directly graphed in VSTAR, the necessary data can be extracted from the tables produced by VSTAR. The results of these are contained in Table 1, and examples are shown in Figure 1. Graphs like these were constructed for all the stars in Table 1, and used to determine the changes and ranges in the period P, the peak-to-peak amplitude A, and the mean magnitude M. The graphs were also used to assess the correlation between the variations.

The values of ΔM (the range in M) cluster between 0.3–0.6 and 0.7–0.9. It is not clear whether the bimodal distribution is significant.

The time scales for the long-term changes in mean magnitude, when quantified in the same way as for the changes in period (section 3.1) and amplitude, give time scales in the range of 20–30 pulsation periods. Since this is also the time scale of amplitude variation, this raises the concern that the mean magnitude variations might be artifacts of the amplitude variations. We also note that ΔM correlates with the mean amplitude A. This may be because both are correlated with some more fundamental parameter, such as temperature. In PRGs, period and amplitude generally increase as temperature

Table 1. Long-term changes in the period, amplitude, and mean magnitude of pulsating red giants, and correlations between these.

Star	*P(days)*	*A*	*ΔP (days)*	*ΔA*	*ΔM*	*σPA*	*σPM*	*σAM*
R And	410	3.05	12.90	0.38	1.34	0	0	0
T And	281	2.07	9.46	0.35	0.36	+	0	0
V And	256	2.12	7.04	0.47	0.44	+	–	–
X And	343	2.52	8.50	0.76	0.33	0	0	0
RR And	331	2.88	4.46	0.49	0.42	0	0	0
RW And	430	2.92	9.65	0.80	0.72	0	0	–
SV And	313	1.97	9.18	0.59	0.53	0	0	–
TU And	313	1.77	10.24	0.45	0.56	(–)	–	0
UW And	237	1.83	5.60	0.53	0.30	+	0	0
YZ And	207	2.15	4.22	0.51	0.64	0	0	0
R Aqr	386	1.74	12.29	1.76	1.03	+	–	–
S Boo	270	1.77	8.60	0.35	0.30	0	0	–
R Car	310	2.33	5.56	0.36	0.23	0	0	0
S Car	151	1.19	2.99	0.45	0.47	+	–	–
R Cas	430	2.60	9.47	0.36	0.72	(–)	(+)	+
S Cas	608	2.26	15.50	0.89	0.94	–	–	+
T Cas	445	1.54	10.82	0.83	0.34	0	(+)	–
U Cas	277	2.71	5.44	0.55	0.33	+	0	–
Z Cas	497	2.03	13.90	0.76	0.47	+	+	+
TY Cas	645	2.11	15.15	0.89	0.58	–	+	–
R Cen	502	0.81	50.40	1.14	0.20	+	0	0
R Oct	405	1.86	13.60	0.73	0.19	(+)	0	0
S Oct	259	2.55	3.93	0.83	1.04	(+)	0	0
T Oct	219	1.65	5.98	0.91	1.00	0	0	0
U Oct	303	2.44	8.10	0.27	0.77	(+)	0	–
R UMi	324	0.43	13.80	0.45	0.31	0	+	–
S UMi	327	1.31	13.10	0.47	0.63	+	0	–
U UMi	325	1.25	12.00	0.42	0.27	+	0	0

Note: P = period; A = amplitude, and M = mean magnitude, and ΔP, ΔA, and ΔM are the total ranges in period, amplitude, and mean magnitude.

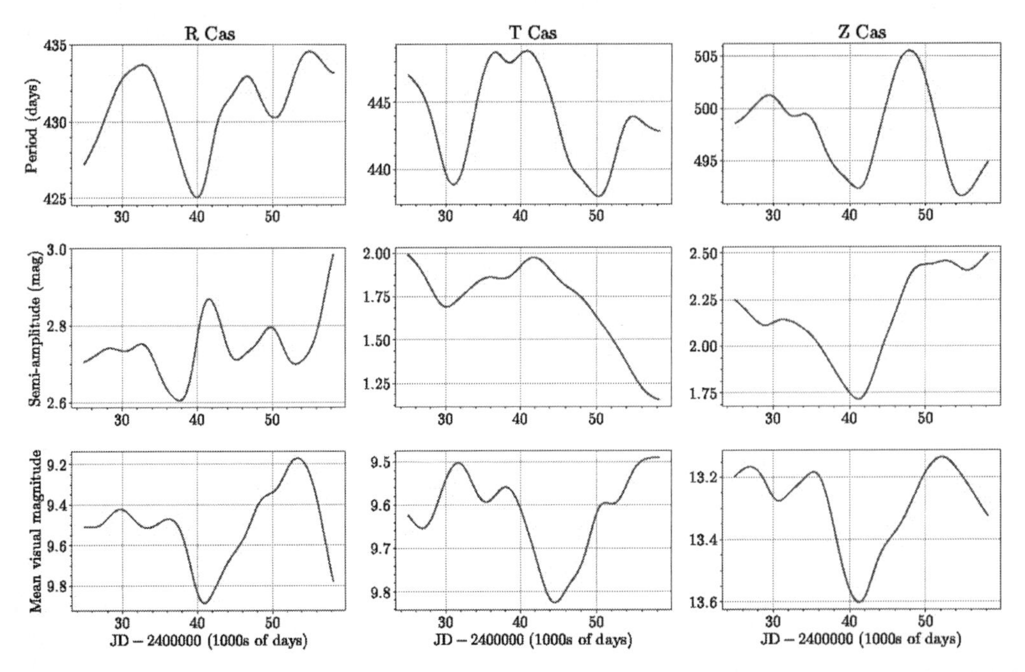

Figure 1. Long-term changes in period (top), semi-amplitude (middle), and mean magnitude (bottom) for three PRGs in Cassiopeia: R Cas (left), T Cas (center), and Z Cas (right). Diagrams such as these were visually and subjectively inspected for correlations between the three parameters such as coincident maxima or minima, as well as their total ranges.

decreases. Large ΔM stars are all long-period stars; smaller ΔM stars occur at all periods.

3.3. Correlations between changing period, amplitude, and mean magnitude?

We compared the long-term changes in period, amplitude, and mean magnitude, and qualitatively assessed, by eye, whether there appeared to be a positive correlation, a negative correlation, or no correlation at all, i.e. whether, with time, they tended to change in the same direction. The changes in period and amplitude were determined using wavelet analysis, and are expressed as the total range in P (ΔP), A (ΔA), and M (ΔM). ΔP increases with P, as might be expected; ΔA increases slightly with P; longer-period stars tend to have larger amplitudes, as is well-known. ΔM does not increase or decrease with P, but lies in the range 0.2 to 0.8. These results are represented by the symbols +, –, and 0, respectively, in Table 1. Correlations involving mean magnitude are less certain than between P and A.

There is a very weak positive correlation between ΔP and ΔA, and a weak negative correlation between ΔA and ΔM. This is discussed further below.

3.4. The advantages of Ursa Minor and Octans

Visual observations such as those in the AID normally contain seasonal gaps, because the star is unavailable for viewing at certain times of the year, depending on its position in the sky. These seasonal gaps produce alias peaks in the Fourier spectrum, due to the one-year periodicity of the times of observations. The alias peaks are frequencies of f ± N / 365.25 where f is the true frequency. The strongest alias peaks are at N = 1. See Percy (2015) for a discussion. For pulsating red giants, the alias peaks can be confused with harmonic or overtone periods.

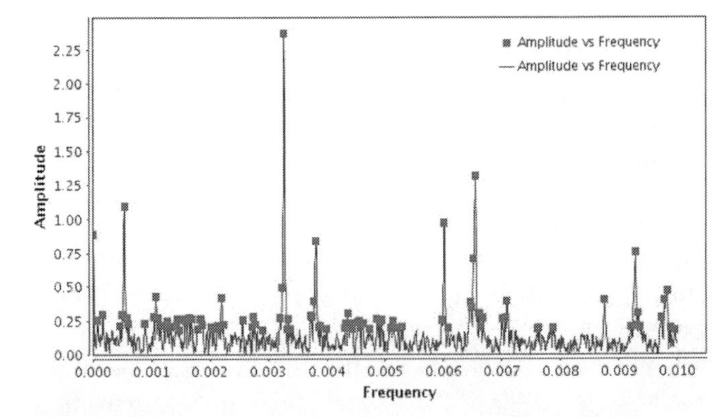

Figure 2. The Fourier spectrum (amplitude in magnitudes versus frequency in cycles / day) of U Cnc, a star near the ecliptic with significant seasonal gaps, and therefore alias peaks in the spectrum. See text for identification of the alias, harmonic, and overtone peaks. The blue line is the Fourier spectrum; the red points are the "top hits" as defined by VSTAR.

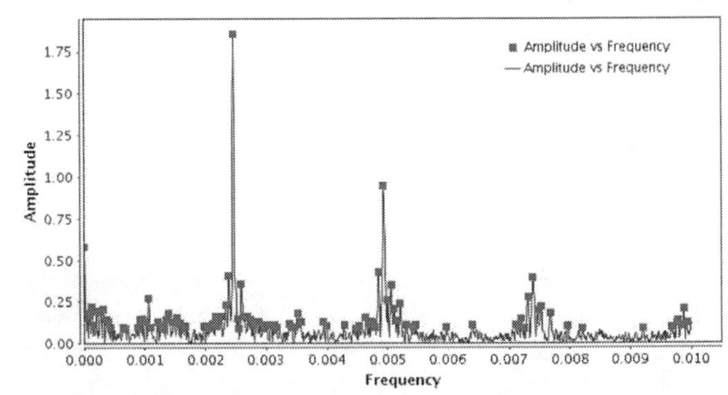

Figure 3. The Fourier spectrum—amplitude versus frequency in cycles / day—of R Oct, with minimal seasonal gaps. Alias peaks are therefore low. The spectrum is dominated by the harmonics which result from the star's non-sinusoidal light curve.

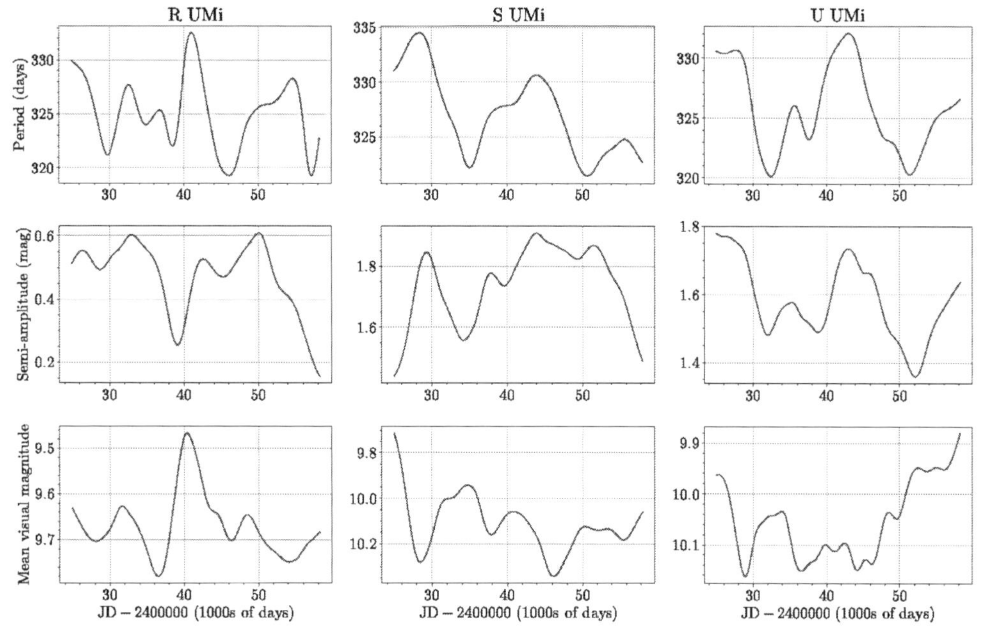

Figure 4. Long-term changes in period (top), semi-amplitude (middle), and mean magnitude (bottom) for three stars in Ursa Minor: R UMi (left), S UMi (center), and U UMi (right). These stars have minimal seasonal gaps to affect the analysis.

Stars near the celestial poles tend to have minimal seasonal gaps, because the stars can be observed all year, without interference from the sun. Ursa Minor and Octans are constellations near the north and south celestial poles, respectively. Figures 2 and 3 compare the DCDFT spectra for U Cnc and R Oct. The solid blue line is the spectrum; the red squares are the "top hits" as defined by VSTAR. U Cnc is near the ecliptic, and shows a complex spectrum with the pulsation frequency 0.003278 cycles per day (cpd) or period 306 days, alias frequencies at 0.000542 and 0.006014 cpd, harmonic frequencies at 0.006554 and 0.009794 cpd, and aliases of the first harmonic at 0.003818 and 0.009296 cpd, and a possible first overtone at 0.007088 cpd. S Oct is near the south celestial pole, and shows only the pulsation frequency 0.002468 cpd (period 405.2 days) and harmonic frequencies at two, three, and four times this.

As noted earlier, the study of these stars was motivated by the concern that interaction between the pulsational variations and the seasonal gaps might produce apparent low-frequency variability in mean magnitude. In fact, the long-term variability (ΔM) of the seven stars in Oct and UMi is similar to that of the other stars, in both total range and time scale. Whether this variability is real, or due to the distribution of the observations over the pulsation cycle or some other observational factor, or a combination of the two, we cannot tell.

As for correlations (Figure 4), there is a tendency for ΔP and ΔA to be positively correlated, ΔA and ΔM to be negatively correlated, and ΔP and ΔM to be uncorrelated. There were similar but weaker correlations among the stars not in Oct or UMi. These correlations are suggestive, but are not present in every star.

3.5. Stars with significant secular period change

The vast majority of PRGs have wandering periods, but a few percent have periods that change secularly and significantly (Templeton *et al.* 2008), probably due to a thermal pulse.

Table 2. Variability properties, their long-term changes, and directions, and correlations between these for PRGs with significant secular period changes.

Star	*P (days)*	*A*	*ΔP (day)*	*ΔA*	*ΔM*	*σPA*	*σPM*	*σAM*
R Aql	282.6	0.83	55↓	0.8↓	0.7↑	+	–	–
R Cen	546.1	0.81	50↓	1.1↓	0.0	+	(–)	(–)
BF Cep	429.3	1.74	14↑	0.4↑	0.1↓	+	–	–
BH Cru	520.6	1.18	35↑	0.5↑	0.3↓	+	–	–
LX Cyg	565.3	1.05	100↑	—	0.8↓	0	0	0
W Dra	279.8	0.93	33↑	0.8↑	1.0↓	0	–	0
R Hya	388.0	1.08	50↓	0.8↓	0.6↑	+	(–)	–
Z Tau	460.0	0.59	40↓	—	1.0↑	0	–	0
T UMi	312.2	0.56	120↓	2.0↓	1.0↑	+	–	–

Because they are so unusual, these stars have previously been studied in some detail; see discussion and references in Templeton *et al.* (2008). For completeness, we list nine of these, in Table 2. The correlations between changes in P, A, and M, as given in the last three columns, are qualitative and based on visual inspection.

There is a generally positive correlation between amplitude and period change, and a negative correlation between mean magnitude change, and period or amplitude change.

3.6. The nature of Z UMi

In the course of undertaking the study of the stars in UMi and Oct, we came upon Z UMi. It is classified in the *General Catalogue of Variable Stars* (Samus *et al.* 2017) as M:, i.e. a possible Mira star, with a period of 475 days. At first inspection, the light curve (Figure 5)—especially the early part—bears some resemblance to a Mira star but, on second inspection, is clearly that of an R Coronae Borealis (RCB) star. Indeed, it was identified as a new RCB star by Benson *et al.* (1994). Fourier analysis gives strongest peaks at "periods" of 1351 and 895 days, but these are just the best fits to the random fadings; they have absolutely no physical significance.

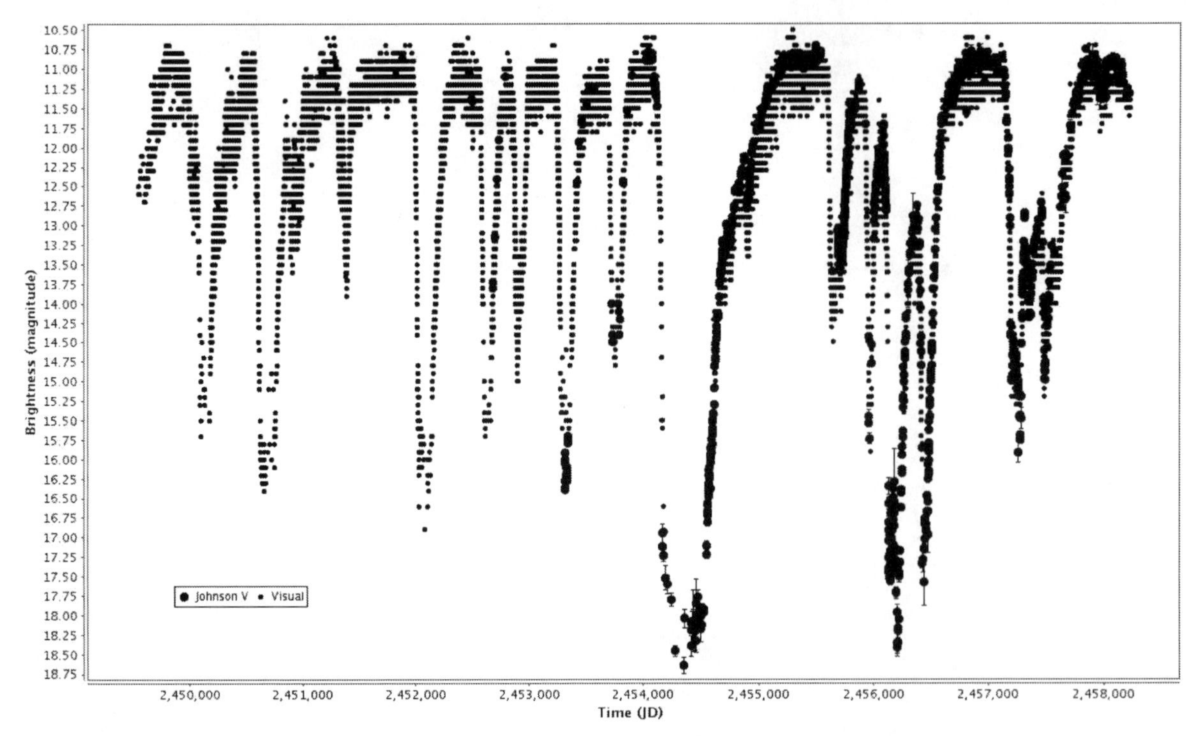

Figure 5. Light curve of Z UMi, from visual observations in the AAVSO International Database. The early variability, especially if sparsely-sampled, bears some resemblance to that of a Mira star.

Table 3. Times of onset of fadings in Z UMi, cycle numbers in the 41.98-day period, and (O–C) analysis of these for a period of 41.98 days and an initial epoch of JD 2450004.

Onset JD	*Cycle*	*(O–C)/P*
2450004	0	0.00
2450557	13	0.17
2451300	31	–0.13
1451977	47	0.00
2452572	61	0.17
2452861	68	0.05
2453239	77	0.06
2453699	88	0.02
2454112	98	–0.14
2455590	133	0.06
2455921	141	–0.05
2456086	145	–0.12
2456394	152	0.22
2457139	170	–0.04
2457937	189	–0.03
2458103	193	–0.07

RCB stars are rare carbon-rich, hydrogen-poor, highly evolved yellow supergiants which undergo fadings of up to 10 magnitudes, then slowly return to normal. The fadings are caused by the nonspherical ejection of carbon-rich dust by the star.

Many RCB stars display low-amplitude pulsations at maximum analogous to the pulsations of Cepheids; see Table 3 in Rao and Lambert (2015) for a list. They give a pulsation period of 130 days for Z UMi, based on very limited photoelectric measurements by Benson *et al.* (1994). We therefore examined AAVSO visual and V observations in the few intervals when the star was near maximum light. No periods with amplitudes greater than 0.05 stand out, though there are some suggestions of low-amplitude variations on time scales of 50–100 days (Figure 6).

We also compiled a list of times of onset of fadings (Table 3), to see if they were "locked" to some period which might be a pulsation period, as has been found in at least five R CrB stars (Crause *et al.* 2007 and references therein). Such a "lock" might imply a causative relation between pulsation and the onset of fading. Indeed, they appear to be locked to a period of 41.98 days—a plausible pulsation period. This period was determined in two ways: (1) dividing the intervals between the fadings by small whole numbers, and looking for commonalities, and (2) calculating a Fourier spectrum of the times of onset of fadings; a peak occurs at 41.98 days. Table 3 lists the times of onset, the cycle numbers of the 41.98-day period, and the values of (O–C) expressed in periods. The average absolute value of the (O–C) is 0.09 cycle, or 4 days. This is similar to a value obtained by Crause *et al.* (2007) in five other R CrB stars whose fadings were locked to a pulsation period.

3.7. Phase curves of pulsating red giants

Most PRGs have reasonably sinusoidal phase curves. We note, however, that PRGs with longer periods tend to have non-sinusoidal phase curves. These include: RW Lyr (503), Z Pup (510), V Cam (523), V Del (528), S Cas (613), and TY Cas (645); the numbers in brackets are the pulsation periods in days. As a way to quantify the non-sinusoidal nature, we used DCDFT in VSTAR to determine the ratio of the first-harmonic amplitude to the fundamental amplitude. The ratio ranges from 0.45 in RW Lyr to 0.78 in V Del, i.e. the phase curves are highly non-sinusoidal. RW Lyr, incidentally, varies in amplitude by a factor of two—unusually large for a Mira star.

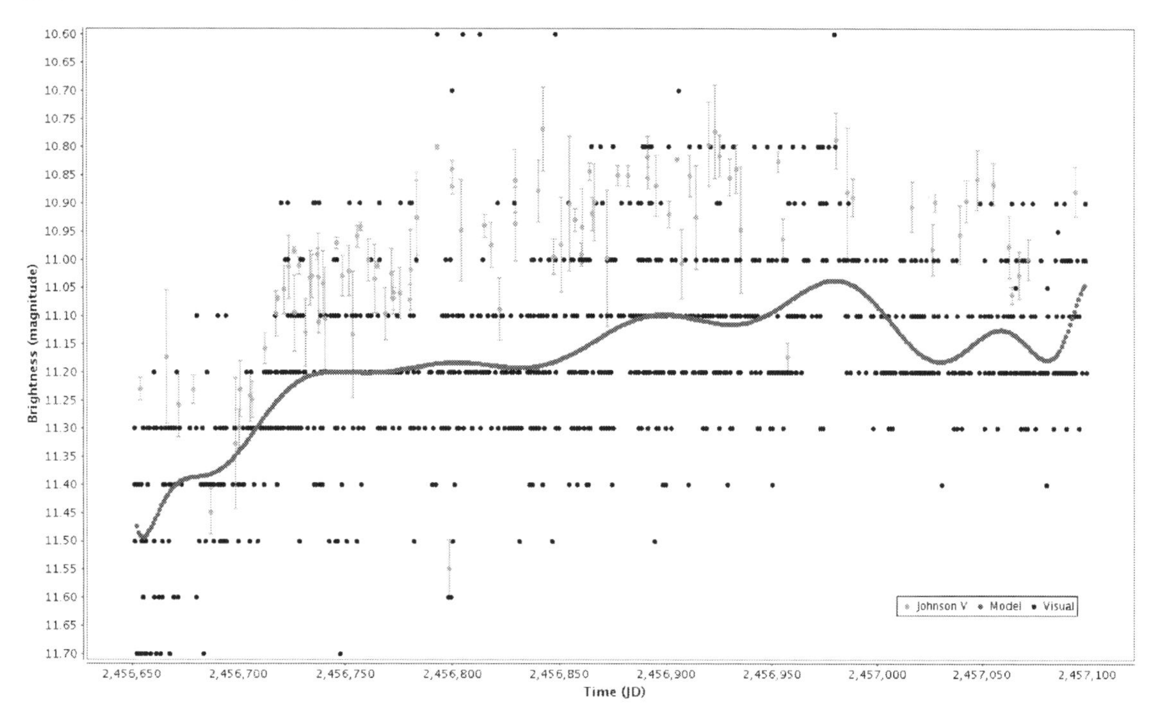

Figure 6. The AAVSO light curve of Z UMi near maximum. The small black points are visual. The solid red line is a polynomial fit to these. The green points with error bars are Johnson V observations. Note the possible low-amplitude variability with a time scale of a few tens of days. The variability is marginal, and there is no evidence for or against periodicity.

We note also that the phase curves of larger-amplitude PRG LSPs are often non-sinusoidal also; see Figures 1 and 4 in Percy and Deibert (2016), for instance, which show the LSP light curves of U Del and Y Lyn. The similarity of the phase curves may be entirely coincidental; there is no evidence otherwise. Or it may be a clue to the nature of the LSP.

4. Discussion

It is not clear whether the long-term variations in mean magnitude of the PRGs are spurious or real, and, if they are real, what the cause is. If spurious, they could arise from the non-random distribution of the observations over the various variability cycles, or changes in the visual observers and their characteristics over time, or be due to changes in the calibration of the visual photometry system (though the AAVSO tries very hard to avoid such changes). If real, they could reflect some long-term variation in the physical properties of the star, perhaps due to the convection process or variations in the amount of obscuring dust around the star. It would be helpful to do a cycle-by-cycle analysis, perhaps using stars with minimal seasonal gaps. It is because of the challenges of studying very long-term changes in mean magnitude using visual data that we and others have not previously carried it out.

The causes of the LSPs and the longer-term variations in period and in amplitude are also not known; see discussion in Percy and Deibert (2016). The variations in period have traditionally been ascribed to random cycle-to-cycle fluctuations, but there is no physical evidence for this. Giant convection cells may somehow be involved in these three types of variations, either through their turnover or through rotational variability. Large granulation cells have recently been imaged on the surface of π^1 Gruis, a PRG (Paladini *et al.* 2018). There have been no explanations proposed, that we know of, for the longterm variations in amplitude.

5. Conclusions

We present some new analyses of PRGs, including long-term changes in mean magnitude, and some interesting possible correlations between the long-term variations in the periods, amplitudes, and mean magnitudes. Since the causes of these long-term variations remain unknown, studies such as this one continue to be useful. Given the complexity of these stars' variations, however, and the limitations of visual data, we cannot say more. A much larger study, possibly with a more quantitative comparison between the variations, might possibly confirm these correlations. Surveys such as the All-Sky Automated Survey for Supernovae have now provided precise photometry of thousands of PRGs over 2000+ days, and these data may eventually help to understand these long-term variations.

Our information on avoiding alias periods (section 3.4) and on the non-sinusoidal phase curves of long-period PRGs may be already obvious or known, but it is useful to point it out here for others who analyze these stars.

As is often the case, we have made an unexpected discovery: a misclassified (at least in the GCVS) RCB star. We have been able to infer a pulsation period of 41.98 days for this star, Z UMi.

And since this project was carried out by an undergraduate student, this paper provides one more example of how such students can develop and apply their science, math, and computing skills by doing (and publishing) real science, with real data. AID and VSTAR are well-suited for such projects.

6. Acknowledgements

We thank the AAVSO observers who made the observations on which this project is based, the AAVSO staff who archived them and made them publicly available, and the developers of the VSTAR package which we used for analysis. We also thank the referee for their useful comments. This paper is based on a research project carried out by undergraduate math and science student ALQ. We acknowledge and thank the University of Toronto Work-Study Program for existing, and for financial support. This project made use of the SIMBAD database, maintained in Strasbourg, France. The Dunlap Institute is funded through an endowment established by the David Dunlap family and the University of Toronto.

References

Benn, D. 2013, VSTAR data analysis software (http://www.aavso.org/vstar-overview).

Benson, P. J., Clayton, G. C., Garnavich, P., and Szkody, P. 1994, *Astron. J.*, **108**, 247.

Crause, L. A., Lawson, W. A., and Henden, A. 2007, *Mon. Not. Roy. Astron. Soc.*, **375**, 301.

Eddington, A. S., and Plakidis, S. 1929, *Mon. Not. Roy. Astron. Soc.*, **90**, 65.

Foster, G. 1996, *Astron. J.*, **112**, 1709.

Kafka, S. 2018, variable star observations from the AAVSO International Database (https://www.aavso.org/aavso-international-database).

Karlsson, T. 2013, *J. Amer. Assoc. Var. Star Obs.*, **41**, 348 (O–C diagrams available at var.astronet.se/mirainfooc.php).

Kiss, L. L., Szatmary, K., Cadmus, R. R. Jr., and Mattei, J. A. 1999, *Astron. Astrophys.*, **346**, 542.

Paladini, C. *et al.* 2018, *Nature*, **553**, 310.

Percy, J. R. 2015, *J. Amer. Assoc. Var. Star Obs.*, **43**, 223.

Percy, J. R., and Abachi, R. 2013, *J. Amer. Assoc. Var. Star Obs.*, **41**, 193.

Percy, J. R., and Colivas, T. 1999, *Publ. Astron. Soc. Pacific*, **111**, 94.

Percy, J. R., and Deibert, E. 2016, *J. Amer. Assoc. Var. Star Obs.*, **44**, 94.

Percy, J. R., and Huang, D. J. 2015, *J. Amer. Assoc. Var. Star Obs.*, **43**, 118.

Percy, J. R., and Kojar, T. 2013, *J. Amer. Assoc. Var. Star Obs.*, **41**, 15.

Percy, J. R., and Laing, J. 2017, *J. Amer. Assoc. Var. Star Obs.*, **45**, 197.

Percy, J. R., and Long, J. 2010, *J. Amer. Assoc. Var. Star Obs.*, **38**, 161.

Percy, J. R., and Tan, P. J. 2013, *J. Amer. Assoc. Var. Star Obs.*, **41**, 75.

Percy, J. R., and Terziev, E. 2011, *J. Amer. Assoc. Var. Star Obs.*, **39**, 1.

Rao, N. K., and Lambert, D. L. 2015, *Mon. Not. Roy. Astron. Soc.*, **447**, 3664.

Samus, N. N., *et al.* 2017, *General Catalogue of Variable Stars*, Sternberg Astronomical Institute, Moscow (http://www.sai.msu.ru/gcvs/gcvs/index.htm).

Templeton, M. R., Willson, L. A., and Foster, G. 2008, *J. Amer. Assoc. Var. Star Obs.*, **36**, 1.

Wood, P. R. 2000, *Publ. Astron. Soc. Australia*, **17**, 18.

Multi-color Photometry of the Hot R Coronae Borealis Star and Proto-planetary Nebula V348 Sagittarii

Arlo U. Landolt
Department of Physics and Astronomy, Louisiana State University, Baton Rouge, LA 70803; landolt@phys.lsu.edu

Visiting astronomer, Kitt Peak National Observatory, National Optical Astronomical Observatory, which is operated by the Association of Universities for Research in Astronomy, Inc., under contract with the National Science Foundation.

Visiting astronomer, Cerro Tololo Inter-American Observatory, which is operated by the Association of Universities for Research in Astronomy, Inc., under contract with the National Science Foundation.

Visiting astronomer; this work makes use of observations obtained at the Las Campanas Observatories.

James L. Clem
Department of Physics and Astronomy, Louisiana State University, Baton Rouge, LA 70803
(Current address: Department of Physics, Grove City College, Grove City, PA 16127); jclem@phys.lsu.edu

Visiting astronomer, Cerro Tololo Inter-American Observatory, which is operated by the Association of Universities for Research in Astronomy, Inc., under contract with the National Science Foundation.

Received April 10, 2019; revised May 17, 2019; accepted May 17, 2019

Abstract A long term program of precision photoelectric UBVRI photometry has been combined with AAVSO archival data for the hot, R CrB-type hydrogen deficient star and proto-planetary nebula, V348 Sgr. CCD data also are described. Since V348 Sgr is one of only four hot R CrB stars, it and other group members deserve continued attention by observers.

1. Introduction

The star now known as V348 Sgr was discovered to be variable in light by Woods (1926), and later, independently, by Schajn (1929). The discovery name assigned by Woods was HV 3976. She found the star's brightness to vary between 11th and fainter than 16.5 magnitude. Woods' discovery note does not state the kind of emulsion utilized, and hence the type of magnitude. (History describing the Harvard College Observatory (HCO) telescopes, leading to an enhanced understanding of the kinds of magnitudes produced by the HCO patrol telescopes may be found at the Digital Access to a Sky Century @ Harvard (DASCH), dasch.rc.fas.harvard.edu/photometry.php, leadingto dasch.rc.fas.harvard.edu/lightcurve.php. Additional insight is located in Laycock *et al.* (2010).)

V348 Sgr appears in the DR2 release of the *Gaia Catalogue* which appears in VizieR in catalogue I/345/gaia2 (Gaia Collab. *et al.* 2016, 2018). V348 Sgr is source number 4079151545960427264 with coordinates R.A. = 18^h 40^m 19.92705^s, Dec. = –22° 54' 29.3880", J2000. It is a member of a small subset of four hot hydrogen-deficient stars. These four stars, MV Sgr, V348 Sgr, DY Cen, and HV 2671, possess the R CrB-type of light curve, that is, they spend the majority of the time at maximum brightness, with occasional excursions to fainter magnitudes (De Marco *et al.* 2002, and references therein).They differ from most R CrB stars in that on average their effective temperatures are 10,000 to 15,000 K hotter. Therefore, these four stars are of special interest, and should continue and remain on observing programs.

V348 Sgr also appears in the literature as AN 21.1929, AAVSO 1834-23, 2MASS J18401992-2254292, and ASAS J184020-2254.5. V348 Sgr does not appear in the UCAC4 catalogue or in the corresponding APASS data release.

A finding chart for V348 Sgr is given in Figure 1. The chart is based on a digitized version of the Palomar Sky Survey I (POSS I) blue survey (Palomar Observatory 1950–1957). The size of the field as presented in the chart is about ten arc minutes on a side.

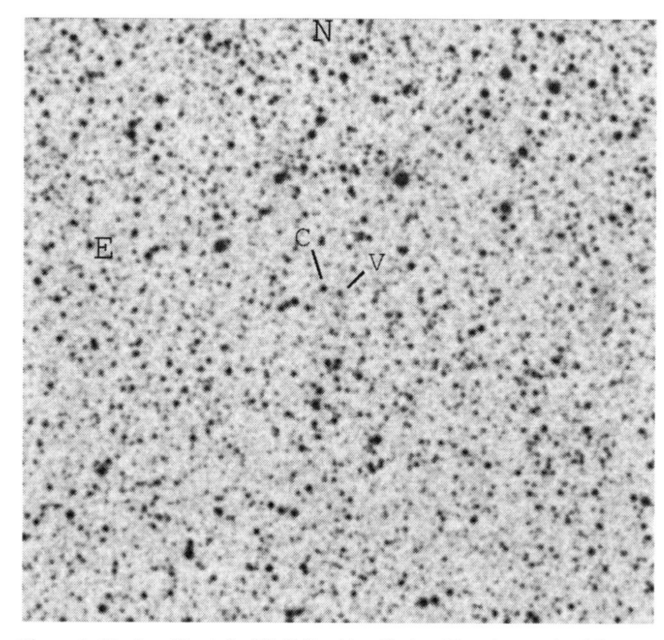

Figure 1. Finding Chart for V348 Sgr identified as V, and a nearby faint star UCAC4 336-170138 identified by C. The field of view is approximately 10 arc minutes on a side.

Excellent and definitive summaries of the characteristics of R CrB stars, including the four stars listed above, have appeared in Clayton (1996, 2012). De Marco *et al.* (2002) thoroughly describe this four-member subset of R CrB stars. They write that these four stars are quite different from each other as evidenced by their spectra. They indicate that the "only common characteristics are their temperatures and light variation." Finally, they found that MV Sgr, V348 Sgr, and DY Cen all exhibit a long-term downward trend in brightness over the time frame under study. Schaefer (2016) has searched archival files and also has discussed the long term behavior of this four-star group of hot R CrB stars. Pollacco *et al.* (1990) showed that the nebulosity surrounding V348 Sgr was an old planetary nebula of extent some 30 arc seconds. Clayton *et al.* (2011) discuss the properties of the dust involved with V348 Sgr via use of Spitzer/IRS spectra. The data in this paper are based on precision photometry in standard bandpasses covering years for which such observations are minimal.

2. Observations

Photoelectric observations of V348 Sgr were taken by AUL in the interval 1982 September 14 to 2001 October 16 ($2445226.53 \leq HJD \leq 2452198.57$), a range of 6,972 days, or 19.1 years. The data were collected at Cerro Tololo Inter-American Observatory's (CTIO's) 0.9-m, 1.0-m (Yale), 1.5-m, and 4.0-m telescopes, and at the Kitt Peak National Observatory 1.3-m telescope.

The dates upon which data were taken, and observatories, telescopes, detectors, and filters utilized all are listed in Table 1. These data were tied into *UBVRI* standard stars as defined in Landolt (1983). All *R* and *I* measures herein are on the Kron-Cousins system. The data, using detectors described in Landolt (1983, 1992), were reduced following precepts outlined in Landolt (2007).

CCD data were taken on 22 nights at the Las Campanas Observatory (LCO) telescopes, 16 nights at the Swope 1.0-m, and 6 at the DuPont 2.5-m. The 1992 October and 1996 August CCD data were obtained at the Swope 1.0-m telescope. The detector was a Texas Instrument (TI#1) 800 × 800 pixel chip whose plate scale was 0.435" pixel^{-1}. The field size was 5.8' on a side. The data were binned 2 × 2. A 2 × 2 inch *UBVRI* filter set borrowed from CTIO meant that the same filter set was used for AUL's CTIO and LCO programs at that time. The composition of the filter set is described in Table 1 in Landolt and Clem (2017). The June 1994 CCD data were obtained at the LCO DuPont 2.5-m telescope, using the same chip and filters as at the Swope telescope.

The CTIO CCD data, calendar years 2008 through 2010, were obtained at the CTIO Yale 1.0-m telescope by JLC, using the Y4KCam CCD. The equipment, data acquisition, and reduction processes were described in Clem and Landolt (2013).

Data were obtained the night of UT 1993 May 11 at the KPNO 0.9-m using the CCDPhot program. This was an IRAF program which used a CCD instead of a photomultiplier as the detector, and apertures defined by software rather than by an aperture wheel. An excellent description of the program and technique was written by Tody and Davis (1992).

The CCDPhot instrumentation included a Tek 2 chip, T5HA, serial number 1115-8-3. For a chip size of 512 × 512 with 27 micron pixels, and a scale of 0.77 arc sec per pixel, the field of view was 6.6 × 6.6 arc min on a side. A more complete description may be found at https://www.noao.edu/noao/noaonews/sep95/art37.html. It was a neat instrumental set-up. A figure illustrating the quantum efficiency of T5HA may be found at https://www.noao.edu/noao/noaonews/jun96/node38.html. Data through the *U* filter did not transform satisfactorily, and are not included herein.

3. Discussion

Data for V348 Sgr in the AAVSO International Database (Kafka 2019) begin on JD 2434917.0, 1954 June 23 UT. We have downloaded data in the interval 1996 April 23 to 2017 November 24 ($2450196.718 \leq 2458082.498$), an interval of 7885.78 days, or 21.6 years, since this subset of data in the AAVSO database is similar in time extent to ours. Visual observations indicating "fainter than" and those taken through filters other than "Johnson *V*" then were eliminated from the listing. The remaining AAVSO observations have been displayed in Figure 2 as black circles.

The photoelectric reduction process recovered the magnitudes and color indices of the standard stars that were observed each night. The rms errors calculated from those recovered magnitude and color indices are listed in Table 2. The first and second columns give the UT date of observation and the corresponding Julian Date, respectively. The telescope at which the data were collected is given in the third column, and the filters through which the data were taken are in the fourth column. The last six columns list the rms errors of the recovered standard stars' magnitude and color indices for that night. The last two lines in Table 2 show that the accuracy of the recovered standard star photometry was one percent or less, except for (*U–B*). When at maximum brightness, V348 Sgr was similar in brightness to the standard stars; when at minimum, it was as much as six magnitudes fainter.

On the night of 2000 May 23 UT, at 08^h 09^m 00^s *UT*, HJD 2451687.83958, V348 Sgr was too faint to measure at the CTIO

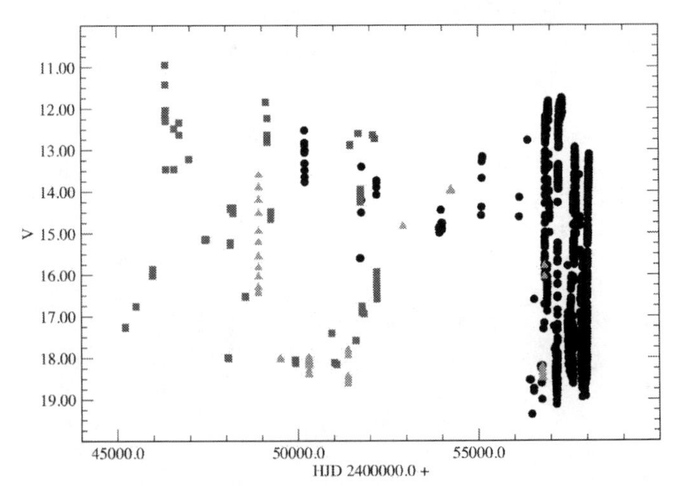

Figure 2. AAVSO V database magnitudes plus *V* photoelectric and CCD magnitudes from this paper for V348 Sgr. Black circles indicate AAVSO data, red squares indicate photoelectric data, and green triangles indicate CCD data.

Table 1. Telescopes, detectors, and filters.

UT (mmddyy)	Observatory Telescope	Detector Set-up	Filter Identification
091482	CTIO 1.5-m	RCA 31034A-02; coldbox 58	(Landolt 1983), Table III
070583	CTIO 1.5-m	RCA 31034A-02; coldbox 59	(Landolt 1983), Table III
100584	CTIO 0.9-m	RCA 31034A-02; coldbox 59	(Landolt 1983), Table III
101184	CTIO 0.9-m	RCA 31034A-02; coldbox 59	(Landolt 1983), Table III
092585	CTIO 1.5-m	RCA 31034A-02; coldbox 59	(Landolt 1983), Table III
093085	CTIO 4.0-m	RCA 31034A-05; coldbox 70	(Landolt 1983), Table III
100385	CTIO 4.0-m	RCA 31034A-05; coldbox 70	(Landolt 1983), Table III
100585	CTIO 1.5-m	RCA 31034A-02; coldbox 59	(Landolt 1983), Table III
100785	CTIO 1.5-m	RCA 31034A-02; coldbox 59	(Landolt 1983), Table III
052486	CTIO 1.5-m	RCA 31034A-02; coldbox 59	(Landolt 1983), Table III
101286	CTIO 1.5-m	RCA 31034A-02; coldbox 59	(Landolt 1983), Table III
070187	KPNO 1.3-m	RCA 31034A-02; coldbox 51	(Landolt 2007), Table 1
091688	CTIO 1.5-m	Hamamatsu R943-02; coldbox 71	(Landolt 1983), Table III
102388	CTIO 1.5-m	Hamamatsu R943-02; coldbox 71	(Landolt 1983), Table III
092689	CTIO 1.5-m	Hamamatsu R943-02; coldbox 71	(Landolt 1983), Table III
060890	CTIO 1.0-m	Hamamatsu R943-02; coldbox 71	(Landolt 1983), Table III
062690	KPNO 1.3-m	RCA 31034A-02; coldbox 51	(Landolt 2007), Table 1
082490	CTIO 1.5-m	Hamamatsu R943-02; coldbox 71	(Landolt 1983), Table III
082690	CTIO 1.5-m	Hamamatsu R943-02; coldbox 71	(Landolt 1983), Table III
110790	CTIO 1.5-m	Hamamatsu R943-02; coldbox 71	(Landolt 1983), Table III
110890	CTIO 1.5-m	Hamamatsu R943-02; coldbox 71	(Landolt 1983), Table III
061791	KPNO 1.3-m	RCA 31034A-02; coldbox 51	(Landolt 2007), Table 1
100791	KPNO 1.3-m	RCA 31034A-02; coldbox 51	(Landolt 2007), Table 1
062992	KPNO 1.3-m	RCA 31034A-02; coldbox 51	(Landolt 2007), Table 1
061593	CTIO 1.5-m	RCA 31034; coldbox 58	(Landolt 1983), Table III
061693	CTIO 1.5-m	RCA 31034; coldbox 58	(Landolt 1983), Table III
061793	CTIO 1.5-m	RCA 31034; coldbox 58	(Landolt 1983), Table III
092493	CTIO 1.0-m	Hamamatsu R943-02; coldbox 50	(Landolt 1983), Table III
092593	CTIO 1.0-m	Hamamatsu R943-02; coldbox 50	(Landolt 1983), Table III
072495	CTIO 1.5-m	Hamamatsu R943-02; coldbox 50	(Landolt 1983), Table III
073195	CTIO 1.0-m	Hamamatsu R943-02; coldbox 50	(Landolt 1983), Table III
092997	CTIO 1.5-m	Burle Industries 31034A-02; coldbox 60	(Landolt 1983), Table III
050898	CTIO 1.5-m	Burle Industries 31034A-02; coldbox 60	(Landolt 1983), Table III
072598	CTIO 1.5-m	Burle Industries 31034A-02; coldbox 60	(Landolt 1983), Table III
092598	CTIO 1.5-m	RCA 31034A-02; coldbox 53	(Landolt 1983), Table III
101099	CTIO 1.5-m	RCA 31034A-02; coldbox 53	(Landolt 1983), Table III
101299	CTIO 1.5-m	RCA 31034A-02; coldbox 53	(Landolt 1983), Table III
031000	CTIO 1.5-m	RCA 31034A-02; coldbox 53	(Landolt 1983), Table III
052300	CTIO 1.5-m	RCA 31034A-02; coldbox 53	(Landolt 1983), Table III
052900	CTIO 1.5-m	RCA 31034A-02; coldbox 53	(Landolt 1983), Table III
071900	CTIO 1.5-m	RCA 31034A-02; coldbox 53	(Landolt 1983), Table III
072000	CTIO 1.5-m	RCA 31034A-02; coldbox 53	(Landolt 1983), Table III
072300	CTIO 1.5-m	RCA 31034A-02; coldbox 53	(Landolt 1983), Table III
072400	CTIO 1.5-m	RCA 31034A-02; coldbox 53	(Landolt 1983), Table III
072500	CTIO 1.5-m	RCA 31034A-02; coldbox 53	(Landolt 1983), Table III
082500	CTIO 1.5-m	RCA 31034A-02; coldbox 53	(Landolt 1983), Table III
082600	CTIO 1.5-m	RCA 31034A-02; coldbox 53	(Landolt 1983), Table III
082700	CTIO 1.5-m	RCA 31034A-02; coldbox 53	(Landolt 1983), Table III
082800	CTIO 1.5-m	RCA 31034A-02; coldbox 53	(Landolt 1983), Table III
082900	CTIO 1.5-m	RCA 31034A-02; coldbox 53	(Landolt 1983), Table III
083000	CTIO 1.5-m	RCA 31034A-02; coldbox 53	(Landolt 1983), Table III
102000	CTIO 1.5-m	RCA 31034A-02; coldbox 53	(Landolt 1983), Table III
102100	CTIO 1.5-m	RCA 31034A-02; coldbox 53	(Landolt 1983), Table III
062801	CTIO 1.5-m	RCA 31034A-02; coldbox 53	(Landolt 1983), Table III
072501	CTIO 1.5-m	RCA 31034A-02; coldbox 53	(Landolt 1983), Table III
082201	CTIO 1.5-m	RCA 31034A-02; coldbox 53	(Landolt 1983), Table III
100701	CTIO 1.5-m	RCA 31034A-02; coldbox 53	(Landolt 1983), Table III
100801	CTIO 1.5-m	RCA 31034A-02; coldbox 53	(Landolt 1983), Table III
100901	CTIO 1.5-m	RCA 31034A-02; coldbox 53	(Landolt 1983), Table III
101001	CTIO 1.5-m	RCA 31034A-02; coldbox 53	(Landolt 1983), Table III
101101	CTIO 1.5-m	RCA 31034A-02; coldbox 53	(Landolt 1983), Table III
101301	CTIO 1.5-m	RCA 31034A-02; coldbox 53	(Landolt 1983), Table III
101501	CTIO 1.5-m	RCA 31034A-02; coldbox 53	(Landolt 1983), Table III
101601	CTIO 1.5-m	RCA 31034A-02; coldbox 53	(Landolt 1983), Table III

Table 2. RMS photometric errors per night recovered from standard stars.

UT (mmddyy)	HJD 2400000.0+	Telescope	Filter	RMS Errors Recovered Standards					
				V	(B–V)	(U–B)	(V–R)	(R–I)	(V–I)
091482	45226.5	CTIO 1.5-m	UBVRI	0.016	0.014	0.050	0.008	0.008	0.008
070583	45520.5	CTIO 1.5-m	UBVRI	0.006	0.007	0.006	0.003	0.004	0.004
100584	45978.5	CTIO 0.9-m	UBVRI	0.010	0.006	0.015	0.008	0.005	0.007
101184	45984.5	CTIO 0.9-m	UBVRI	0.016	0.005	0.027	0.005	0.003	0.004
092585	46333.5	CTIO 1.5-m	UBVRI	0.012	0.011	0.032	0.014	0.011	0.018
093085	46338.5	CTIO 4.0-m	UBVRI	0.012	0.020	0.100	0.007	0.012	0.017
100385	46341.5	CTIO 4.0-m	UBVRI	0.037	0.040	0.054	0.035	0.043	0.068
100585	46343.5	CTIO 1.5-m	UBVRI	0.010	0.007	0.043	0.003	0.006	0.008
100785	46345.5	CTIO 1.5-m	UBVRI	0.007	0.011	0.052	0.009	0.005	0.012
052486	46574.5	CTIO 1.5-m	UBVRI	0.004	0.008	0.025	0.006	0.017	0.017
101286	46715.5	CTIO 1.5-m	UBVRI	0.010	0.014	0.036	0.007	0.011	0.009
070187	46977.5	KPNO 1.3-m	UBVRI	0.019	0.022	0.020	0.016	0.019	0.021
091688	47420.5	CTIO 1.5-m	UBVRI	0.004	0.010	0.032	0.006	0.006	0.007
102388	47457.5	CTIO 1.5-m	UBVRI	0.009	0.010	0.042	0.008	0.006	0.009
092689	47795.5	CTIO 1.5-m	UBVRI	0.009	0.011	0.041	0.005	0.006	0.010
060890	48050.5	CTIO 1.0-m	UBVRI	0.006	0.011	0.028	0.006	0.010	0.012
062690	48068.5	KPNO 1.3-m	UBVRI	0.020	0.007	0.017	0.005	0.009	0.013
082490	48127.5	CTIO 1.5-m	UBVRI	0.011	0.009	0.032	0.006	0.005	0.006
082690	48129.5	CTIO 1.5-m	UBVRI	0.010	0.008	0.029	0.005	0.008	0.009
110790	48202.5	CTIO 1.5-m	UBVRI	0.007	0.008	0.034	0.007	0.008	0.011
110890	48203.5	CTIO 1.5-m	UBVRI	0.008	0.014	0.038	0.006	0.007	0.010
061791	48424.5	KPNO 1.3-m	UBVRI	0.006	0.006	0.023	0.005	0.006	0.008
100791	48536.5	KPNO 1.3-m	UBVRI	0.012	0.007	0.022	0.007	0.008	0.013
062992	48802.5	KPNO 1.3-m	UBVRI	0.008	0.007	0.014	0.007	0.003	0.006
051193	49118.5	KPNO 0.9-m	UBVRI	0.015	0.016	0.051	0.023	0.007	0.022
061593	49153.5	CTIO 1.5-m	UBVRI	0.008	0.009	0.022	0.004	0.014	0.016
061693	49154.5	CTIO 1.5-m	UBVRI	0.007	0.004	0.016	0.005	0.009	0.011
061793	49155.5	CTIO 1.5-m	UBVRI	0.008	0.007	0.028	0.006	0.005	0.009
092493	49254.4	CTIO 1.0-m	UBVRI	0.009	0.010	0.028	0.008	0.032	—
092593	49255.5	CTIO 1.0-m	UBVRI	0.010	0.008	0.031	0.011	0.024	—
072495	49922.5	CTIO 1.5-m	UBVRI	0.007	0.009	0.020	0.004	0.010	0.011
073195	49929.5	CTIO 1.0-m	UBV	0.004	0.008	0.020	—	—	—
092997	50720.5	CTIO 1.5-m	UBVRI	0.006	0.009	0.031	0.004	0.007	0.008
050898	50941.5	CTIO 1.5-m	UBVRI	0.008	0.009	0.014	0.004	0.005	0.004
072598	51019.5	CTIO 1.5-m	UBVRI	0.015	0.014	0.020	0.006	0.011	0.014
092598	51081.5	CTIO 1.5-m	UBVRI	0.008	0.009	0.032	0.007	0.011	0.014
101099	51461.5	CTIO 1.5-m	UBVRI	0.004	0.010	0.033	0.003	0.007	0.005
101299	51463.5	CTIO 1.5-m	UBVRI	0.005	0.006	0.033	0.004	0.008	0.008
031000	51613.5	CTIO 1.5-m	UBVRI	0.010	0.008	0.016	0.006	0.005	0.006
052300	51687.5	CTIO 1.5-m	UBVRI	0.008	0.010	0.019	0.006	0.012	0.015
052900	51693.5	CTIO 1.5-m	UBVRI	0.008	0.012	0.023	0.004	0.006	0.007
071900	51744.5	CTIO 1.5-m	UBVRI	0.007	0.007	0.020	0.004	0.004	0.007
072000	51745.5	CTIO 1.5-m	UBVRI	0.006	0.009	0.026	0.003	0.010	0.011
072300	51748.5	CTIO 1.5-m	UBVRI	0.004	0.008	0.016	0.004	0.005	0.006
072400	51749.5	CTIO 1.5-m	UBVRI	0.007	0.007	0.015	0.004	0.006	0.008
072500	51750.5	CTIO 1.5-m	UBVRI	0.005	0.008	0.020	0.003	0.004	0.006
082500	51781.5	CTIO 1.5-m	UBVRI	0.010	0.011	0.036	0.005	0.008	0.009
082600	51782.5	CTIO 1.5-m	UBVRI	0.006	0.009	0.031	0.004	0.007	0.008
082700	51783.5	CTIO 1.5-m	UBVRI	0.009	0.010	0.033	0.003	0.011	0.010
082800	51784.5	CTIO 1.5-m	UBVRI	0.006	0.010	0.036	0.003	0.006	0.005
082900	51785.5	CTIO 1.5-m	UBVRI	0.007	0.008	0.031	0.004	0.006	0.007
083000	51786.5	CTIO 1.5-m	UBVRI	0.010	0.009	0.014	0.004	0.002	0.005
102000	51837.5	CTIO 1.5-m	UBVRI	0.008	0.009	0.033	0.003	0.005	0.006
102100	51838.5	CTIO 1.5-m	UBVRI	0.007	0.010	0.031	0.003	0.006	0.006
062801	52088.5	CTIO 1.5-m	UBVRI	0.007	0.011	0.022	0.006	0.004	0.008
072501	52115.5	CTIO 1.5-m	UBVRI	0.007	0.007	0.023	0.004	0.008	0.010
082201	52143.5	CTIO 1.5-m	UBVRI	0.008	0.011	0.031	0.005	0.008	0.011
100701	52189.5	CTIO 1.5-m	UBVRI	0.010	0.010	0.034	0.005	0.014	0.015
100801	52190.5	CTIO 1.5-m	UBVRI	0.009	0.013	0.033	0.004	0.016	0.017
100901	52191.5	CTIO 1.5-m	UBVRI	0.009	0.013	0.035	0.005	0.005	0.007
101001	52192.5	CTIO 1.5-m	UBVRI	0.010	0.008	0.031	0.005	0.009	0.011
101101	52193.5	CTIO 1.5-m	UBVRI	0.007	0.011	0.033	0.005	0.011	0.012
101301	52195.5	CTIO 1.5-m	UBVRI	0.005	0.010	0.033	0.005	0.006	0.007
101501	52197.5	CTIO 1.5-m	UBVRI	0.007	0.010	0.037	0.007	0.029	0.031
101601	52198.5	CTIO 1.5-m	UBVRI	0.009	0.012	0.040	0.007	0.012	0.016
			ave.	0.009	0.010	0.030	0.006	0.009	0.011
			±	0.005	0.005	0.013	0.005	0.007	0.009

Table 3. V348 Sgr photoelectric data.

HJD	*V*	*(B–V)*	*(U–B)*	*(V–R)*	*(R–I)*	*(V–I)*
2445226.53218	12.744	+0.559	–0.375	+0.398	+0.363	+0.761
2445226.53466	12.726	+0.562	–0.370	+0.409	+0.363	+0.772
2445520.76341	13.238	+0.664	–0.249	+0.459	+0.444	+0.903
2445978.59159	13.977	+0.911	–0.033	+0.645	+0.566	+1.213
2445984.54833	14.143	+1.161	–0.185	+0.596	+0.569	+1.167
2446333.52951	19.053	+0.272	–1.532	+2.078	+0.743	+2.856
2446333.53762	17.825	+0.532	–0.960	+1.211	+0.340	+1.574
2446338.51876	18.583	+1.090	–0.426	+1.104	+0.746	+1.854
2446341.53179	17.702	+0.613	–0.912	+0.952	+0.512	+1.491
2446341.53922	17.755	+0.634	–0.989	+0.915	+0.607	+1.541
2446343.54113	17.967	+0.056	–1.026	+1.697	+0.763	+2.475
2446345.52463	16.532	+0.822	–0.595	+1.041	+0.659	+1.707
2446574.87950	17.519	+0.197	—	—	—	—
2446574.88446	16.538	+1.426	—	—	—	—
2446715.52914	17.367	+2.075	—	—	—	—
2446715.53758	17.662	+0.855	–0.560	+1.189	+0.702	+1.895
2446977.79809	16.778	+0.577	–0.757	+1.072	+0.536	+1.622
2447420.52371	14.837	+1.256	+0.850	+0.711	+0.694	+1.408
2447420.52916	14.862	+1.189	+1.036	+0.726	+0.677	+1.400
2447457.54620	14.857	+1.426	—	+0.682	+0.667	+1.344
2448050.71173	12.011	+0.362	–0.650	+0.317	+0.281	+0.594
2448068.81702	11.992	+0.394	–0.603	+0.300	+0.297	+0.596
2448127.65392	14.722	+0.756	–0.260	+0.575	+0.497	+1.070
2448127.65874	14.776	+0.720	–0.284	+0.621	+0.468	+1.086
2448129.67057	15.614	+0.685	–0.448	+0.697	+0.547	+1.239
2448129.67842	15.582	+0.687	–0.473	+0.677	+0.551	+1.222
2448202.52816	15.616	+0.937	–0.236	+0.825	+0.653	+1.474
2448203.53091	15.482	+0.994	–0.190	+0.620	+0.793	+1.403
2448536.61322	13.478	+0.811	–0.028	+0.545	+0.500	+1.046
2448536.61628	13.463	+0.825	–0.113	+0.558	+0.452	+1.013
2449118.97954	18.13	–0.20	—	+1.86	+0.79	+2.62
2449153.78668	17.183	+0.594	–0.773	—	—	—
2449153.79625	17.282	+0.618	–0.850	—	—	—
2449154.75816	17.273	+0.456	–0.781	+1.176	+0.695	+1.864
2449154.77087	17.274	+0.441	–0.763	+1.070	+0.529	+1.592
2449155.80644	17.375	+0.555	–0.964	+1.060	+0.492	+1.551
2449155.81863	17.772	+0.375	–1.067	+1.528	+0.598	+2.125
2449254.56876	15.49	+0.8	—	—	—	—
2449254.57160	15.34	+1.6	—	—	—	—
2449255.50394	15.52	+0.9	—	—	—	—
2449922.67670	11.951	+0.353	–0.668	+0.284	+0.281	+0.564
2449922.68047	11.948	+0.365	–0.704	+0.276	+0.296	+0.571
2449929.50884	11.874	+0.323	–0.694	—	—	—
2450941.82178	12.590	+0.529	–0.402	+0.401	+0.399	+0.799
2451019.74143	11.885	+0.401	–0.608	+0.280	+0.298	+0.578
2451081.53852	11.845	+0.337	–0.695	+0.268	+0.266	+0.534
2451461.53119	17.113	+0.517	–0.827	+0.933	+0.309	+1.247
2451463.54158	17.133	+0.365	–0.673	+0.943	+0.532	+1.476
2451613.86655	12.421	+0.490	–0.438	+0.366	+0.331	+0.698
2451693.77032	17.407	+0.321	–0.895	+0.922	+0.105	+1.027
2451744.73326	15.752	+1.547	+0.087	+0.981	+0.833	+1.819
2451745.75336	15.811	+1.256	+0.172	+0.881	+0.728	+1.610
2451748.64006	15.976	+1.006	–0.457	+0.943	+0.714	+1.658
2451748.65226	16.056	+1.119	–0.428	+0.936	+0.846	+1.780
2451749.66877	15.997	+0.965	–0.474	+0.965	+0.673	+1.637
2451750.71958	15.901	+1.053	–0.456	+0.941	+0.778	+1.721
2451781.52578	13.240	+0.717	–0.217	+0.470	+0.465	+0.937
2451782.54509	13.173	+0.691	–0.230	+0.451	+0.453	+0.905
2451783.61339	13.164	+0.665	–0.218	+0.456	+0.440	+0.892
2451784.59330	13.098	+0.665	–0.241	+0.461	+0.442	+0.900
2451785.57684	13.113	+0.658	–0.232	+0.459	+0.433	+0.892
2451786.61392	13.134	+0.652	–0.244	+0.454	+0.446	+0.900
2451837.53474	13.071	+0.644	–0.225	+0.464	+0.443	+0.912
2451838.54138	13.059	+0.650	–0.253	+0.458	+0.441	+0.894
2452088.79054	17.370	+0.575	–0.696	+0.918	+0.099	+1.019
2452143.56922	17.277	+0.494	–0.827	+1.197	+0.844	+2.042
2452189.53989	14.072	+0.915	+0.026	+0.606	+0.487	+1.089
2452190.55294	13.896	+0.918	–0.050	+0.614	+0.566	+1.182
2452191.57834	13.782	+0.882	+0.007	+0.587	+0.510	+1.105
2452192.57987	13.731	+0.864	–0.010	+0.580	+0.510	+1.085
2452193.57090	13.647	+0.857	–0.011	+0.566	+0.511	+1.093
2452195.57484	13.587	+0.817	–0.128	+0.509	+0.488	+1.017
2452197.57594	13.428	+0.773	–0.125	+0.518	+0.487	+0.989
2452198.57746	13.420	+0.736	–0.140	+0.507	+0.419	+0.928

1.5-m telescope. Also on the night of 2001 July 25, at $03^h\,22^m\,00^s$ *UT*, HJD 2452115.64028, V348 Sgr was barely visible, and too faint to measure. In each of these two instances, the observing log indicates that it was estimated that $V \sim$ 16th magnitude.

Johnson *V* magnitude photoelectric data from the observations reported in this manuscript, Table 3, then were overlayed in Figure 2 onto the AAVSO database observations. Our photoelectric observations are plotted in red. One is reminded that the AAVSO database observations are in Julian Days (JDs), whereas the authors' are in Heliocentric Julian Days (HJDs).

CCD data for V348 Sgr, from Table 4 and plotted with green symbols in Figure 2, were obtained by JLC at the CTIO Yale 1.0-m telescope in the interval 2008 June 29 to 2010 May 13 *UT* (2454646.7 < *UT* < 2455329.7). Figure 2, therefore, is a composite of the *V* data with the AAVSO data shown in black, the photoelectric data in red, and the CCD data in green.

Figure 3 is the result of 33 images taken over an eleven-night run at the LCO Swope 1.0-m telescope in the time interval UT 1992 October 5 through 1992 October 15 (2448901 < HJD < 2448911). V348 Sgr serendipitously was caught brightening some three magnitudes over these eleven nights (Landolt and Uomoto 1992). These data are in Table 4.

Figure 4 finds V348 Sgr more or less constant near 18th *V* magnitude over a six-night interval, UT 1996 August through 1996 August 11 (2450301 < HJD < 2450306), from data also taken at the Swope telescope. The scale of the figure matches that of Figure 3 for ease of comparison. These data are in Table 4.

Figures 5 and 6 illustrate the behavior of the *UBVRI* photoelectric color indices as a function of Heliocentric Julian Day (HJD), using the same HJD scale as in Figure 2. These data are in Table 3. Except for *(R–I)*, each color index exhibits a maximum change of two magnitudes. These differences arise since at maximum brightness, the hot R CrB star dominates, whereas at minimum light, the planetary nebulosity dominates.

Figures 7 (60 images), 8 (81 images), and 9 (141 images) present the *V* filter CCD data obtained by JLC at the CTIO Yale 1.0-m telescope on the successive nights of UT 2007 May 21 and 22. These data are in Table 4. Several of the data points in Figure 7 exhibit larger error bars which resulted from intermittent clouds at that point in the night. The purpose of Figure 9 is to show V348 Sgr's behavior near maximum brightness on successive nights. Although not periodic, real variations through the *V* filter are visible at the three percent level. Percy and Dembski (2018) note that "most or all RCrB

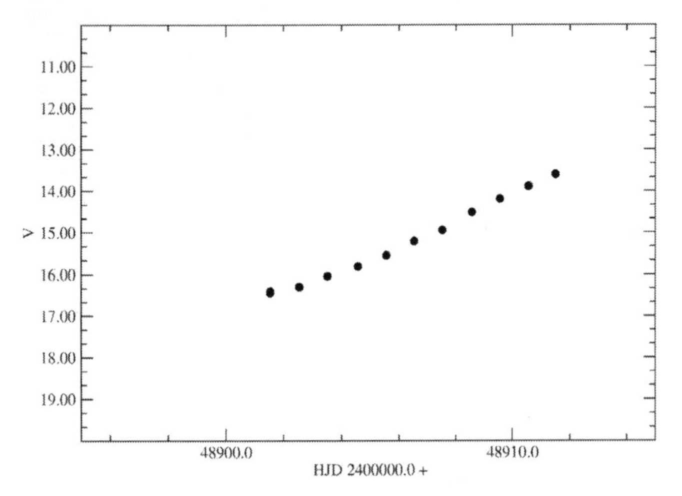

Figure 3. CCD data from the Las Campanas Observatory's Swope 1.0-m telescope for eleven nights in the interval 2448901 ≤ HJD ≤ 2448911.

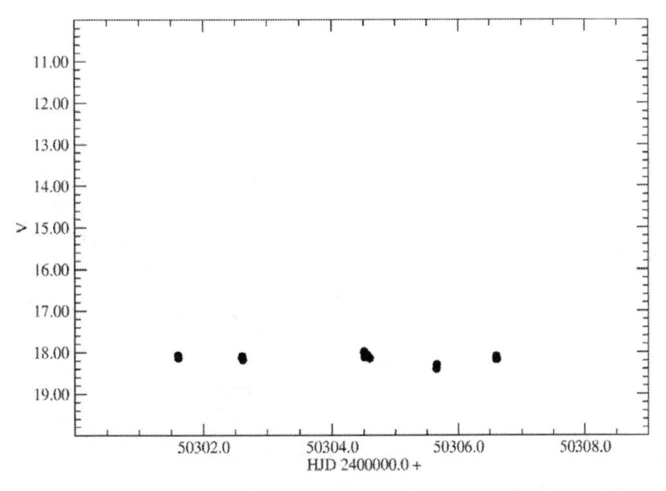

Figure 4. CCD data from the Las Campanas Observatory's Swope 1.0-m telescope for six nights in the interval 2450301 ≤ HJD ≤ 2450306.

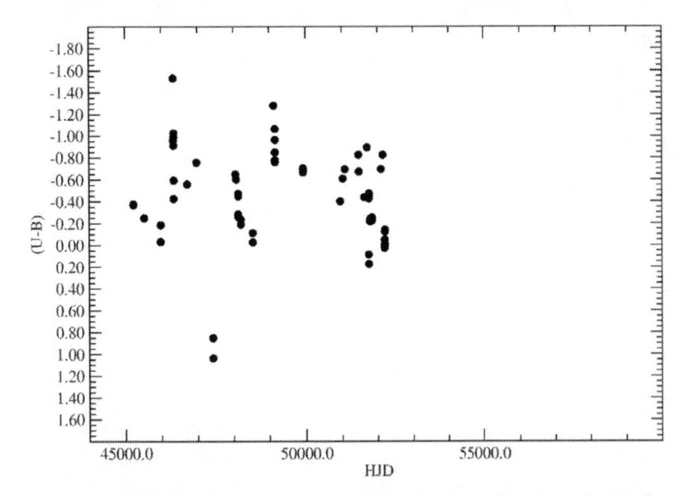

Figure 5. Photoelectric (*U–B*) color index data as a function of HJD for V348 Sgr from this paper.

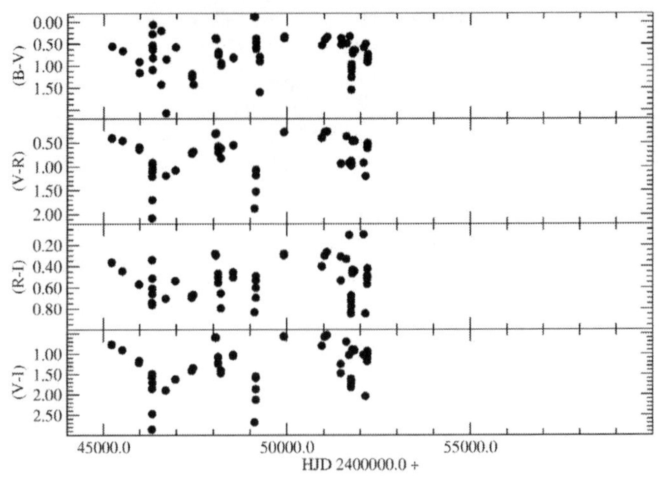

Figure 6. Photoelectric (*B–V*), (*V–R*), (*R–I*), and (*V–I*) color index data as a function of HJD for V348 Sgr from this paper.

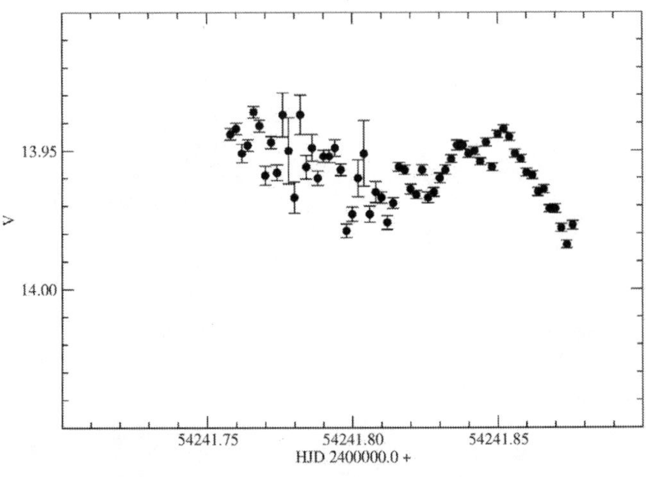

Figure 7. CCD data from the CTIO Yale 1.0-m telescope for UT 2007 May 21 (HJD 2454241).

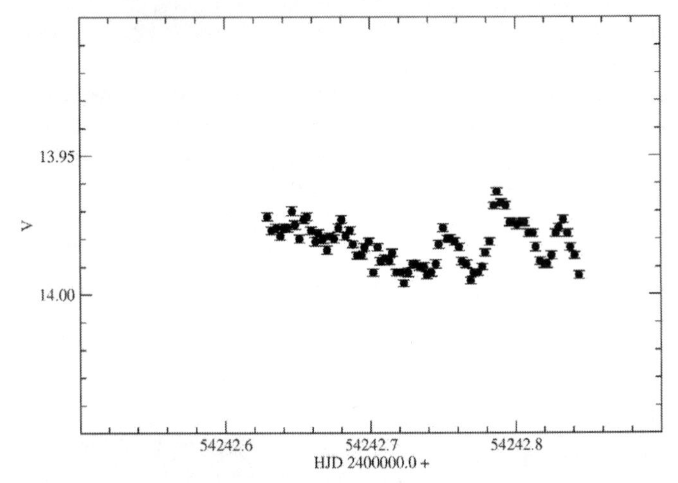

Figure 8. CCD data from the CTIO Yale 1.0-m telescope for UT 2007 May 22 (HJD 2454242).

Table 4. V348 Sgr CCD data.

HJD	*V*	*SDev*
2448901.547356	16.401	0.0138
2448901.548930	16.435	0.0168
2448901.550446	16.445	0.0136
2448902.551609	16.292	0.0088
2448902.553136	16.302	0.0100
2448902.554652	16.301	0.0093
2448903.516791	16.040	0.0104
2448903.518341	16.046	0.0108
2448903.519915	16.036	0.0101
2448904.562916	15.813	0.0094
2448904.564467	15.812	0.0104
2448904.565994	15.810	0.0136
2448905.557783	15.550	0.0070
2448905.559288	15.540	0.0067
2448905.560804	15.557	0.0086
2448906.535650	15.213	0.0072
2448906.537201	15.208	0.0063
2448906.538728	15.200	0.0052
2448907.523631	14.938	0.0047
2448907.525147	14.941	0.0038
2448907.526663	14.947	0.0041
2448908.567627	14.510	0.0071
2448908.569143	14.515	0.0080
2448908.570659	14.512	0.0072
2448909.574543	14.196	0.0084
2448909.576059	14.191	0.0067
2448909.577633	14.193	0.0090
2448910.568671	13.885	0.0021
2448910.570187	13.889	0.0022
2448910.571692	13.898	0.0020
2448911.490260	13.590	0.0019
2448911.491787	13.593	0.0021
2448911.493303	13.603	0.0022
2449506.815125	17.995	0.0155
2449506.816201	18.042	0.0173
2450301.606179	18.067	0.0286
2450301.611734	18.137	0.0328
2450302.602382	18.126	0.0340
2450302.607127	18.091	0.0317
2450302.610171	18.154	0.0314
2450302.613226	18.101	0.0334
2450302.620876	18.182	0.0364
2450304.500927	18.011	0.0238
2450304.506204	17.977	0.0209
2450304.511169	18.119	0.0323
2450304.521065	18.115	0.0310
2450304.525428	18.065	0.0270
2450304.558736	18.064	0.0243
2450304.562497	18.080	0.0244
2450304.591616	18.150	0.0299
2450304.595632	18.145	0.0321
2450305.647652	18.403	0.0704
2450305.650696	18.377	0.0701
2450305.653635	18.299	0.0562
2450305.656991	18.297	0.0621
2450306.604062	18.168	0.0336
2450306.607985	18.090	0.0284
2450306.617510	18.166	0.0346
2451390.685563	18.627	0.0382
2451390.687126	18.619	0.0410
2451391.654418	18.484	0.0328
2451391.693443	18.556	0.0386
2451391.695110	18.550	0.0364
2451392.689621	18.453	0.0382
2451392.691288	18.436	0.0379
2451393.690630	17.952	0.0289
2451393.692123	17.946	0.0315
2451393.780509	17.848	0.0196
2451393.781991	17.802	0.0162
2451394.692395	17.924	0.0241
2452918.485326	14.830	0.0055
2452918.486622	14.835	0.0051
2454241.758161	13.944	0.0021
2454241.760140	13.942	0.0021
2454241.762122	13.951	0.0034
2454241.764096	13.948	0.0021
2454241.766075	13.936	0.0021
2454241.768054	13.941	0.0022
2454241.770034	13.959	0.0034
2454241.772013	13.947	0.0022
2454241.773988	13.958	0.0029
2454241.775970	13.937	0.0079
2454241.777949	13.950	0.0120
2454241.779927	13.967	0.0056
2454241.781907	13.937	0.0071
2454241.783887	13.956	0.0043
2454241.785862	13.949	0.0049
2454241.787841	13.960	0.0026
2454241.789819	13.952	0.0022
2454241.791799	13.952	0.0024
2454241.793773	13.949	0.0030
2454241.795753	13.957	0.0021
2454241.797731	13.979	0.0024
2454241.799711	13.973	0.0026
2454241.801689	13.960	0.0067
2454241.803670	13.951	0.0119
2454241.805650	13.973	0.0029
2454241.807630	13.965	0.0038
2454241.809610	13.967	0.0020
2454241.811591	13.976	0.0025
2454241.813570	13.969	0.0020
2454241.815549	13.956	0.0016
2454241.818235	13.957	0.0017
2454241.820215	13.964	0.0018
2454241.822195	13.966	0.0015
2454241.824173	13.957	0.0018
2454241.826152	13.967	0.0019
2454241.828131	13.965	0.0016
2454241.830110	13.960	0.0018
2454241.832084	13.957	0.0017
2454241.834063	13.953	0.0015
2454241.836039	13.948	0.0018
2454241.838019	13.948	0.0014
2454241.839999	13.951	0.0015
2454241.841977	13.950	0.0014
2454241.843955	13.954	0.0013
2454241.845930	13.947	0.0015
2454241.847910	13.956	0.0014
2454241.849886	13.944	0.0013
2454241.851865	13.942	0.0013
2454241.853844	13.945	0.0014
2454241.855824	13.951	0.0015
2454241.857804	13.953	0.0014
2454241.859784	13.958	0.0014
2454241.861764	13.959	0.0014
2454241.863741	13.965	0.0015
2454241.865715	13.964	0.0015
2454241.867698	13.971	0.0013
2454241.869673	13.971	0.0016
2454241.871654	13.978	0.0015
2454241.873629	13.984	0.0014
2454241.875611	13.977	0.0016
2454242.629041	13.972	0.0015
2454242.632351	13.977	0.0013
2454242.635026	13.976	0.0014
2454242.637699	13.979	0.0014
2454242.640372	13.976	0.0015
2454242.643045	13.976	0.0014
2454242.645717	13.970	0.0016
2454242.648391	13.975	0.0015
2454242.651062	13.980	0.0014
2454242.653735	13.973	0.0014
2454242.656407	13.972	0.0014
2454242.659079	13.977	0.0013
2454242.661752	13.981	0.0015
2454242.664424	13.978	0.0013
2454242.667097	13.980	0.0015
2454242.669770	13.984	0.0015
2454242.672442	13.979	0.0012
2454242.675115	13.980	0.0014
2454242.677788	13.976	0.0014
2454242.680462	13.973	0.0013
2454242.683137	13.979	0.0014
2454242.685810	13.977	0.0013
2454242.688483	13.982	0.0012
2454242.691154	13.986	0.0013
2454242.693826	13.986	0.0013
2454242.696498	13.983	0.0013
2454242.699171	13.981	0.0013
2454242.701844	13.992	0.0013
2454242.704516	13.983	0.0013
2454242.707190	13.988	0.0013
2454242.709863	13.987	0.0012
2454242.712602	13.988	0.0012
2454242.715273	13.985	0.0013
2454242.717948	13.992	0.0012
2454242.720620	13.992	0.0012
2454242.723293	13.996	0.0012
2454242.725964	13.992	0.0015
2454242.728635	13.989	0.0012
2454242.731309	13.989	0.0012
2454242.733983	13.990	0.0012
2454242.736654	13.990	0.0014
2454242.739329	13.993	0.0013
2454242.742000	13.992	0.0014
2454242.744677	13.989	0.0015
2454242.747347	13.982	0.0015
2454242.750024	13.976	0.0014
2454242.752699	13.980	0.0012
2454242.755370	13.980	0.0014
2454242.758043	13.981	0.0013
2454242.760722	13.983	0.0013
2454242.763400	13.988	0.0013
2454242.766083	13.989	0.0013
2454242.768760	13.995	0.0013
2454242.771430	13.992	0.0013
2454242.774107	13.992	0.0011
2454242.776777	13.990	0.0011
2454242.779453	13.985	0.0013
2454242.782125	13.981	0.0013
2454242.784798	13.968	0.0012
2454242.787470	13.963	0.0013
2454242.790142	13.967	0.0013
2454242.792919	13.968	0.0014
2454242.795594	13.974	0.0014
2454242.798267	13.974	0.0013
2454242.800942	13.975	0.0015
2454242.803615	13.974	0.0013
2454242.806288	13.974	0.0014
2454242.808960	13.978	0.0014
2454242.811634	13.978	0.0013
2454242.814307	13.983	0.0016
2454242.816980	13.988	0.0013
2454242.819781	13.989	0.0014
2454242.822440	13.989	0.0014
2454242.825117	13.986	0.0014
2454242.827790	13.978	0.0013
2454242.830463	13.976	0.0014
2454242.833141	13.973	0.0013
2454242.835814	13.978	0.0013
2454242.838489	13.983	0.0013
2454242.841163	13.986	0.0013
2454242.843837	13.993	0.0014
2456774.977437	18.212	0.0262
2456774.983091	18.272	0.0267
2456774.987899	18.353	0.0287
2456775.984039	18.367	0.0287
2456779.979752	18.492	0.0402
2456779.982046	18.460	0.0466
2456780.968839	18.434	0.0462
2456838.835146	16.047	0.0056
2456839.817621	15.787	0.0027

stars undergo small amplitude pulsations with periods of a few weeks." However, Figures 7, 8, and 9 illustrate that light variations also occur on the time scale of tens of minutes, say in the range of 0.01 to 0.1 day. Intensive monitoring should be filter-defined with integration times short enough to resolve short time variations, but long enough to obtain adequate signal to noise. Well calibrated observations through the Johnson *V* filter are recommended, thereby permitting easier comparison to the majority of the photometric data in the literature.

It should be noted that the short timescale variations discovered herein contrast with those of the cooler R CrB stars (Clayton 1996). Many of the cooler R CrB stars have pulsation periods on the order of 40 to 100 days. It could be, of course, that such periods, mostly dependent upon observations in databases such as the AAVSO's, are more the result of the observing technique, a measurement per night over days and weeks. Intensive well-calibrated short timescale observations of the cooler R CrB stars also might be fruitful.

Looking at recent data displayed in the AAVSO database for V348 Sgr, in the time interval 2457100 < HJD < 2458400 (2008 December 8 to 2018 September 25), one notes the simultaneous decline in brightness in the *V* and *I* photometric passbands as V348 Sgr approaches minimum brightness. The decline in the *B* passband is less. The *U* photometric band data in this paper are the only such data known to the authors for V348 Sgr in this time frame.

Figures 10, 11, 12, 13, and 14 illustrate the behavior of the (*U–B*), (*B–V*), (*V–R*), (*R–I*), and (*V–I*) color indices as a function of the *V* magnitude. The ordinate scale is the same for these figures to better illustrate the photometric behavior of V348 Sgr. As V348 Sgr fades, (*U–B*) initially reddens and then, during the final five magnitudes of decline, becomes more blue.

Two points in Figure 10 stand out. The two reddest points are from 1988 September 16 UT. The data points taken at the CTIO 1.5-m on HJD 2447420.52371 and 2447420.52916 are at *V* = 14.837, (*U–B*) = +0.850 and *V* = 14.862, (*U–B*) = +1.036. Those data were taken through a 1.4-mm diaphragm (14 arc seconds), as were the other photoelectric data. The observing log indicated raw data errors of 2% in *V*, 4.8% in (*B–V*), 13% in (*U–B*), 1% in (*V–R*), 2% in (*R–I*), and 0.6% in (*V–I*), as support for the validity of the plotted data points for this night's data. The two measures were taken 7.5 minutes of time apart. The standard star photometry errors were less than one percent for that night, except 3% for (*U–B*). The sky was clear all night, with seeing between 2.5^s and 3^s of arc. Additional precise and accurate data taken when V348 Sgr is faint are needed.

Whereas the (*U–B*) color index data points stand out in Figure 10, for V348 Sgr on UT 1988 September 16, measures in the (*B–V*), (*V–R*), (*R–I*), and (*V–I*) color indices do not in Figures 11, 12, 13, and 14, respectively. The AAVSO database observations of this date do not provide aid in interpretation. However, perusal of recent AAVSO database multicolor data in the time interval 2457100 < HJD < 2458400 show color indices for a magnitude of *V* ~ 14.85 to be similar to those found herein. There are no comparable (*U–B*) data points in the AAVSO database. An interpretation is that the measured (*U–B*) color index on 1988 September 16 results from the planetary nebula which surrounds V348 Sgr, not a satisfactory statement.

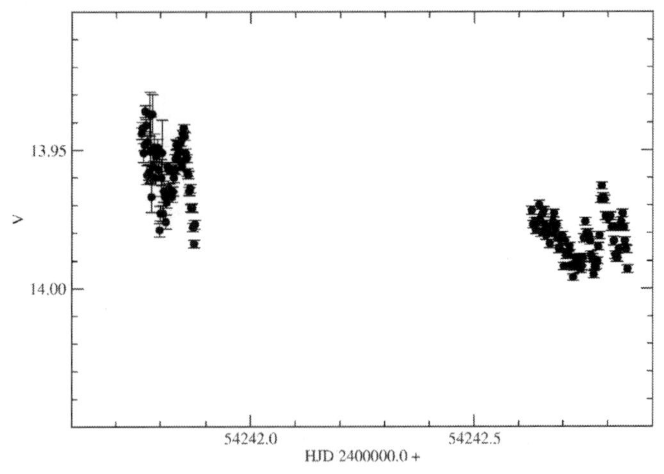

Figure 9. CCD data from the CTIO Yale 1.0-m telescope for UT 2007 May 21 and 22 (HJD 2454241 and 2454242).

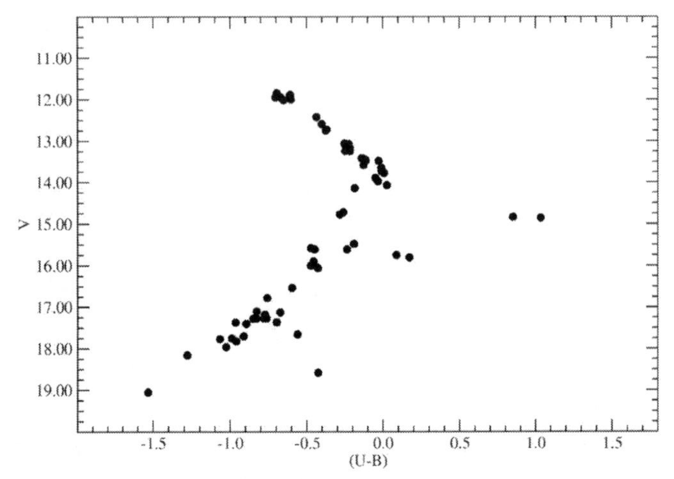

Figure 10. *V* magnitudes vs (*U–B*) color index for the photoelectric data for V348 Sgr in this paper.

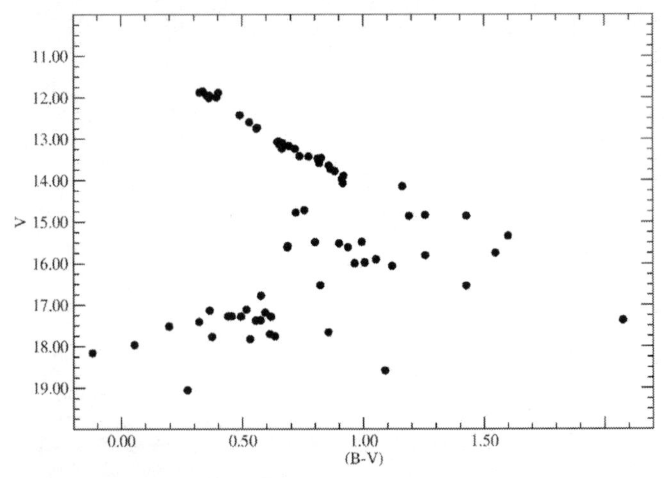

Figure 11. *V* magnitudes vs (*B–V*) color index for the photoelectric data for V348 Sgr in this paper.

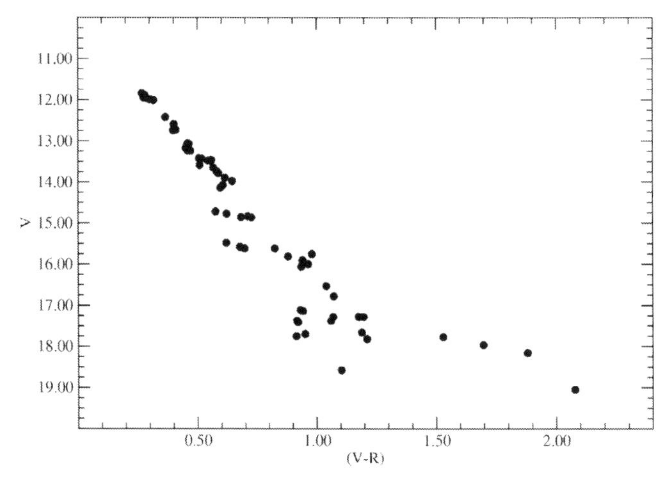

Figure 12. *V* magnitudes vs (*V–R*) color index for the photoelectric data for V348 Sgr in this paper.

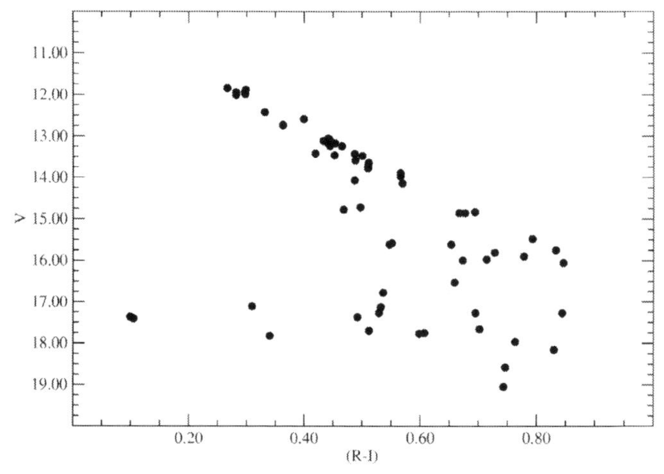

Figure 13. *V* magnitudes vs (*R–I*) color index for the photoelectric data for V348 Sgr in this paper.

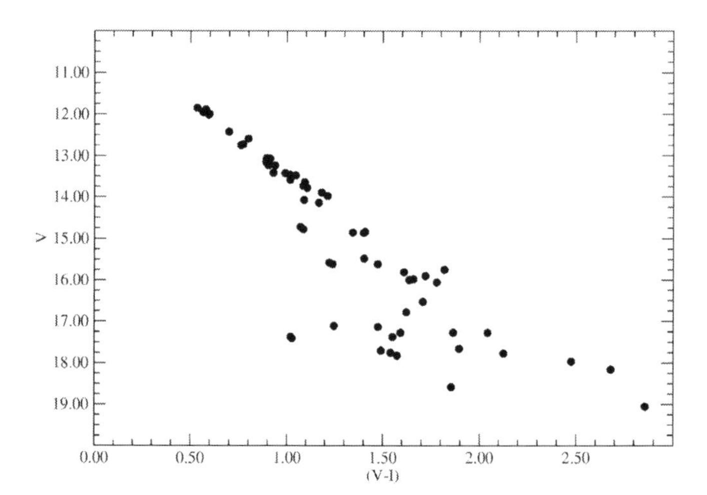

Figure 14. *V* magnitudes vs (*V–I*) color index for the photoelectric data for V348 Sgr in this paper.

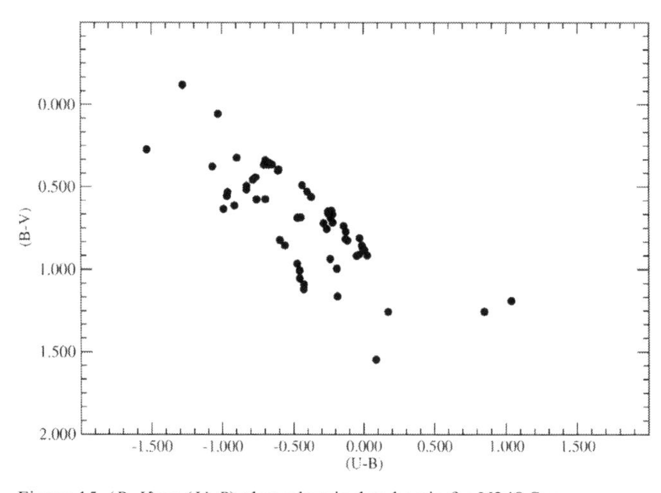

Figure 15. (*B–V*) vs (*U–B*) photoelectric data herein for V348 Sgr.

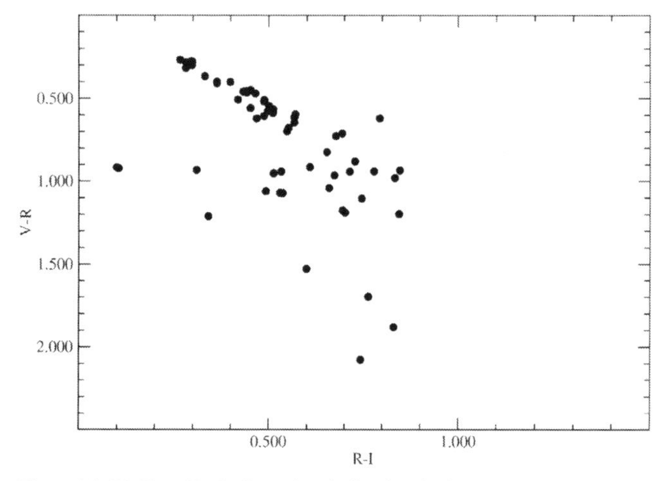

Figure 16. (*V–R*) vs (*R–I*) photoelectric data herein for V348 Sgr.

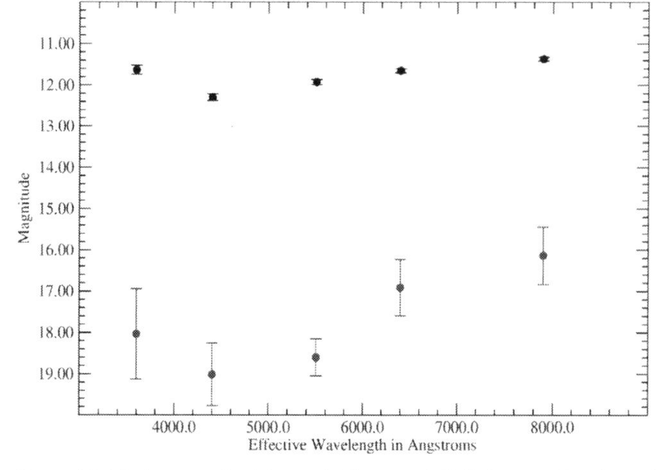

Figure 17. Maximum and minimum brightness of V348 Sgr as a function of effective wavelengths of UBVRI filters.

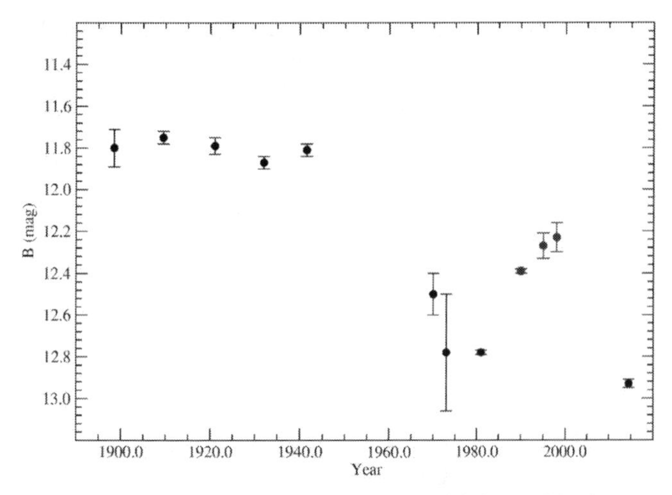

Figure 18. Behavior of B magnitude near maximum light for V348 Sgr between 1896 and 1998; black circles from Schaefer (2016) and blue circles from data herein.

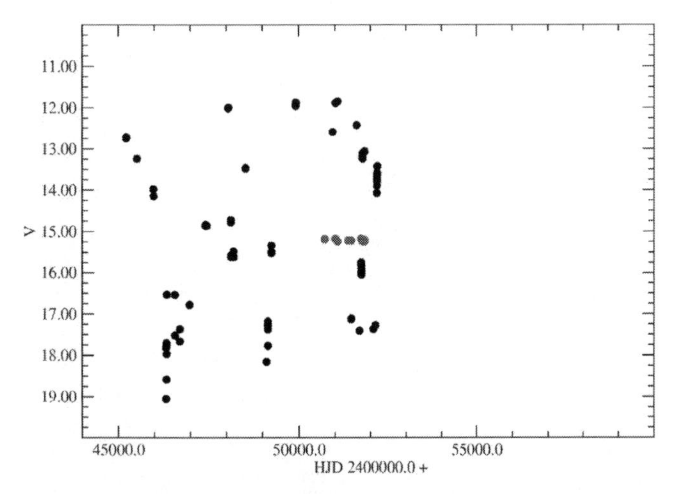

Figure 19. Behavior of V magnitude of star C (red filled circles) on nights when it was observed along with V348 Sgr (black filled circles).

Figure 15 and Figure 16 show that as the shorter wavelength color index becomes more red, so does the longer wavelength color index. Figure 15 presents a definite but broad relationship between $(B–V)$ as a function of $(U–B)$ as both become more red. The relationship is tight in Figure 16 in the color index interval ($+0.3 < (V–R) < +0.7$; $+0.3 < (R–I) < +0.6$), after which scatter increases due to the faintness of the star. As the referee wrote, "The bewildering pattern of data points in Figures 15 and 16 suggests a path for future investigation." What is needed is a series of well calibrated observations, particularly when V348 Sgr is faint.

The CCDphot data from UT 1993 May 11 are included in Table 3 at HJD 2449118.97954. Since these data were obtained at KPNO's 0.9-m telescope at an air mass of 1.8, the magnitude and color indices of a nearby star, C (see description of star C below), were adjusted to match that star's average magnitude and color indices as determined at CTIO, where the star was high in the sky. Such determined differences then were applied to the measured magnitude and color indices of V348 Sgr taken at KPNO, resulting in the values to be found in Table 3. The nearest in time observation, ten days earlier in the AAVSO database, to which this photometry may be compared is an observation by Thomas Cragg where he determined a visual observation of fainter than 15.5 on JD 2449108.2.

Figure 17 is a summary, based on the data herein, of the brightness of V348 Sgr at maximum and minimum brightness as a function of wavelength through the U, B, V , R, and I Johnson Kron filters. The maxima were taken from dates where $V \approx 12$th magnitude or brighter. The minima were taken from dates where $V \approx 18$th magnitude or fainter. The error bars at minima light are larger just because the photometry is less accurate. Nevertheless, V348 Sgr is fainter at the B wavelength at both maxima and minima when compared to the other filters. And V348 Sgr brightens at both maxima and minima as one proceeds from the V to the R to the I filter. This follows from the discussion in Clayton *et al.* (2011).

Schaefer (2016) described a long term decline in the average B magnitude of V348 Sgr. He illustrated this decline with archival data (his Table 2) and displayed in his Figure 2. These same data are presented as black filled circles in Figure 18 herein. Similar data from Table 3 herein are displayed as blue filled circles. The current data confirm the long term, but indicate a less steep decline. Actually the current photoelectric data show the maximum B magnitude to have brightened somewhat. At least a partial explanation lies in the difficulty in identifying a time interval of maximum brightness.

A star, identified in Figure 1 as C, located to the north and east of V348 Sgr, at $\Delta\alpha = +1.075^s$ and $\Delta\delta = +5.53''$, was used as a comparison star. It was intermediate in brightness between the bright and faint limits of the light variations of V348 Sgr. This star appears in the UCAC4 Catalogue as UCAC4 336-170138 (Zacharias *et al.* 2013). Its coordinates from the Gaia proper motion catalogue, VizieR's catalogue I/343/gps1, are R.A. = $18^h\,40^m\,21.02934^s$, Dec. = $-22°\,54'\,24.1221''$, J2000. This same catalogue lists this star's proper motion as $\mu_\alpha = -5.586 \pm 1.322$ and $\mu_\delta = +9.709 \pm 1.532$ mas yr^{-1}. The star labeled C herein is identified as star 12 in Figures 3a and 3b in Heck *et al.* (1985).

Ten photoelectric observations of star C, all taken at the CTIO 1.5-m telescope, over a three-year interval provided an average magnitude and color indices of $V = 14.788 \pm 0.024$, $(B–V) = +1.231 \pm 0.056$, $(U–B) = +0.872 \pm 0.054$, $(V–R) = +0.668 \pm 0.031$, $(R–I) = +0.653 \pm 0.019$, and $(V–I) = +1.321 \pm 0.044$. Figure 19 illustrates the behavior of star C on nights when it was observed along with V348 Sgr itself.

As a byproduct of the CCD observations of V348 Sgr, 219 data points were obtained of star C through a Johnson V filter on 31 nights. The resulting magnitude was $V = 14.791 \pm 0.013$, in good agreement with the photoelectric results. The corresponding color indices from star C's CCD data are $(B–V) = +1.224 \pm 0.002$, $(U–B) = +0.880 \pm 0.006$, $(V–R) = +0.685 \pm 0.002$, $(R–I) = +0.655 \pm 0.0025$, and $(V–I) = +1.340 \pm 0.002$.

This is particularly gratifying since star C is faint for photoelectric measurements at a 1.5-m telescope, especially in as crowded a field as is evidenced in Figure 1. Identifying the same spot for a photoelectrically-based sky background reading consistently night to night over years is tricky. That is why CCDs excel in crowded fields, as if additional evidence is needed.

4. Summary

Calibrated photometric photoelectric, CCDphot, and CCD data of the hot R CrB star V348 Sgr have been obtained by the authors over an interval of 21.6 years. The current data confirm a long term decline in brightness, but with a smaller slope than heretofore determined. These accurate multicolor photometric data aid in the zero point determination of data in databases and in the definition of the long-term photometric behavior of the light and color curves for V348 Sgr. Intensive monitoring is crucial for understanding the apparent short time variations. These data should be calibrated and filter defined.

5. Acknowledgements

It is a pleasure to thank the staffs of CTIO, KPNO, and LCO for their help in making the observing runs a success. The authors note with appreciation G. Clayton's comments, and recognize with gratitude the long term observation efforts of the AAVSO community. The authors thank the referee for helpful comments.

The data reported in this paper came from observing runs supported by AFOSR grant 82-0192, Space Telescope Science Institute grant STScI CW-0004-85, and NSF grants AST 9114457, 9313868, 9528177, 0097895, and 0803158.

References

Clayton, G. C. 1996, *Publ. Astron. Soc. Pacific*, **108**, 225.
Clayton, G. C. 2012, *J. Amer. Assoc. Var. Star Obs.*, **40**, 539.
Clayton, G. C., *et al.* 2011, *Astron. J.*, **142**, 54.
Clem, J. L., and Landolt, A. U. 2013, *Astron. J.*, **146**, 88.
De Marco, O., Clayton, G. C., Herwig, F., Pollacco, D. L., Clark, J. S., and Kilkenny, D. 2002, *Astron. J.*, **123**, 3387.
Gaia Collaboration, *et al.* 2016, *Astron. Astrophys.*, **595A**, 1.
Gaia Collaboration, *et al.* 2018, *Astron. Astrophys.*, **616A**, 1.
Heck, A., Houziaux, L., Manfroid, J., Jones, D. H. P., and Andrews, P. J. 1985, *Astron. Astrophys., Suppl. Ser.*, **61**, 375.
Kafka, S. 2019, variable star observations from the AAVSO International Database (https://www.aavso.org/aavso-international-database).
Landolt, A. U. 1983, *Astron. J.*, **88**, 439.
Landolt, A. U. 1992, *Astron. J.*, **104**, 340.
Landolt, A. U. 2007, in *The Future of Photometric, Spectrophotometric, and Polarimetric Standardization*, ed., C. Sterken, ASP Conf. Ser. 364, Astronomical Society of the Pacific, San Francisco, 27.
Landolt, A. U., and Clem, J. L. 2017, *J. Amer. Assoc. Var. Star Obs.*, **45**, 159.
Landolt, A. U., and Uomoto, A. K. 1992, *IAU Circ.*, No. 5640, 2.
Laycock, S., Tang, S., Grindlay, J., Los, E., Simcoe, R., and Mink, D. 2010, *Astron. J.*, **140**, 1062.
Percy, J. R., and Dembski, K. H. 2018, *J. Amer. Assoc. Var. Star Obs.*, **46**, 127.
Pollacco, D. L., Tadhunter, C. N., and Hill, P. W. 1990 *Mon. Not. Roy. Astron. Soc.*, **245**, 204.
Schaefer, B. E. 2016, *Mon. Not. Roy. Astron. Soc.*, **460**, 1233.
Schajn, P. 1929, *Astron. Nachr.*, **235**, 417.
Tody, D., and Davis, L. E. 1992, in *Astronomical Data Analysis Software and Systems I*, eds. D. M. Worrall, C. Biemesderfer, J. Barnes, ASP Conf. Ser. 25, Astronomical Society of the Pacific, San Francisco, 484.
Woods, I. E. 1926, *Bull. Harvard Coll. Obs.*, No. 838, 11.
Zacharias, N., Finch, C. T., Girard, T. M., Henden, A., Bartlett, J. L., Monet, D. G., and Zacharias, M. I. 2013, *Astron. J.*, **145**, 44.

Sky Brightness at Zenith During the January 2019 Total Lunar Eclipse

Jennifer J. Birriel
J. Kevin Adkins
Department of Mathematics and Physics, Morehead State University, 150 University Boulevard, Morehead, KY 40351; j.birriel@moreheadstate.edu; jkadkins@moreheadstate.edu

Received April 9, 2019; revised April 24, 29, 2019; accepted April 30, 2019

Abstract Lunar eclipses occur during the full moon phase when the moon is obscured by Earth's shadow. During these events, the night sky brightness changes as the full moon rises and then passes first into the penumbral and then the umbral shadow. We acquired sky brightness data at zenith using a Unihedron Sky Quality Meter during the 20–21 January 2019 total lunar eclipse as seen from Morehead, Kentucky. The resulting sky brightness curve shows an obvious signature when the moon enters the umbral (partial) eclipse phases and the total eclipse phase. During the total eclipse phase, the brightness curve is flat and measures 19.1 ± 0.1 mag / arcsec^2. The observed brightness at totality is close to typical new moon in January night at our location, which measures 19.3 ± 0.1 mag / arcsec^2. The partial eclipse phase is symmetric on either side of totality. The penumbral phase is more difficult to identify in the plot, without comparison to a typical full moon night. There is a clear asymmetry in the curve just before and just after the umbral phase. This asymmetry is probably due to changes in terrestrial atmospheric conditions, such as high altitude clouds.

1. Introduction

Photometric studies of sky brightness during solar eclipses are common (e.g. Pramudya and Arkanuddin 2016 and references therein), while those examining the evolution of sky brightness during a lunar eclipse are extremely rare. During the 6 July 1982 lunar eclipse, Morton (1983) monitored changes in the brightness and color of the night sky using the 31-inch reflector at Lowell Observatory in Arizona. He positioned the telescope 20 degrees due north of the moon and tracked at lunar speed. Two decades later, Dvorak (2005) serendipitously obtained eclipse sky brightness data while making CCD observations of the eclipsing binary QQ Cas during the 24–25 October 2004 lunar eclipse. He produced a plot the sky brightness in ADU versus time. We will compare our results to those of Morton's and Dvorak's.

The entire total lunar eclipse of January 2019 was visible across all of North and South America, most of Europe, and western Africa (e.g. https://www.timeanddate.com/eclipse/lunar/2019-january-21) . On the East Coast of the United States, the eclipse began around 9:30 p.m. while on the West Coast the eclipse began around 6:30 p.m. All across the zone of totality, the duration of the eclipse was 5 hours, 11 minutes, and 33 seconds with totality lasting 61 minutes and 58 seconds.

2. Instrumentation and observations

Night sky photometry can be performed quite easily using Unihedron Sky Quality Meters (Unihedron 2019; Cinzano 2005). There are several models of this device but all contain the same photodiode sensor (the TAOS TSL237S) and the same infrared blocking filter (a HOYA CM-500). Each SQM model is designed to measure visible light at zenith. SQMs contain an onboard temperature sensor and provide temperature corrected sky brightness readings in magnitudes per square arcsecond (mpsas). The measurement uncertainty of each device is ±0.1 mpsas.

Our device is a Sky Quality Meter fitted with a lens and enabled with Ethernet connectivity, hereafter, SQM-LE. The lens reduces the field of view of the SQM-LE to a 20-degree cone centered at zenith. Our device is located in weatherproof housing on the roof of a four-story building on the campus of Morehead State University in Morehead, Kentucky. The geographic coordinates of our location are 38° 11' 2.23" N and 83° 25' 57.67" W, 225 meters above sea level. The SQM-LE is controlled by a personal computer with SQM READER PRO software by KnightWare (http://www.knightware.biz/sqm/).

On the night of the eclipse, the temperature started at 15 degrees Fahrenheit and dropped steadily to 7 degrees near the end of the eclipse. In the early part of the night there were some passing clouds and winds averaged 7 miles per hour. Reported visibility was 10 miles the entire night. Astronomical seeing was good to very good and transparency between 4 and 5 (Astronomical League 2019). After midnight, winds dropped to zero and skies remained mainly clear throughout the remainder of the eclipse. Weather conditions on the night of the eclipse were obtained from timeanddate.com and are provided by CustomWeather, Inc. (2019). The SQM-LE took readings every two minutes beginning at sunset and ending at sunrise with all data logged to a text file. The predicted times for eclipse stages and lunar altitude at our location are provided in Table 1.

Table 1. Predicted times for eclipse stage.

Eclipse Stage	*Time (EST)*	*Lunar Altitude (°)*
Moon Enters Penumbra	9:26 p.m.	45.7
Moon Enters Umbra	10:33 p.m.	56.2
Moon Enters Totality	11:41 p.m.	66.8
Middle of Eclipse	12:12 a.m.	70.2
Moon Exits Totality	12:43 a.m.	71.9
Moon Exits Umbra	1:50 a.m.	67.9
Moon Exits Penumbra	2:48 a.m.	59.3

Note: Data from Thorsen (1995–2019), copyright © Time and Date AS 1995–2019. All rights reserved. Used by permission.

3. Results and analysis

The sky brightness during the night of the eclipse (20 January to 21 January 2019) are displayed in Figure 1. The plot displays mpsas versus local (Eastern Standard) time. Recall that magnitude is an inverse scale, with brighter values indicated by smaller numerical values.

How does the night of the eclipse compare to a new moon night or a full moon night? We have historical plots that serve as a good comparison of sky brightness. These data are measurements of night sky brightness at zenith taken with the same SQM-LE at the same location. Note that a clear new moon night, Figure 2, displays a constant night sky brightness after astronomical twilight. On the other hand, a clear full moon night, Figure 3, displays a steady increase in brightness until the moon reaches maximum altitude in the sky and then decreases as the full moon sets.

On the night of 20 January 2019, astronomical twilight began at 6:43 p.m. The full moon rose in the East and the sky began to brighten at zenith as it would on a normal full moon night. In Figure 1, it is difficult to see when the Moon enters the penumbral shadow; we will examine this later. The signature of the umbral phase of the eclipse is clear around 10:33 p.m., as indicated by the steady increase in mpsas, indicating a darkening of the sky. This corresponds to the first partial eclipse phase. Between roughly 11:43 p.m. and 12:43 a.m., the sky brightness reached a constant value of 19.1 ± 0.1 mpsas at totality. From just after 12:43 a.m. until about 1:43 a.m., the sky steadily brightened as the moon entered the second partial phase. There is a flattening in the brightness curve between 1:43 and 2:43 a.m., when the moon was in the penumbral shadow. The jagged feature just after 2:43 a.m. is a passing cloud; clouds have been shown to amplify night sky brightness (Kyba *et al.* 2011). After 2:43 a.m., the moon exited Earth's shadow. The night sky brightness following this decreased as it would on a clear, full moon night.

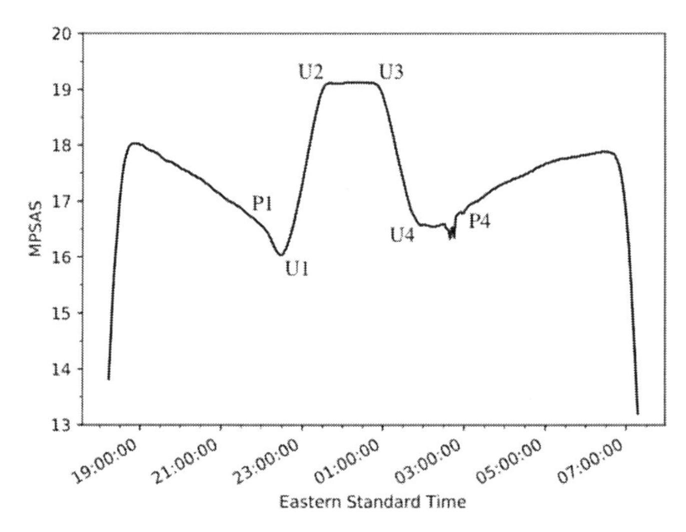

Figure 1. Measured night sky brightness in magnitudes per square arcsecond (mpsas) versus local (Eastern Standard) time during the January 2019 total lunar eclipse as observed from Morehead, Kentucky. This plot was created with the Matplotlib library in PYTHON (Hunter 2007).

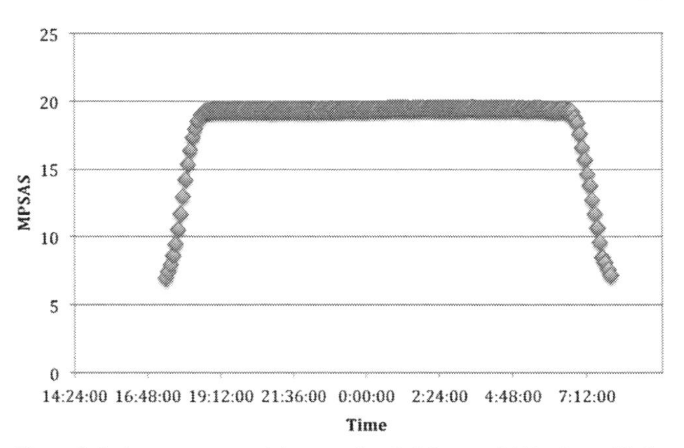

Figure 2. A clear new moon night occurring 7–8 January 2013 taken with the same SQM-LE at the same location.

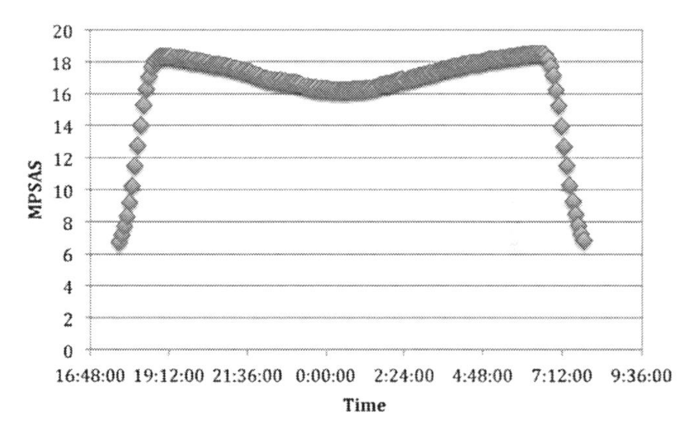

Figure 3. A clear full moon night occurring 26–27 January 2013 taken with the same SQM-LE at the same location.

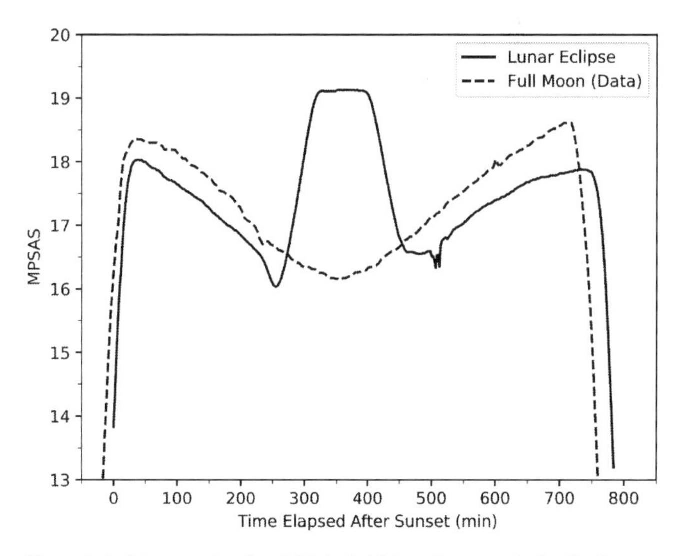

Figure 4. A plot comparing the night sky brightness in mpsas during the January 2019 lunar eclipse to a January 2013 full moon night with no lunar eclipse versus the time elapsed after sunset. The systematic difference of about 0.5 mpsas is attributed to recent construction on campus as noted in the text. This plot was created with the Matplotlib library in PYTHON (Hunter 2007).

To help identify when the Moon enters Earth's penumbral shadow, in Figure 4 we plot the data presented in Figure 3 collected on a full moon night in January 2013, without an eclipse, on top of the lunar eclipse data. Unfortunately, a hard drive failure resulted in the loss of the majority of our historical data, leaving only plots from previous talks and analyses. The full moon data presented in Figure 4 were extracted from Figure 3 with the WebPlotDigitizer (Rohatgi 2019) automatic extraction feature. In Figure 4, it is clear that the full moon night was on average about 0.5 mpsas darker than the lunar eclipse night. Because of the large lapse in time between datasets, we at first attributed this offset to the recent grand opening of a newly constructed student union directly adjacent to the building where our SQM-LE collects data. However, a colleague (Cool 2019) pointed out that the lunar altitude at our location on January 26, 2013 was 4.77 degrees higher than on the night of the eclipse. Further investigation revealed that the moon was 11% closer on the night of the eclipse as compared to January 26, 2013 (timeanddate.com): as a result, the moon was 23% brighter on the night of the lunar eclipse. These two astronomical factors would far outweigh the effects of local construction. In Figure 4, it is easy to see the concave-up shape of the full moon sky brightness curve. The brightness asymmetry before entering umbral phase (towards U1) and the exiting penumbral phase (towards P2) most likely results from changes in atmospheric conditions (Dvorak 2005). Interestingly, just before entering the umbral phase, there is a small increase in brightness, for which we have no explanation.

During a total lunar eclipse the surface of the moon is not completely dark: sunlight is refracted and transmitted by the terrestrial atmosphere onto the lunar surface (Keenan 1929; Danjon 1985). Full disk lunar photometry observations during lunar eclipses are numerous (e.g. Di Giovanni 2018, references therein) and such studies are used to probe the structure of the terrestrial atmosphere. The sky brightness at totality was 19.1 ± 0.1, which is consistent with a typical clear, new moon night in Morehead, Kentucky, which averages 19.3 ± 0.1 mpsas. The small difference is insignificant given that even on a clear night atmospheric extinction varies with season, dust, and air pollution (Krisciunas *et al.* 1987). Our measurements indicate that night sky brightness during totality is similar to that of a new moon night, in agreement with the results of Morton (1983).

Our SQM-LE sky brightness at zenith qualitatively agrees with those of Morton (1983) and Dvorak (2005) despite the use of distinctly different measurement methods. In all three observations, the umbral and totality phases are qualitatively similar in shape. During totality, we observed a nearly constant slope of zero, consistent with the observations of Dvorak. The eclipse data of Morton (1983) exhibit a zero slope during totality with the exception of a small positive "bump" in the second half of totality. This bump might be due to atmospheric conditions or the result of volcanic dust as mentioned by Morton.

The penumbral phases of our data are clearly asymmetric. The 1982 eclipse observed by Morton exhibits no asymmetry in the first and second penumbral phases. However, the October 2004 eclipse reported on by Dvorak exhibits penumbral phase asymmetry. In our data, the first penumbral phase exhibits an increase in brightness just before entering the first umbral phase, in qualitative agreement with the observations of Dvorak. On the other hand, Dvorak recorded an immediate, gradual increase in brightness as the moon exited the umbral phase. During the 1982 eclipse, the moon traversed through the middle of Earth's shadow. The 2019 and 2004 eclipses had similar geometry, with the moon traversing the upper third of the Earth's shadow. However, it is difficult to see how the lunar path would result in the observed asymmetries, especially given how closely the concave upward shape of our eclipse data matches that of a full moon, non-eclipse night. We posit that the asymmetry is due to atmospheric changes. Neither Morton nor Dvorak reported atmospheric conditions or weather. However, full disk photometry of the lunar surface during eclipses indicate that atmospheric conditions such as high altitude clouds, aerosols, and volcanic dust affect lunar brightness during eclipses (e.g. Muñoz and Pallé 2011). It is reasonable to assume that this might also affect night sky brightness. It is of note that Schaude (1999) observed unexplained increases in brightness just before and after the start of penumbral eclipses, just as we observe in our data.

4. Conclusions

The observed times for the phases of the lunar eclipse are consistent with predictions to within ±2 minutes, which is not surprising given the time resolution of our observations. The partial and total phases of the lunar eclipse can be identified by visual inspection. The measured brightness at totality is consistent with that of a new moon night. On the other hand, determination of the penumbral phases is more difficult and requires additional data from a full moon clear night for comparison. We posit that the origin of the brightness asymmetry in the two penumbral phases is due to atmospheric effects.

This study represents a tentative step towards filling a gap in sky brightness photometry during lunar eclipses. It is also the first study of night sky brightness during a lunar eclipse using inexpensive equipment, the meter, housing, and computer costing just under $900 US. This study is complementary to a daytime study recently done during the 2016 total solar eclipse (Pramudya and Arkanuddin 2016) using the same device. We suggest that coordinated observing campaigns of future lunar eclipses using a network of SQM devices might prove useful to further examine the origin of the observed asymmetry in the penumbral phase, particularly if paired with simultaneous observations of full disk lunar eclipse photometry.

5. Acknowledgements

The authors would like to thank Robert Dick for his many useful comments that improved the clarity and quality of this paper. We would also like to thank Duane Skaggs, our former Department chair, who spent the spring of 2018 helping J. Birriel get the SQM-LE back online with a dedicated IP address after our original IP address was de-activated as a result of a campus wide system update.

References

The Astronomical League. 2019, Seeing and Transparency Guide (https://www.astroleague.org/content/seeing-and-transparency-guide).

Cinzano, P. 2005, *ISTIL Internal Report No.9*, **1.4**, 1.

Cool, A. 2019, private communication (20 May).

CustomWeather, Inc. 2019, weather information (https://customweather.com/ (via https://www.timeanddate.com/weather)).

Danjon, A. 1985, *Griffith Obs.*, **49**, 9.

Di Giovanni, G. 2018, *J. Br. Astron. Assoc.*, **128**, 10.

Dvorak, S. 2005, *J. Amer. Assoc. Var. Star Obs.*, **34**, 72.

Hunter, J. D. 2007, *Comput. Sci. Eng.*, **9**, 90.

Keenan, P. C. 1929, *Publ. Astron. Soc. Pacific*, **41**, 297.

Krisciunas, K., *et al.* 1987, *Publ. Astron. Soc. Pacific*, **99**, 887.

Kyba, C. C. M., Ruhtz, T., Fischer, J., and Hölker, F. 2011, *PLoS One*, **6**, e17307 (https://journals.plos.org/plosone/article?id=10.1371/journal.pone.0017307).

Morton, J. C. 1983, *Obs.*, **103**, 24.

Muñoz, G. A., and Pallé, E. 2011, *J. Quant. Spectrosc. Radiat. Transfer*, **112**, 1609.

Pramudya, Y., and Arkanuddin, M. 2016, *J. Phys., Conf. Ser.*, **771**, e012013.

Rohatgi, A. 2019, WebPlotDigitizer (https://apps.automeris.io/wpd).

Schaude, R. W., Jr. 1999, *Int. Amat.-Professional Photoelectric Photom. Commun.*, No. 78, 3.

Thorsen, S. 1995–2019, Time and Date AS (https://www.timeanddate.com).

Unihedron. 2019, Unihedron Sky Quality Meter (http://unihedron.com/projects/sqm-le).

Visual Times of Maxima for Short Period Pulsating Stars V

Gerard Samolyk
P. O. Box 20677, Greenfield, WI 53220; gsamolyk@wi.rr.com

Received January 9, 2019; accepted January 9, 2019

Abstract This compilation contains 503 times of maxima of 8 short period pulsating stars (primarily RR Lyrae type): RR Leo, SS Leo, TV Leo, WW Leo, SZ Lyn, RZ Lyr, AV Peg, RV UMa. These were reduced from a portion of the visual observations made from 1966 to 2014 that are included in the AAVSO International Database.

1. Observations

This is the fifth in a series of papers to publish of times of maxima derived from visual observations reported to the AAVSO International Database as part of the RR Lyr committee legacy program. The goal of this project is to fill some historical gaps in the O–C history for these stars. This list contains times of maxima for RR Lyr stars located in the constellations Leo, Lynx, Lyra, Pegasus, and Ursa Major. This list will be web-archived and made available through the AAVSO ftp site at ftp://ftp.aavso.org/public/datasets/gsamj471vismax5.txt.

These observations were reduced by the writer using the PERANSO program (Vanmunster 2007). The linear elements in the *General Catalogue of Variable Stars* (Kholopov *et al.* 1985) were used to compute the O–C values for all stars.

Figures 1, 2, and 3 are O–C plots for three of the stars included in Table 1. These plots include the visual times of maxima listed in this paper plus more recent times of maxima observed with CCDs. The circled CCD times of maxima on the plots were previously published in *JAAVSO* (Samolyk 2010–2018).

References

Kholopov, P. N., *et al.* 1985, *General Catalogue of Variable Stars*, 4th ed., Moscow.
Samolyk, G. 2010, *J. Amer. Assoc. Var. Star Obs.*, **38**, 12.
Samolyk, G. 2011, *J. Amer. Assoc. Var. Star Obs.*, **39**, 23.
Samolyk, G. 2012, *J. Amer. Assoc. Var. Star Obs.*, **40**, 923.
Samolyk, G. 2013, *J. Amer. Assoc. Var. Star Obs.*, **41**, 85.
Samolyk, G. 2014, *J. Amer. Assoc. Var. Star Obs.*, **42**, 124.
Samolyk, G. 2015, *J. Amer. Assoc. Var. Star Obs.*, **43**, 74.
Samolyk, G. 2016, *J. Amer. Assoc. Var. Star Obs.*, **44**, 66.
Samolyk, G. 2017, *J. Amer. Assoc. Var. Star Obs.*, **45**, 116.
Samolyk, G. 2018, *J. Amer. Assoc. Var. Star Obs.*, **46**, 74.
Vanmunster, T. 2007, PERANSO period analysis software (http://www.peranso.com).

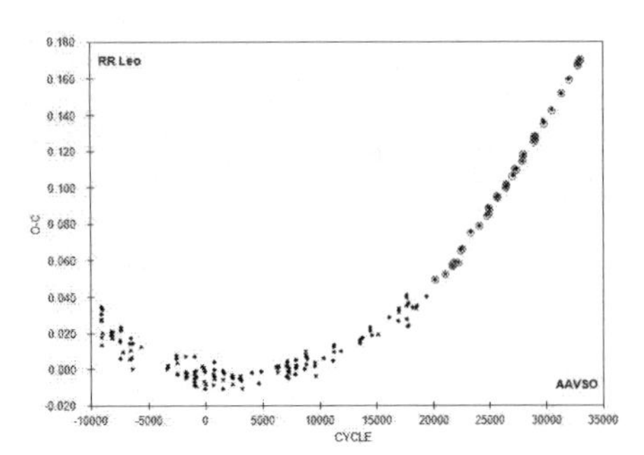

Figure 1. O–C plot for RR Leo. The fundamental period of this star has been increasing since 1966.

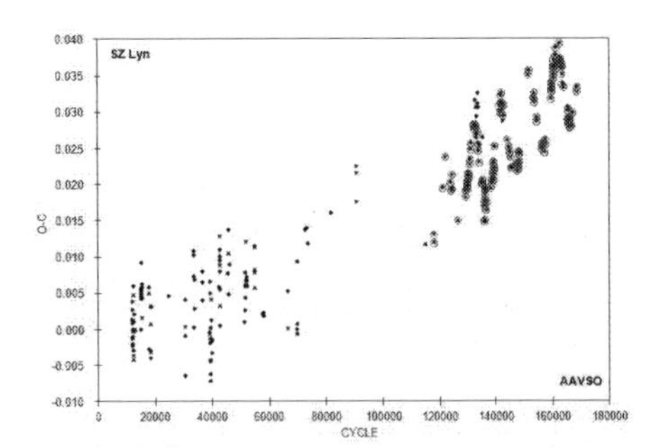

Figure 2. O–C plot for SZ Lyn. The fundamental period of this star increased in 1973. SZ Lyn is part of a binary system with an orbital period of about 3.3 years. The oscillation in O–C is caused by the changing light travel time due to this orbit.

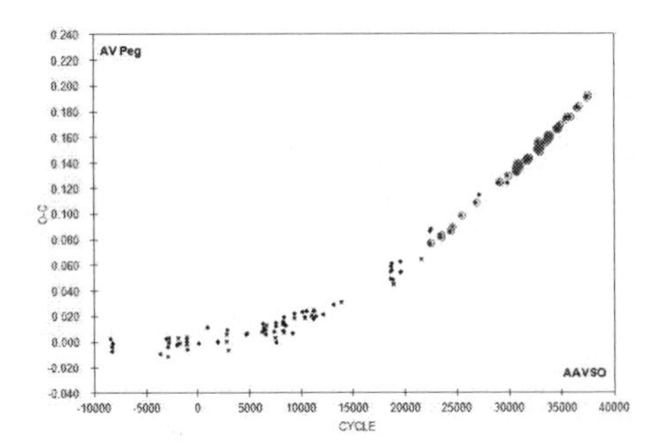

Figure 3. O–C plot for AV Peg. The fundamental period of this star has been increasing since 1969.

Table 1. Recent times of minima of stars in the AAVSO short period pulsator program.

Star	JD (max) Hel. 2400000 +	Cycle	O–C (day)	Observer	Error (day)
RR Leo	39144.721	–9175	0.028	M. Baldwin	0.003
RR Leo	39148.793	–9166	0.028	M. Baldwin	0.006
RR Leo	39168.698	–9122	0.028	M. Baldwin	0.004
RR Leo	39169.606	–9120	0.031	M. Baldwin	0.005
RR Leo	39173.665	–9111	0.018	M. Baldwin	0.004
RR Leo	39174.565	–9109	0.014	M. Baldwin	0.005
RR Leo	39178.658	–9100	0.035	M. Baldwin	0.005
RR Leo	39182.715	–9091	0.020	M. Baldwin	0.005
RR Leo	39197.657	–9058	0.034	M. Baldwin	0.004
RR Leo	39528.797	–8326	0.022	M. Baldwin	0.003
RR Leo	39530.604	–8322	0.019	M. Baldwin	0.008
RR Leo	39534.676	–8313	0.020	M. Baldwin	0.003
RR Leo	39558.653	–8260	0.020	M. Baldwin	0.003
RR Leo	39567.699	–8240	0.018	M. Baldwin	0.002
RR Leo	39595.751	–8178	0.021	M. Baldwin	0.003
RR Leo	39890.706	–7526	0.016	M. Baldwin	0.004
RR Leo	39894.768	–7517	0.006	M. Baldwin	0.004
RR Leo	39895.683	–7515	0.017	M. Baldwin	0.005
RR Leo	39917.856	–7466	0.022	M. Baldwin	0.004
RR Leo	39918.762	–7464	0.024	M. Baldwin	0.004
RR Leo	39975.750	–7338	0.010	M. Baldwin	0.005
RR Leo	40270.715	–6686	0.015	M. Baldwin	0.005
RR Leo	40274.778	–6677	0.006	L. Hazel	0.004
RR Leo	40280.671	–6664	0.018	M. Baldwin	0.008
RR Leo	40294.688	–6633	0.011	M. Baldwin	0.003
RR Leo	40327.709	–6560	0.007	M. Baldwin	0.004
RR Leo	40347.608	–6516	0.001	L. Hazel	0.004
RR Leo	40350.789	–6509	0.015	T. Cragg	0.004
RR Leo	40702.749	–5731	0.013	T. Cragg	0.003
RR Leo	41751.384	–3413	0.000	M. Baldwin	0.004
RR Leo	41765.410	–3382	0.002	M. Baldwin	0.003
RR Leo	41766.315	–3380	0.002	M. Baldwin	0.004
RR Leo	42105.615	–2630	0.007	M. Baldwin	0.003
RR Leo	42124.616	–2588	0.008	M. Baldwin	0.005
RR Leo	42129.591	–2577	0.007	M. Baldwin	0.004
RR Leo	42148.589	–2535	0.004	M. Baldwin	0.004
RR Leo	42157.631	–2515	–0.002	M. Baldwin	0.005
RR Leo	42160.797	–2508	–0.003	G. E. Underhay	0.006
RR Leo	42489.697	–1781	0.007	M. Baldwin	0.003
RR Leo	42504.616	–1748	–0.003	M. Baldwin	0.002
RR Leo	42507.781	–1741	–0.004	M. Baldwin	0.003
RR Leo	42508.689	–1739	–0.001	M. Baldwin	0.004
RR Leo	42509.593	–1737	–0.002	M. Baldwin	0.005
RR Leo	42541.711	–1666	–0.004	T. Cragg	0.005
RR Leo	42832.601	–1023	–0.003	M. Baldwin	0.003
RR Leo	42835.778	–1016	0.008	M. Baldwin	0.008
RR Leo	42836.674	–1014	–0.001	M. Baldwin	0.003
RR Leo	42837.573	–1012	–0.007	M. Baldwin	0.003
RR Leo	42844.810	–996	–0.008	M. Baldwin	0.004
RR Leo	42845.718	–994	–0.005	M. Baldwin	0.006
RR Leo	42874.672	–930	–0.004	M. Baldwin	0.003
RR Leo	42898.644	–877	–0.009	M. Baldwin	0.004
RR Leo	42903.629	–866	0.000	M. Baldwin	0.003
RR Leo	42908.602	–855	–0.004	M. Baldwin	0.004
RR Leo	43144.757	–333	0.002	M. Baldwin	0.004
RR Leo	43211.703	–185	–0.006	M. Baldwin	0.003
RR Leo	43226.639	–152	0.001	M. Baldwin	0.005
RR Leo	43244.733	–112	–0.001	M. Baldwin	0.004
RR Leo	43248.795	–103	–0.010	G. E. Underhay	0.007
RR Leo	43277.751	–39	–0.008	G. E. Underhay	0.005
RR Leo	43578.592	626	–0.008	M. Baldwin	0.004
RR Leo	43606.646	688	–0.003	M. Baldwin	0.004
RR Leo	43610.715	697	–0.005	M. Baldwin	0.003
RR Leo	43625.649	730	0.000	M. Baldwin	0.004
RR Leo	43630.630	741	0.005	M. Baldwin	0.004
RR Leo	43639.670	761	–0.003	M. Baldwin	0.006
RR Leo	43967.653	1486	–0.005	M. Baldwin	0.004
RR Leo	43981.682	1517	–0.001	M. Baldwin	0.005
RR Leo	43984.839	1524	–0.010	G. E. Underhay	0.005
RR Leo	44010.634	1581	–0.002	M. Baldwin	0.004
RR Leo	44314.641	2253	–0.003	M. Baldwin	0.005
RR Leo	44341.780	2313	–0.008	G. E. Underhay	0.005
RR Leo	44342.688	2315	–0.004	M. Baldwin	0.003
RR Leo	44351.737	2335	–0.003	M. Baldwin	0.003
RR Leo	44630.862	2952	–0.005	G. Hanson	0.003
RR Leo	44640.816	2974	–0.004	G. Hanson	0.003
RR Leo	44704.601	3115	–0.006	M. Baldwin	0.004
RR Leo	44717.716	3144	–0.011	G. E. Underhay	0.006
RR Leo	45084.616	3955	–0.002	G. Chaple	0.004
RR Leo	45378.666	4605	–0.007	G. Chaple	0.003
RR Leo	45464.627	4795	–0.001	M. Baldwin	0.003
RR Leo	46057.717	6106	0.002	M. Baldwin	0.007
RR Leo	46114.719	6232	0.002	M. Baldwin	0.004
RR Leo	46142.764	6294	–0.001	M. Baldwin	0.005
RR Leo	46143.671	6296	0.001	M. Baldwin	0.005
RR Leo	46181.673	6380	0.002	M. Baldwin	0.004
RR Leo	46490.652	7063	–0.004	M. Baldwin	0.004
RR Leo	46490.658	7063	0.002	M. Heifner	0.005
RR Leo	46495.629	7074	–0.003	M. Heifner	0.003
RR Leo	46514.636	7116	0.003	M. Baldwin	0.004
RR Leo	46518.709	7125	0.005	M. Baldwin	0.002
RR Leo	46523.685	7136	0.004	M. Baldwin	0.006
RR Leo	46527.751	7145	–0.001	M. Baldwin	0.005
RR Leo	46531.819	7154	–0.005	M. Baldwin	0.006
RR Leo	46532.730	7156	0.002	M. Baldwin	0.002
RR Leo	46561.680	7220	–0.002	R. Hill	0.004
RR Leo	46570.730	7240	0.001	R. Hill	0.004
RR Leo	46831.758	7817	–0.002	M. Baldwin	0.002
RR Leo	46850.762	7859	0.001	M. Baldwin	0.003
RR Leo	46852.576	7863	0.005	M. Baldwin	0.005
RR Leo	46860.716	7881	0.002	M. Baldwin	0.005
RR Leo	47198.652	8628	0.001	M. Baldwin	0.004
RR Leo	47227.615	8692	0.010	M. Baldwin	0.005
RR Leo	47241.637	8723	0.008	M. Baldwin	0.008
RR Leo	47255.660	8754	0.007	M. Baldwin	0.007
RR Leo	47264.708	8774	0.007	M. Baldwin	0.003
RR Leo	47278.728	8805	0.003	R. Hill	0.005
RR Leo	47293.660	8838	0.006	M. Baldwin	0.004
RR Leo	47597.666	9510	0.004	M. Baldwin	0.003
RR Leo	47621.641	9563	0.002	M. Baldwin	0.004
RR Leo	47640.636	9605	–0.004	G. Samolyk	0.004
RR Leo	47948.726	10286	0.007	M. Baldwin	0.004
RR Leo	48318.782	11104	0.005	R. Hill	0.006
RR Leo	48348.648	11170	0.013	G. Samolyk	0.004
RR Leo	48357.693	11190	0.010	M. Baldwin	0.004
RR Leo	48362.673	11201	0.014	M. Baldwin	0.005
RR Leo	48636.820	11807	0.010	R. Hill	0.005
RR Leo	49389.609	13471	0.017	M. Baldwin	0.008
RR Leo	49398.655	13491	0.015	M. Baldwin	0.003
RR Leo	49488.684	13690	0.018	M. Baldwin	0.004
RR Leo	49801.746	14382	0.024	M. Baldwin	0.004
RR Leo	49802.649	14384	0.022	M. Baldwin	0.006
RR Leo	49866.434	14525	0.019	R. Papini	0.003
RR Leo	50123.846	15094	0.020	R. Hill	0.004
RR Leo	50548.653	16033	0.029	M. Baldwin	0.003
RR Leo	50914.637	16842	0.027	G. Chaple	0.006
RR Leo	50923.691	16862	0.033	M. Baldwin	0.004
RR Leo	50928.666	16873	0.032	M. Baldwin	0.004
RR Leo	50952.645	16926	0.034	M. Baldwin	0.003
RR Leo	51256.655	17598	0.036	M. Baldwin	0.004
RR Leo	51256.659	17598	0.040	R. Berg	0.005
RR Leo	51259.814	17605	0.028	M. Baldwin	0.003

Table continued on following pages

Table 1. Recent times of minima of stars in the AAVSO short period pulsator program, cont.

Star	JD (max) Hel. 2400000+	Cycle	O–C (day)	Observer	Error (day)	Star	JD (max) Hel. 2400000+	Cycle	O–C (day)	Observer	Error (day)
RR Leo	51261.637	17609	0.041	R. Berg	0.006	WW Leo	43245.675	13873	0.009	M. Baldwin	0.004
RR Leo	51275.644	17640	0.024	M. Baldwin	0.003	WW Leo	43626.640	14505	–0.025	M. Baldwin	0.003
RR Leo	51298.729	17691	0.037	R. Berg	0.004	WW Leo	44696.703	16280	–0.012	M. Baldwin	0.005
RR Leo	51308.669	17713	0.024	M. Baldwin	0.003	WW Leo	46114.611	18632	0.004	M. Baldwin	0.004
RR Leo	51501.851	18140	0.035	M. Baldwin	0.003	WW Leo	46143.561	18680	0.017	M. Baldwin	0.008
RR Leo	51611.782	18383	0.034	R. Berg	0.007	WW Leo	46495.629	19264	0.024	M. Heifner	0.002
RR Leo	51669.690	18511	0.036	R. Berg	0.004	WW Leo	47231.680	20485	0.001	M. Baldwin	0.004
RR Leo	52049.705	19351	0.040	R. Berg	0.005	WW Leo	47260.622	20533	0.006	M. Baldwin	0.003
RR Leo	54585.417	24956	0.088	S. Swierczynski	0.006	WW Leo	47266.645	20543	0.001	M. Baldwin	0.005
RR Leo	56748.359	29737	0.137	M. Rzepka	0.006	WW Leo	47557.823	21026	0.004	M. Baldwin	0.004
SS Leo	46142.650	6963	0.007	M. Baldwin	0.004	WW Leo	47586.740	21074	–0.015	M. Baldwin	0.005
SS Leo	46888.623	8154	0.004	M. Baldwin	0.008	WW Leo	47594.610	21087	0.018	M. Baldwin	0.005
SS Leo	47260.683	8748	0.016	M. Baldwin	0.008	WW Leo	47597.611	21092	0.005	M. Baldwin	0.004
SS Leo	47270.692	8764	0.003	M. Baldwin	0.006	WW Leo	47644.631	21170	0.003	M. Baldwin	0.008
SS Leo	47615.806	9315	0.002	R. Hill	0.005	WW Leo	47650.652	21180	–0.005	M. Baldwin	0.005
SS Leo	47674.683	9409	0.002	M. Baldwin	0.005	WW Leo	47914.720	21618	0.017	M. Baldwin	0.008
SS Leo	47999.753	9928	0.000	M. Baldwin	0.005	WW Leo	47940.620	21661	–0.005	M. Baldwin	0.003
SS Leo	48004.761	9936	–0.003	M. Baldwin	0.004	WW Leo	47943.637	21666	–0.003	M. Baldwin	0.004
SS Leo	48683.724	11020	0.003	M. Baldwin	0.005	WW Leo	47955.721	21686	0.024	M. Baldwin	0.006
SS Leo	48718.789	11076	–0.007	M. Baldwin	0.004	WW Leo	47976.805	21721	0.009	M. Baldwin	0.008
SS Leo	48745.729	11119	0.000	M. Baldwin	0.004	WW Leo	47999.703	21759	–0.001	M. Baldwin	0.006
SS Leo	49134.698	11740	0.009	M. Baldwin	0.008	WW Leo	48673.683	22877	–0.002	M. Baldwin	0.005
SS Leo	49801.736	12805	–0.009	M. Baldwin	0.005	WW Leo	48682.722	22892	–0.006	M. Baldwin	0.004
SS Leo	49843.713	12872	0.003	M. Baldwin	0.005	WW Leo	49095.682	23577	0.005	M. Baldwin	0.004
SS Leo	50190.689	13426	–0.016	M. Baldwin	0.004	WW Leo	49423.629	24121	0.004	M. Baldwin	0.003
SS Leo	50222.646	13477	–0.002	M. Baldwin	0.003	WW Leo	49450.773	24166	0.020	M. Baldwin	0.008
SS Leo	50545.824	13993	–0.018	M. Baldwin	0.003	WW Leo	49743.741	24652	0.005	M. Baldwin	0.007
SS Leo	50564.633	14023	0.001	M. Baldwin	0.004	WW Leo	49778.694	24710	–0.007	M. Baldwin	0.004
SS Leo	50579.655	14047	–0.010	M. Baldwin	0.004	WW Leo	49787.741	24725	–0.002	M. Baldwin	0.004
SS Leo	51259.857	15133	–0.017	M. Baldwin	0.004	WW Leo	49810.650	24763	–0.001	M. Baldwin	0.006
TV Leo	39173.678	3169	0.018	M. Baldwin	0.007	WW Leo	49813.675	24768	0.009	M. Baldwin	0.004
TV Leo	39917.838	4275	0.003	M. Baldwin	0.007	WW Leo	50138.616	25307	0.017	M. Baldwin	0.003
TV Leo	41801.805	7075	–0.017	T. Cragg	0.004	WW Leo	50153.671	25332	0.001	M. Baldwin	0.004
TV Leo	46532.681	14106	0.032	M. Baldwin	0.006	WW Leo	50165.733	25352	0.006	M. Baldwin	0.002
TV Leo	46534.678	14109	0.011	M. Baldwin	0.005	WW Leo	50191.662	25395	0.012	M. Baldwin	0.005
TV Leo	47231.765	15145	0.022	M. Baldwin	0.006	WW Leo	50514.773	25931	–0.002	M. Baldwin	0.005
TV Leo	47976.633	16252	0.043	M. Baldwin	0.006	WW Leo	50540.697	25974	0.000	M. Baldwin	0.004
TV Leo	47978.640	16255	0.031	M. Baldwin	0.005	WW Leo	50546.723	25984	–0.002	M. Baldwin	0.006
TV Leo	48379.665	16851	0.036	M. Baldwin	0.004	WW Leo	52310.672	28910	0.021	M. Baldwin	0.003
TV Leo	48741.669	17389	0.045	M. Baldwin	0.005	WW Leo	52319.724	28925	0.031	M. Baldwin	0.003
TV Leo	48743.693	17392	0.051	M. Baldwin	0.008	WW Leo	52380.595	29026	0.014	M. Baldwin	0.004
TV Leo	48745.712	17395	0.051	M. Baldwin	0.006	SZ Lyn	39495.605	11376	0.002	M. Baldwin	0.003
TV Leo	49018.891	17801	0.052	M. Baldwin	0.005	SZ Lyn	39495.725	11377	0.001	M. Baldwin	0.002
TV Leo	49835.740	19015	0.058	M. Baldwin	0.004	SZ Lyn	39499.579	11409	–0.002	M. Baldwin	0.001
TV Leo	50158.723	19495	0.072	M. Baldwin	0.005	SZ Lyn	39500.665	11418	–0.001	M. Baldwin	0.002
TV Leo	50191.682	19544	0.061	M. Baldwin	0.004	SZ Lyn	39505.606	11459	–0.002	M. Baldwin	0.001
TV Leo	50553.687	20082	0.071	M. Baldwin	0.003	SZ Lyn	39506.571	11467	–0.001	M. Baldwin	0.001
TV Leo	50896.824	20592	0.053	R. Hill	0.006	SZ Lyn	39526.581	11633	0.000	M. Baldwin	0.002
WW Leo	39168.617	7110	–0.007	M. Baldwin	0.004	SZ Lyn	39528.634	11650	0.004	M. Baldwin	0.004
WW Leo	39171.630	7115	–0.008	M. Baldwin	0.006	SZ Lyn	39530.682	11667	0.003	M. Baldwin	0.003
WW Leo	39174.652	7120	0.000	M. Baldwin	0.006	SZ Lyn	39533.578	11691	0.006	M. Baldwin	0.003
WW Leo	39180.673	7130	–0.008	M. Baldwin	0.007	SZ Lyn	39533.689	11692	–0.004	M. Baldwin	0.001
WW Leo	39532.715	7714	–0.027	M. Baldwin	0.008	SZ Lyn	39534.654	11700	–0.003	M. Baldwin	0.002
WW Leo	39558.666	7757	0.001	M. Baldwin	0.005	SZ Lyn	39537.670	11725	0.000	M. Baldwin	0.001
WW Leo	39567.683	7772	–0.024	M. Baldwin	0.004	SZ Lyn	39556.590	11882	–0.004	M. Baldwin	0.001
WW Leo	39916.745	8351	–0.010	M. Baldwin	0.003	SZ Lyn	39556.716	11883	0.001	M. Baldwin	0.001
WW Leo	40294.729	8978	–0.010	M. Baldwin	0.004	SZ Lyn	39558.648	11899	0.005	M. Baldwin	0.001
WW Leo	42477.630	12599	–0.012	M. Baldwin	0.004	SZ Lyn	39558.765	11900	0.001	M. Baldwin	0.004
WW Leo	42509.588	12652	–0.004	M. Baldwin	0.004	SZ Lyn	39598.663	12231	0.002	M. Baldwin	0.002
WW Leo	42832.716	13188	–0.001	M. Baldwin	0.005	SZ Lyn	39612.643	12347	0.000	M. Baldwin	0.002
WW Leo	42835.735	13193	0.003	M. Baldwin	0.005	SZ Lyn	39884.695	14604	0.005	M. Baldwin	0.002
WW Leo	42844.779	13208	0.005	M. Baldwin	0.004	SZ Lyn	39886.628	14620	0.009	M. Baldwin	0.003
WW Leo	42861.651	13236	–0.003	M. Baldwin	0.004	SZ Lyn	39886.745	14621	0.006	M. Baldwin	0.003
WW Leo	42873.707	13256	–0.004	M. Baldwin	0.005	SZ Lyn	39890.717	14654	0.000	M. Baldwin	0.004
WW Leo	43219.739	13830	–0.005	M. Baldwin	0.004	SZ Lyn	39892.650	14670	0.004	M. Baldwin	0.003
WW Leo	43242.636	13868	–0.016	M. Baldwin	0.004	SZ Lyn	39892.771	14671	0.005	M. Baldwin	0.005

Table continued on following pages

Table 1. Recent times of minima of stars in the AAVSO short period pulsator program, cont.

Star	JD (max) Hel. 2400000+	Cycle	O–C (day)	Observer	Error (day)
SZ Lyn	39893.736	14679	0.006	M. Baldwin	0.003
SZ Lyn	39894.700	14687	0.005	M. Baldwin	0.001
SZ Lyn	39895.664	14695	0.005	M. Baldwin	0.003
SZ Lyn	39896.629	14703	0.006	M. Baldwin	0.001
SZ Lyn	39896.748	14704	0.004	M. Baldwin	0.002
SZ Lyn	39907.717	14795	0.005	M. Baldwin	0.004
SZ Lyn	39912.656	14836	0.002	M. Baldwin	0.004
SZ Lyn	39915.674	14861	0.006	M. Baldwin	0.005
SZ Lyn	39918.687	14886	0.006	M. Baldwin	0.002
SZ Lyn	40208.694	17292	0.006	M. Baldwin	0.001
SZ Lyn	40208.806	17293	–0.003	M. Baldwin	0.003
SZ Lyn	40211.827	17318	0.005	M. Baldwin	0.002
SZ Lyn	40278.716	17873	–0.003	M. Baldwin	0.003
SZ Lyn	40292.697	17989	–0.004	M. Baldwin	0.002
SZ Lyn	40293.666	17997	0.001	M. Baldwin	0.003
SZ Lyn	40293.789	17998	0.003	M. Baldwin	0.002
SZ Lyn	40294.747	18006	–0.003	M. Baldwin	0.003
SZ Lyn	40321.753	18230	0.003	M. Baldwin	0.001
SZ Lyn	41065.696	24402	0.005	P. Atwood	0.002
SZ Lyn	41765.397	30207	0.000	M. Baldwin	0.001
SZ Lyn	41766.360	30215	–0.001	M. Baldwin	0.002
SZ Lyn	41766.475	30216	–0.006	M. Baldwin	0.001
SZ Lyn	41773.356	30273	0.004	M. Baldwin	0.005
SZ Lyn	42124.591	33187	0.000	M. Baldwin	0.003
SZ Lyn	42129.540	33228	0.007	M. Baldwin	0.002
SZ Lyn	42129.664	33229	0.011	M. Baldwin	0.002
SZ Lyn	42133.641	33262	0.010	M. Baldwin	0.003
SZ Lyn	42148.580	33386	0.003	M. Baldwin	0.002
SZ Lyn	42155.575	33444	0.007	M. Baldwin	0.003
SZ Lyn	42491.626	36232	0.007	M. Baldwin	0.001
SZ Lyn	42491.748	36233	0.008	M. Baldwin	0.003
SZ Lyn	42508.619	36373	0.004	M. Baldwin	0.003
SZ Lyn	42782.711	38647	0.000	M. Baldwin	0.003
SZ Lyn	42802.599	38812	–0.001	M. Baldwin	0.002
SZ Lyn	42828.631	39028	–0.004	M. Baldwin	0.002
SZ Lyn	42832.611	39061	–0.002	M. Baldwin	0.003
SZ Lyn	42832.740	39062	0.007	M. Baldwin	0.003
SZ Lyn	42835.619	39086	–0.007	M. Baldwin	0.002
SZ Lyn	42835.747	39087	0.000	M. Baldwin	0.002
SZ Lyn	42836.586	39094	–0.004	M. Baldwin	0.003
SZ Lyn	42837.674	39103	–0.001	M. Baldwin	0.002
SZ Lyn	42842.611	39144	–0.006	M. Baldwin	0.002
SZ Lyn	42843.707	39153	0.005	M. Baldwin	0.001
SZ Lyn	42861.663	39302	0.001	M. Baldwin	0.005
SZ Lyn	42863.715	39319	0.004	M. Baldwin	0.003
SZ Lyn	42886.611	39509	–0.001	M. Baldwin	0.004
SZ Lyn	42887.694	39518	–0.003	M. Baldwin	0.003
SZ Lyn	43211.701	42206	0.006	M. Baldwin	0.002
SZ Lyn	43219.660	42272	0.010	M. Baldwin	0.003
SZ Lyn	43223.637	42305	0.009	M. Baldwin	0.002
SZ Lyn	43226.642	42330	0.001	M. Baldwin	0.002
SZ Lyn	43228.578	42346	0.008	M. Baldwin	0.001
SZ Lyn	43228.696	42347	0.006	M. Baldwin	0.001
SZ Lyn	43241.591	42454	0.003	M. Baldwin	0.003
SZ Lyn	43243.647	42471	0.010	M. Baldwin	0.006
SZ Lyn	43244.614	42479	0.013	M. Baldwin	0.004
SZ Lyn	43245.697	42488	0.011	M. Baldwin	0.002
SZ Lyn	43247.624	42504	0.010	M. Baldwin	0.003
SZ Lyn	43548.598	45001	0.008	M. Baldwin	0.001
SZ Lyn	43600.675	45433	0.014	M. Baldwin	0.005
SZ Lyn	43606.578	45482	0.011	M. Baldwin	0.002
SZ Lyn	43606.693	45483	0.005	M. Baldwin	0.005
SZ Lyn	43625.621	45640	0.009	M. Baldwin	0.002
SZ Lyn	44253.600	50850	0.001	M. Baldwin	0.004
SZ Lyn	44313.628	51348	0.003	M. Baldwin	0.004
SZ Lyn	44314.594	51356	0.004	M. Baldwin	0.003
SZ Lyn	44314.718	51357	0.008	M. Baldwin	0.005
SZ Lyn	44317.609	51381	0.006	M. Baldwin	0.003
SZ Lyn	44340.632	51572	0.007	M. Baldwin	0.007
SZ Lyn	44351.600	51663	0.006	M. Baldwin	0.003
SZ Lyn	44351.721	51664	0.007	M. Baldwin	0.004
SZ Lyn	44353.655	51680	0.012	M. Baldwin	0.002
SZ Lyn	44367.631	51796	0.006	M. Baldwin	0.003
SZ Lyn	44368.596	51804	0.007	M. Baldwin	0.005
SZ Lyn	44373.659	51846	0.007	M. Baldwin	0.004
SZ Lyn	44374.622	51854	0.006	M. Baldwin	0.002
SZ Lyn	44701.633	54567	0.006	M. Baldwin	0.004
SZ Lyn	44701.756	54568	0.008	M. Baldwin	0.002
SZ Lyn	44702.603	54575	0.012	M. Baldwin	0.001
SZ Lyn	44702.720	54576	0.008	M. Baldwin	0.001
SZ Lyn	44704.652	54592	0.011	M. Baldwin	0.003
SZ Lyn	45029.605	57288	0.002	G. Chaple	0.002
SZ Lyn	45058.654	57529	0.002	G. Chaple	0.003
SZ Lyn	45082.640	57728	0.002	G. Chaple	0.004
SZ Lyn	46123.583	66364	0.005	M. Baldwin	0.002
SZ Lyn	46125.627	66381	0.000	M. Baldwin	0.004
SZ Lyn	46501.576	69500	0.001	M. Baldwin	0.004
SZ Lyn	46511.579	69583	–0.001	M. Baldwin	0.003
SZ Lyn	46514.593	69608	0.000	M. Baldwin	0.001
SZ Lyn	46518.580	69641	0.009	M. Baldwin	0.001
SZ Lyn	46862.591	72495	0.014	M. Baldwin	0.002
SZ Lyn	46911.649	72902	0.014	M. Baldwin	0.002
SZ Lyn	46939.611	73134	0.012	M. Baldwin	0.001
SZ Lyn	47948.613	81505	0.016	M. Baldwin	0.002
SZ Lyn	49013.420	90339	0.018	M. Martignoni	0.003
SZ Lyn	49020.415	90397	0.022	M. Martignoni	0.004
SZ Lyn	49032.590	90498	0.023	M. Martignoni	0.005
SZ Lyn	51943.377	114647	0.012	S. Foglia	0.005
SZ Lyn	54064.691	132246	0.032	R. Harvan	0.002
SZ Lyn	54111.578	132635	0.031	R. Harvan	0.002
SZ Lyn	54118.565	132693	0.027	R. Harvan	0.003
SZ Lyn	54148.581	132942	0.029	R. Harvan	0.001
SZ Lyn	54168.593	133108	0.033	R. Harvan	0.001
SZ Lyn	54178.596	133191	0.031	R. Harvan	0.001
SZ Lyn	54211.622	133465	0.031	R. Harvan	0.001
SZ Lyn	54234.639	133656	0.025	R. Harvan	0.002
SZ Lyn	54394.831	134985	0.027	R. Harvan	0.001
SZ Lyn	55286.430	142382	0.029	J. Starzomski	0.006
RZ Lyr	45645.566	8728	0.017	M. Heifner	0.004
RZ Lyr	47677.756	12703	0.019	R. Hill	0.006
RZ Lyr	47678.774	12705	0.015	R. Hill	0.005
RZ Lyr	49991.626	17229	0.006	R. Hill	0.006
RZ Lyr	49993.659	17233	–0.006	R. Hill	0.005
RZ Lyr	49996.736	17239	0.004	R. Hill	0.007
RZ Lyr	52539.642	22213	–0.009	R. Berg	0.004
RZ Lyr	52541.684	22217	–0.012	R. Hill	0.006
RZ Lyr	52543.728	22221	–0.013	R. Hill	0.006
RZ Lyr	52543.732	22221	–0.009	R. Hill	0.007
RZ Lyr	52794.750	22712	–0.011	R. Hill	0.008
RZ Lyr	52796.795	22716	–0.011	R. Hill	0.007
RZ Lyr	54003.324	25076	–0.014	S. Swierczynski	0.004
RZ Lyr	54004.357	25078	–0.003	S. Swierczynski	0.005
RZ Lyr	54005.375	25080	–0.008	S. Swierczynski	0.002
RZ Lyr	54271.743	25601	0.003	P. Soron	0.004
RZ Lyr	54274.813	25607	0.005	P. Soron	0.007
RZ Lyr	54356.603	25767	–0.003	R. Harvan	0.002
AV Peg	40471.744	–8501	0.003	M. Baldwin	0.006
AV Peg	40512.727	–8396	–0.003	M. Baldwin	0.003
AV Peg	40520.531	–8376	–0.007	L. Hazel	0.008
AV Peg	40562.698	–8268	0.000	M. Baldwin	0.003
AV Peg	42343.579	–3706	–0.008	M. Baldwin	0.003
AV Peg	42570.789	–3124	0.004	M. Baldwin	0.005

Table continued on next page

Table 1. Recent times of minima of stars in the AAVSO short period pulsator program, cont.

Star	*JD (max) Hel. 2400000+*	*Cycle*	*O–C (day)*	*Observer*	*Error (day)*
AV Peg	42631.681	–2968	–0.003	M. Baldwin	0.004
AV Peg	42652.754	–2914	–0.010	M. Baldwin	0.006
AV Peg	42663.699	–2886	0.004	M. Baldwin	0.004
AV Peg	42692.583	–2812	0.001	M. Baldwin	0.004
AV Peg	42994.732	–2038	0.000	M. Baldwin	0.003
AV Peg	43021.667	–1969	–0.001	M. Baldwin	0.004
AV Peg	43028.699	–1951	0.004	M. Baldwin	0.003
AV Peg	43096.620	–1777	0.000	M. Baldwin	0.003
AV Peg	43350.756	–1126	0.002	M. Baldwin	0.004
AV Peg	43357.780	–1108	–0.001	M. Baldwin	0.005
AV Peg	43375.743	–1062	0.005	M. Baldwin	0.005
AV Peg	43404.621	–988	–0.005	M. Baldwin	0.003
AV Peg	43817.642	70	0.000	M. Baldwin	0.004
AV Peg	44141.666	900	0.013	M. Baldwin	0.002
AV Peg	44546.473	1937	0.001	B. Wingate	0.004
AV Peg	44554.670	1958	0.000	M. Heifner	0.001
AV Peg	44871.655	2770	0.001	M. Baldwin	0.003
AV Peg	44874.784	2778	0.007	L. Cook	0.003
AV Peg	44896.648	2834	0.010	M. Heifner	0.002
AV Peg	44946.600	2962	–0.006	M. Heifner	0.002
AV Peg	45608.689	4658	0.008	M. Heifner	0.003
AV Peg	45633.672	4722	0.007	M. Heifner	0.003
AV Peg	46211.819	6203	0.009	M. Baldwin	0.005
AV Peg	46254.767	6313	0.016	M. Baldwin	0.005
AV Peg	46263.742	6336	0.012	M. Baldwin	0.006
AV Peg	46329.714	6505	0.011	M. Baldwin	0.004
AV Peg	46354.694	6569	0.007	M. Baldwin	0.004
AV Peg	46365.632	6597	0.014	M. Heifner	0.003
AV Peg	46671.681	7381	0.009	M. Baldwin	0.002
AV Peg	46714.617	7491	0.004	M. Baldwin	0.003
AV Peg	46723.592	7514	0.001	M. Baldwin	0.005
AV Peg	46725.559	7519	0.016	M. Baldwin	0.003
AV Peg	46732.583	7537	0.013	M. Baldwin	0.003
AV Peg	46979.687	8170	0.010	M. Baldwin	0.002
AV Peg	46995.699	8211	0.016	M. Baldwin	0.004
AV Peg	47002.724	8229	0.015	M. Baldwin	0.002
AV Peg	47023.810	8283	0.020	M. Baldwin	0.002
AV Peg	47029.661	8298	0.016	M. Baldwin	0.003
AV Peg	47038.632	8321	0.008	M. Baldwin	0.004
AV Peg	47081.579	8431	0.014	M. Baldwin	0.004
AV Peg	47358.739	9141	0.008	M. Baldwin	0.003
AV Peg	47410.674	9274	0.023	M. Baldwin	0.004
AV Peg	47419.649	9297	0.019	M. Baldwin	0.005
AV Peg	47721.804	10071	0.024	M. Baldwin	0.005
AV Peg	47814.709	10309	0.020	R. Hill	0.007
AV Peg	47823.687	10332	0.020	R. Hill	0.003
AV Peg	47861.559	10429	0.025	M. Baldwin	0.003
AV Peg	48065.721	10952	0.021	M. Baldwin	0.004
AV Peg	48149.656	11167	0.026	M. Baldwin	0.005
AV Peg	48151.601	11172	0.019	M. Baldwin	0.005
AV Peg	48158.634	11190	0.025	M. Baldwin	0.004
AV Peg	48233.582	11382	0.021	M. Baldwin	0.004
AV Peg	48507.626	12084	0.022	M. Baldwin	0.003
AV Peg	48898.789	13086	0.030	M. Baldwin	0.003
AV Peg	49194.695	13844	0.032	M. Baldwin	0.003
AV Peg	51037.672	18565	0.050	R. Berg	0.005
AV Peg	51047.832	18591	0.060	M. Baldwin	0.003
AV Peg	51058.758	18619	0.055	R. Berg	0.004
AV Peg	51067.738	18642	0.057	R. Berg	0.005
AV Peg	51085.700	18688	0.062	R. Berg	0.005
AV Peg	51110.672	18752	0.050	R. Berg	0.005
AV Peg	51132.529	18808	0.046	R. Berg	0.006
AV Peg	51144.631	18839	0.046	R. Berg	0.006
AV Peg	51418.683	19541	0.055	R. Berg	0.007
AV Peg	51420.643	19546	0.063	R. Berg	0.007
AV Peg	51436.641	19587	0.056	R. Berg	0.008
AV Peg	52200.614	21544	0.065	R. Berg	0.006
AV Peg	52513.716	22346	0.087	R. Berg	0.008
AV Peg	52540.654	22415	0.089	R. Berg	0.005
AV Peg	54388.714	27149	0.115	P. Soron	0.002
AV Peg	55422.436	29797	0.125	J. Starzomski	0.005
RV UMa	44630.856	–950	0.002	G. Hanson	0.003
RV UMa	46193.710	2389	0.004	M. Baldwin	0.004
RV UMa	46273.755	2560	0.010	M. Baldwin	0.005
RV UMa	46274.689	2562	0.008	M. Baldwin	0.006
RV UMa	46531.652	3111	0.006	M. Baldwin	0.004
RV UMa	46553.652	3158	0.008	M. Baldwin	0.003
RV UMa	46560.678	3173	0.013	M. Baldwin	0.003
RV UMa	46612.622	3284	0.002	G. Chaple	0.005
RV UMa	46850.860	3793	–0.003	M. Baldwin	0.006
RV UMa	46858.820	3810	0.000	M. Baldwin	0.006
RV UMa	46861.647	3816	0.019	M. Baldwin	0.004
RV UMa	46912.660	3925	0.014	M. Baldwin	0.004
RV UMa	46939.799	3983	0.005	M. Baldwin	0.005
RV UMa	46948.702	4002	0.015	M. Baldwin	0.002
RV UMa	47022.661	4160	0.020	M. Baldwin	0.005
RV UMa	47271.649	4692	0.000	M. Baldwin	0.004
RV UMa	47293.656	4739	0.009	M. Baldwin	0.003
RV UMa	49118.639	8638	0.026	M. Baldwin	0.005
RV UMa	49154.679	8715	0.025	M. Baldwin	0.004
RV UMa	49160.778	8728	0.039	M. Baldwin	0.004
RV UMa	49161.717	8730	0.042	M. Baldwin	0.003
RV UMa	49572.673	9608	0.042	M. Baldwin	0.003
RV UMa	49580.630	9625	0.041	M. Baldwin	0.003
RV UMa	49587.638	9640	0.029	M. Baldwin	0.005
RV UMa	49843.681	10187	0.043	M. Baldwin	0.004
RV UMa	49857.722	10217	0.042	M. Baldwin	0.006
RV UMa	49901.718	10311	0.040	M. Baldwin	0.003
RV UMa	50290.680	11142	0.044	M. Baldwin	0.005
RV UMa	50540.639	11676	0.059	M. Baldwin	0.004
RV UMa	50921.641	12490	0.061	M. Baldwin	0.002
RV UMa	50928.653	12505	0.052	M. Baldwin	0.003
RV UMa	50950.665	12552	0.065	M. Baldwin	0.003
RV UMa	51068.613	12804	0.062	M. Baldwin	0.002
RV UMa	51667.747	14084	0.079	R. Berg	0.006
RV UMa	52049.696	14900	0.091	R. Berg	0.005
RV UMa	54624.515	20401	0.112	S. Swierczynski	0.006

Recent Maxima of 85 Short Period Pulsating Stars

Gerard Samolyk
P.O. Box 20677, Greenfield, WI 53220; gsamolyk@wi.rr.com

Received January 23, 2019; accepted January 23, 2019

Abstract This paper contains times of maxima for 85 short period pulsating stars (primarily RR Lyrae and δ Scuti stars). This represents the CCD observations received by the AAVSO's Short Period Pulsator (SPP) Section in 2017.

1. Recent observations

This accompanying list contains times of maxima calculated from CCD observations made by participants in the AAVSO's Short Period Pulsator (SPP) Section. This list will be web-archived and made available through the AAVSO ftp site at ftp:ftp.aavso.org/public/datasets/gsamj471spp85.txt. The error estimate is included. RR Lyr stars in this list, along with data from earlier AAVSO publications, are included in the GEOS database at: http://rr-lyr.irap.omp.eu/dbrr/. This database does not include δ Scuti stars. All observations were reduced by the writer using the PERANSO program (Vanmunster 2007). Column F indicates the filter used. A "C" indicates a clear filter.

The linear elements in the *General Catalogue of Variable Stars* (Kholopov *et al.* 1985) were used to compute the O–C values for most stars. For a few exceptions where the GCVS elements are missing or are in significant error, light elements from another source are used: VY CrB (Antipin 1996); RZ Cap and DG Hya (Samolyk 2010); V2416 Cyg (Samolyk 2018); V2771 Cyg (AAVSO VSX site; Watson *et al.* 2014); CV Peg and FR Psc (Le Borgne 2000–2017); and GW UMa (Hintz *et al.* 2001).

References

Antipin, S. V. 1996, *Inf. Bull. Var. Stars*, No. 4343, 1.

Hintz, E. G., Bush, T. C., and Rose, M. B. 2005, *Astron. J.*, **130**, 2876.

Kholopov, P. N., *et al.* 1985, *General Catalogue of Variable Stars*, 4th ed., Moscow.

Le Borgne, J. F., ed. 2000–2017, GEOS database (http://rr-lyr.irap.omp.eu/dbrr/index.php).

Samolyk, G. 2010, *J. Amer. Assoc. Var. Stars*, **38**, 12.

Samolyk, G. 2018, *J. Amer. Assoc. Var. Stars*, **46**, 74.

Vanmunster, T. 2007, PERANSO period analysis software (http://www.peranso.com).

Watson, C., Henden, A. A., and Price, C. A. 2014, AAVSO International Variable Star Index (https://www.aavso.org/vsx).

Table 1. Recent times of minima of stars in the AAVSO short period pulsator program.

Star	JD (max) Hel. 2400000 +	Cycle	O–C (day)	F	Observer	Error (day)
SW And	58317.8056	90860	–0.4971	V	G. Samolyk	0.0007
SW And	58372.6468	90984	–0.4985	V	G. Samolyk	0.0008
SW And	58381.4929	91004	–0.4980	V	T. Arranz	0.0008
SW And	58389.4511	91022	–0.5008	V	T. Arranz	0.0007
SW And	58400.5087	91047	–0.5002	V	T. Arranz	0.0008
SW And	58408.4704	91065	–0.4996	V	T. Arranz	0.0009
SW And	58409.8055	91068	–0.4913	V	R. Sabo	0.0029
SW And	58436.3315	91128	–0.5021	V	T. Arranz	0.0007
SW And	58440.3127	91137	–0.5014	V	T. Arranz	0.0009
SW And	58455.3469	91171	–0.5047	V	T. Arranz	0.0009
XX And	58330.8652	26625	0.2853	V	G. Samolyk	0.0018
XX And	58346.7650	26647	0.2847	V	G. Samolyk	0.0017
XX And	58377.8433	26690	0.2848	V	K. Menzies	0.0015
XX And	58399.5246	26720	0.2837	V	T. Arranz	0.0013
XX And	58409.6334	26734	0.2740	V	R. Sabo	0.0035
XX And	58425.5440	26756	0.2842	V	T. Arranz	0.0011
XX And	58449.3932	26789	0.2828	V	T. Arranz	0.0013
XX And	58483.3661	26836	0.2865	V	T. Arranz	0.0013
AC And	58317.7653	13803	0.3666	V	G. Samolyk	0.0017
AC And	58322.8939	13810	0.5165	V	G. Samolyk	0.0017
AC And	58360.7044	13863	0.6313	V	G. Samolyk	0.0035
AC And	58397.5258	13915	0.4682	V	T. Arranz	0.0015
AC And	58402.4109	13922	0.3746	V	T. Arranz	0.0025
AT And	58299.9144	25865	–0.0061	V	T. Polakis	0.0019
AT And	58385.6662	26004	–0.0055	V	G. Samolyk	0.0016
AT And	58392.4477	26015	–0.0100	V	T. Arranz	0.0014
AT And	58395.5355	26020	–0.0068	V	T. Arranz	0.0015
AT And	58413.4229	26049	–0.0099	V	T. Arranz	0.0018
AT And	58426.3802	26070	–0.0078	V	T. Arranz	0.0014
AT And	58461.5459	26127	–0.0063	V	G. Samolyk	0.0014
DY And	58300.8707	36858	–0.1759	V	T. Polakis	0.0019
GM And	58340.8492	46267	0.0437	V	K. Menzies	0.0016
SW Aqr	58355.5192	72410	–0.0012	V	T. Arranz	0.0008
SW Aqr	58356.4389	72412	–0.0001	V	T. Arranz	0.0009
TZ Aqr	58343.8116	36406	0.0119	V	G. Samolyk	0.0021
YZ Aqr	58371.7931	41685	0.0828	V	G. Samolyk	0.0015
AA Aqr	58360.7646	61772	–0.1797	V	G. Samolyk	0.0019
AA Aqr	58390.6018	61821	–0.1781	V	G. Samolyk	0.0015
BO Aqr	58407.7046	24187	0.2247	V	G. Samolyk	0.0022
BR Aqr	58410.7406	43112	–0.2243	V	G. Samolyk	0.0011
CY Aqr	58390.3259	394537	0.0157	V	T. Arranz	0.0003
CY Aqr	58390.3866	394538	0.0153	V	T. Arranz	0.0003
CY Aqr	58391.3023	394553	0.0155	V	T. Arranz	0.0003
CY Aqr	58391.3635	394554	0.0156	V	T. Arranz	0.0003
CY Aqr	58391.4244	394555	0.0155	V	T. Arranz	0.0003
CY Aqr	58407.5386	394819	0.0156	V	G. Samolyk	0.0004
CY Aqr	58407.5995	394820	0.0155	V	G. Samolyk	0.0003
CY Aqr	58407.6606	394821	0.0155	V	G. Samolyk	0.0004
CY Aqr	58407.7220	394822	0.0159	V	G. Samolyk	0.0003
CY Aqr	58407.7829	394823	0.0158	V	G. Samolyk	0.0005
SY Ari	58337.8828	39233	–0.0785	V	R. Sabo	0.0012
TZ Aur	58135.7656	97615	0.0157	V	R. Sabo	0.0006

Table continued on following pages

Table 1. Recent times of minima of stars in the AAVSO short period pulsator program, cont.

Star	JD (max) Hel. 2400000+	Cycle	O–C (day)	F	Observer	Error (day)
TZ Aur	58384.8717	98251	0.0167	V	G. Samolyk	0.0011
TZ Aur	58415.8137	98330	0.0164	V	N. Simmons	0.0008
BH Aur	58181.6241	33831	0.0081	V	G. Samolyk	0.0012
BH Aur	58413.7759	34340	0.0102	V	G. Samolyk	0.0011
BH Aur	58462.5771	34447	0.0098	V	G. Samolyk	0.0011
BH Aur	58463.4892	34449	0.0097	V	T. Arranz	0.0013
BH Aur	58485.3822	34497	0.0104	V	T. Arranz	0.0009
RS Boo	58193.7692	43524	–0.0217	V	G. Samolyk	0.0009
RS Boo	58214.5218	43579	–0.0227	V	T. Arranz	0.0006
RS Boo	58215.6532	43582	–0.0234	V	G. Samolyk	0.0006
RS Boo	58225.4661	43608	–0.0213	V	T. Arranz	0.0007
RS Boo	58242.4452	43653	–0.0224	V	T. Arranz	0.0009
RS Boo	58253.3860	43682	–0.0245	V	T. Arranz	0.0007
RS Boo	58255.6491	43688	–0.0254	V	N. Simmons	0.0007
ST Boo	58199.8288	62701	0.0944	V	G. Samolyk	0.0011
ST Boo	58226.5997	62744	0.1068	V	T. Arranz	0.0009
ST Boo	58251.4971	62784	0.1126	V	T. Arranz	0.0007
ST Boo	58284.4792	62837	0.1133	V	T. Arranz	0.0006
ST Boo	58294.4337	62853	0.1111	V	T. Arranz	0.0009
SW Boo	58191.7598	30418	0.5021	V	G. Samolyk	0.0011
SW Boo	58243.6299	30519	0.5058	V	T. Arranz	0.0011
SW Boo	58302.6905	30634	0.5107	V	T. Polakis	0.0009
SZ Boo	58181.8078	58417	0.0125	V	G. Samolyk	0.0011
SZ Boo	58234.6125	58518	0.0124	V	T. Arranz	0.0008
TV Boo	58193.8251	107449	0.1194	C	G. Samolyk	0.0019
TV Boo	58199.7434	107468	0.0991	V	G. Samolyk	0.0012
TV Boo	58209.7517	107500	0.1055	V	G. Samolyk	0.0021
TV Boo	58234.7701	107580	0.1192	V	R. Sabo	0.0026
TV Boo	58291.6593	107762	0.1225	V	K. Menzies	0.0015
TV Boo	58300.7326	107791	0.1316	V	T. Polakis	0.0013
TW Boo	58228.7511	58875	–0.0986	V	G. Samolyk	0.0009
TW Boo	58246.8470	58909	–0.1000	V	K. Menzies	0.0009
UU Boo	58154.9107	48302	0.3267	V	G. Samolyk	0.0009
UU Boo	58187.8109	48374	0.3286	V	G. Samolyk	0.0009
UU Boo	58227.5637	48461	0.3293	V	T. Arranz	0.0007
UY Boo	58218.7545	25171	0.8550	V	G. Samolyk	0.0011
UY Boo	58243.4582	25209	0.8269	V	T. Arranz	0.0012
UY Cam	58195.6441	84745	–0.0980	V	G. Samolyk	0.0024
UY Cam	58470.7059	85775	–0.0898	V	G. Samolyk	0.0029
RW Cnc	58132.8486	33948	0.2229	V	K. Menzies	0.0008
RW Cnc	58227.5173	34121	0.2262	V	T. Arranz	0.0009
RW Cnc	58463.9053	34553	0.2243	V	G. Samolyk	0.0011
TT Cnc	58181.6715	32367	0.1378	V	G. Samolyk	0.0019
TT Cnc	58467.9020	32875	0.1360	V	G. Samolyk	0.0011
VZ Cnc	58191.6794	102567	0.0248	C	G. Samolyk	0.0009
VZ Cnc	58216.6449	102707	0.0194	V	N. Simmons	0.0019
SS CVn	58215.7487	38856	–0.3683	V	N. Simmons	0.0008
SS CVn	58215.7490	38856	–0.3680	V	G. Samolyk	0.0009
RV Cap	58322.7990	54584	–0.1220	V	G. Samolyk	0.0011
RV Cap	58388.6285	54731	–0.1109	V	G. Samolyk	0.0014
RZ Cap	58342.7064	17351	0.0027	V	G. Samolyk	0.0019
VW Cap	58341.8114	104792	0.2775	V	G. Samolyk	0.0032
YZ Cap	58373.6414	53551	0.0421	V	G. Samolyk	0.0013
AN Cap	58360.6773	7722	0.0009	V	G. Samolyk	0.0026
RR Cet	58376.8308	45559	0.0178	V	G. Samolyk	0.0013
RR Cet	58451.4894	45694	0.0176	V	T. Arranz	0.0009
RU Cet	58375.8601	31639	0.1362	V	G. Samolyk	0.0009
RU Cet	58462.6201	31787	0.1267	V	G. Samolyk	0.0011
RV Cet	58404.8539	30945	0.2851	V	G. Samolyk	0.0022
RX Cet	58390.7535	31837	0.3287	V	G. Samolyk	0.0014
RX Cet	58459.6005	31957	0.3326	V	G. Samolyk	0.0015
RZ Cet	58384.8459	47939	–0.2144	V	G. Samolyk	0.0014
TY Cet	58158.5976	20797	–0.0154	V	G. Samolyk	0.0026
TY Cet	58407.8321	21567	–0.0145	V	G. Samolyk	0.0021
UU Cet	58373.8244	28322	–0.1777	V	G. Samolyk	0.0021
TU Com	58299.7426	63543	0.5270	V	T. Polakis	0.0021
VY CrB	58278.8080	35646	–0.1673	V	K. Menzies	0.0015
VY CrB	58298.7130	35689	–0.1694	V	T. Polakis	0.0014
VY CrB	58303.8031	35700	–0.1718	V	T. Polakis	0.0013
XX Cyg	58199.9070	101913	0.0042	V	G. Samolyk	0.0007
XX Cyg	58246.7056	102260	0.0046	V	G. Samolyk	0.0007
XX Cyg	58246.8408	102261	0.0050	V	G. Samolyk	0.0008
XX Cyg	58295.7958	102624	0.0039	V	K. Menzies	0.0005
XX Cyg	58307.7996	102713	0.0047	V	G. Samolyk	0.0006
XZ Cyg	58234.7756	30240	–2.6724	V	G. Samolyk	0.0011
XZ Cyg	58254.8374	30283	–2.6787	V	G. Samolyk	0.0009
XZ Cyg	58261.8355	30298	–2.6811	V	G. Samolyk	0.0008
XZ Cyg	58263.7019	30302	–2.6815	V	G. Samolyk	0.0007
XZ Cyg	58271.6282	30319	–2.6891	V	G. Samolyk	0.0007
XZ Cyg	58297.7637	30375	–2.6888	V	G. Samolyk	0.0009
XZ Cyg	58302.8987	30386	–2.6875	V	T. Polakis	0.0009
XZ Cyg	58305.7011	30392	–2.6853	V	G. Samolyk	0.0007
XZ Cyg	58317.8285	30418	–2.6921	V	G. Samolyk	0.0007
XZ Cyg	58322.4892	30428	–2.6984	V	T. Arranz	0.0008
XZ Cyg	58324.8211	30433	–2.7000	V	G. Samolyk	0.0009
XZ Cyg	58328.5488	30441	–2.7059	V	T. Arranz	0.0008
XZ Cyg	58329.4808	30443	–2.7073	V	T. Arranz	0.0008
XZ Cyg	58336.4798	30458	–2.7088	V	T. Arranz	0.0009
XZ Cyg	58351.4283	30490	–2.6947	V	T. Arranz	0.0007
XZ Cyg	58388.7435	30570	–2.7155	V	H. Smith	0.0009
DM Cyg	58271.8259	37368	0.0914	V	G. Samolyk	0.0008
DM Cyg	58300.7939	37437	0.0891	V	K. Menzies	0.0009
DM Cyg	58300.7941	37437	0.0893	V	T. Polakis	0.0009
DM Cyg	58308.7719	37456	0.0897	V	G. Samolyk	0.0014
DM Cyg	58323.8901	37492	0.0930	V	R. Sabo	0.0009
DM Cyg	58329.7649	37506	0.0897	V	R. Sabo	0.0008
DM Cyg	58339.4230	37529	0.0911	V	T. Arranz	0.0007
DM Cyg	58354.5383	37565	0.0914	V	T. Arranz	0.0009
DM Cyg	58375.5345	37615	0.0946	V	T. Arranz	0.0009
DM Cyg	58376.3735	37617	0.0939	V	T. Arranz	0.0009
DM Cyg	58402.4022	37679	0.0913	V	T. Arranz	0.0008
DM Cyg	58426.3342	37736	0.0912	V	T. Arranz	0.0008
V2416 Cyg	58199.8584	82077	0.0008	V	G. Samolyk	0.0014
V2416 Cyg	58199.9143	82078	0.0008	V	G. Samolyk	0.0015
V2416 Cyg	58246.6371	82914	–0.0002	V	G. Samolyk	0.0020
V2416 Cyg	58246.6932	82915	0.0000	V	G. Samolyk	0.0019
V2416 Cyg	58246.7487	82916	–0.0003	V	G. Samolyk	0.0024
V2416 Cyg	58246.8030	82917	–0.0019	V	G. Samolyk	0.0017
V2416 Cyg	58295.8196	83794	–0.0006	V	K. Menzies	0.0012
V2416 Cyg	58307.8332	84009	–0.0033	V	G. Samolyk	0.0013
V2771 Cyg	58350.7061	26616	0.0733	TG	G. Conrad	0.0027
V2771 Cyg	58358.7720	26670	0.0703	TG	G. Conrad	0.0026
V2771 Cyg	58359.6722	26676	0.0739	TG	G. Conrad	0.0026
RW Dra	58148.8957	42381	0.2393	V	G. Samolyk	0.0010
RW Dra	58231.7244	42568	0.2425	V	G. Samolyk	0.0009
RW Dra	58255.6794	42622	0.2800	V	G. Samolyk	0.0009
RW Dra	58284.4394	42687	0.2504	V	T. Arranz	0.0008
RW Dra	58291.5459	42703	0.2702	V	T. Arranz	0.0008
RW Dra	58303.5086	42730	0.2742	V	T. Arranz	0.0007
RW Dra	58317.6457	42762	0.2379	V	G. Samolyk	0.0014
RW Dra	58319.4168	42766	0.2374	V	T. Arranz	0.0014
RW Dra	58327.4096	42784	0.2577	V	T. Arranz	0.0011
XZ Dra	58215.8730	34182	–0.1215	V	G. Samolyk	0.0013
XZ Dra	58294.4974	34347	–0.1191	V	T. Arranz	0.0009
XZ Dra	58296.8760	34352	–0.1229	V	T. Polakis	0.0011
RX Eri	57334.7917	60694	–0.0094	V	G. Samolyk	0.0019
RX Eri	58124.6376	62039	–0.0096	V	G. Samolyk	0.0012
RX Eri	58425.8981	62552	–0.0065	V	G. Samolyk	0.0015
RR Gem	58137.7882	42237	–0.6246	V	G. Samolyk	0.0007
RR Gem	58173.5445	42327	–0.6263	V	T. Arranz	0.0007
RR Gem	58175.5296	42332	–0.6277	V	T. Arranz	0.0005
RR Gem	58407.9435	42917	–0.6405	V	R. Sabo	0.0015

Table continued on next page

Table 1. Recent times of minima of stars in the AAVSO short period pulsator program, cont.

Star	*JD (max) Hel. 2400000 +*	*Cycle*	*O–C (day)*	*F*	*Observer*	*Error (day)*
GQ Gem	58162.8168	48787	–0.2088	V	R. Sabo	0.0014
TW Her	58261.6727	91883	–0.0177	V	G. Samolyk	0.0008
TW Her	58271.6614	91908	–0.0190	V	G. Samolyk	0.0006
TW Her	58336.3973	92070	–0.0183	V	T. Arranz	0.0005
TW Her	58342.3910	92085	–0.0186	V	T. Arranz	0.0005
VX Her	58227.5895	80104	–0.0776	V	T. Arranz	0.0006
VZ Her	58218.8632	48668	0.0874	V	G. Samolyk	0.0007
VZ Her	58237.7980	48711	0.0882	V	G. Samolyk	0.0008
VZ Her	58275.6671	48797	0.0891	V	G. Samolyk	0.0011
VZ Her	58296.8015	48845	0.0877	V	T. Polakis	0.0007
AR Her	58192.8816	35614	–1.0426	V	G. Samolyk	0.0009
AR Her	58216.8115	35665	–1.0841	V	G. Samolyk	0.0002
AR Her	58218.7143	35669	–1.0614	V	G. Samolyk	0.0015
AR Her	58229.5343	35692	–1.0521	V	T. Arranz	0.0008
AR Her	58234.7017	35703	–1.0550	V	G. Samolyk	0.0009
AR Her	58243.5953	35722	–1.0919	V	G. Samolyk	0.0013
AR Her	58254.4467	35745	–1.0512	V	T. Arranz	0.0009
AR Her	58255.8735	35748	–1.0344	V	G. Samolyk	0.0015
AR Her	58265.7194	35769	–1.0591	V	G. Samolyk	0.0008
AR Her	58285.4690	35811	–1.0507	V	T. Arranz	0.0009
AR Her	58292.5166	35826	–1.0535	V	T. Arranz	0.0016
AR Her	58293.4545	35828	–1.0557	V	T. Arranz	0.0007
AR Her	58297.6782	35837	–1.0622	V	G. Samolyk	0.0008
AR Her	58305.6448	35854	–1.0861	V	G. Samolyk	0.0009
AR Her	58309.3960	35862	–1.0951	V	T. Arranz	0.0009
AR Her	58329.6431	35905	–1.0592	V	K. Menzies	0.0012
DL Her	58210.8195	33809	0.0450	V	G. Samolyk	0.0019
DL Her	58286.5563	33937	0.0534	V	T. Arranz	0.0011
DL Her	58292.4826	33947	0.0634	V	T. Arranz	0.0014
DL Her	58297.8108	33956	0.0670	V	T. Polakis	0.0014
DL Her	58305.4926	33969	0.0576	V	T. Arranz	0.0009
DL Her	58324.4246	34001	0.0575	V	T. Arranz	0.0011
DY Her	58209.9089	166657	–0.0330	V	G. Samolyk	0.0007
DY Her	58263.7132	167019	–0.0332	V	G. Samolyk	0.0006
DY Her	58263.8615	167020	–0.0336	V	G. Samolyk	0.0006
DY Her	58269.8064	167060	–0.0339	V	K. Menzies	0.0005
DY Her	58289.7233	167194	–0.0336	V	K. Menzies	0.0006
DY Her	58298.7890	167255	–0.0344	V	T. Polakis	0.0009
DY Her	58302.6544	167281	–0.0335	V	G. Samolyk	0.0007
DY Her	58302.8031	167282	–0.0334	V	G. Samolyk	0.0007
DY Her	58303.6948	167288	–0.0335	V	T. Polakis	0.0007
DY Her	58316.6258	167375	–0.0334	V	G. Samolyk	0.0007
DY Her	58319.7474	167396	–0.0331	V	R. Sabo	0.0008
LS Her	58265.6578	131108	–0.0264	V	G. Samolyk	0.0021
LS Her	58318.7551	131338	–0.0149	V	R. Sabo	0.0021
SZ Hya	58124.9441	32473	–0.2696	V	G. Samolyk	0.0011
SZ Hya	58137.7944	32497	–0.3130	V	G. Samolyk	0.0018
SZ Hya	58187.7738	32590	–0.2970	V	G. Samolyk	0.0039
SZ Hya	58192.6380	32599	–0.2679	V	G. Samolyk	0.0018
UU Hya	58123.9523	35591	0.0091	V	G. Samolyk	0.0015
UU Hya	58195.7267	35728	0.0135	V	G. Samolyk	0.0016
UU Hya	58216.6769	35768	0.0090	V	N. Simmons	0.0011
DG Hya	58137.7588	7641	0.0253	V	G. Samolyk	0.0021
DH Hya	58132.8680	55126	0.1092	V	G. Samolyk	0.0011
DH Hya	58157.8076	55177	0.1099	V	G. Samolyk	0.0010
RR Leo	58193.7885	32932	0.1703	V	G. Samolyk	0.0007
RR Leo	58205.5511	32958	0.1707	V	T. Arranz	0.0007
RR Leo	58467.9457	33538	0.1772	V	G. Samolyk	0.0007
SS Leo	58187.7564	26194	–0.1100	V	G. Samolyk	0.0014
SS Leo	58228.4693	26259	–0.1094	V	T. Arranz	0.0009
ST Leo	58195.7574	63334	–0.0183	V	G. Samolyk	0.0011
ST Leo	58263.6283	63476	–0.0211	V	K. Menzies	0.0014
TV Leo	58200.7145	31447	0.1288	V	G. Samolyk	0.0016
TV Leo	58240.4143	31506	0.1303	V	T. Arranz	0.0010
WW Leo	58161.9174	38616	0.0505	V	G. Samolyk	0.0015
AA Leo	58136.9107	30970	–0.1113	V	G. Samolyk	0.0013
U Lep	58468.7220	29436	0.0425	V	G. Samolyk	0.0015
SZ Lyn	58136.5980	166028	0.0281	V	G. Samolyk	0.0007
SZ Lyn	58136.7193	166029	0.0288	V	G. Samolyk	0.0007
SZ Lyn	58136.8401	166030	0.0291	V	G. Samolyk	0.0006
SZ Lyn	58136.9599	166031	0.0284	V	G. Samolyk	0.0008
SZ Lyn	58209.7628	166635	0.0282	V	G. Samolyk	0.0009
SZ Lyn	58231.5814	166816	0.0299	V	K. Menzies	0.0009
SZ Lyn	58411.7840	168311	0.0328	V	G. Samolyk	0.0007
SZ Lyn	58411.9052	168312	0.0335	V	G. Samolyk	0.0007
SZ Lyn	58461.6869	168725	0.0343	V	N. Simmons	0.0008
SZ Lyn	58461.8076	168726	0.0344	V	N. Simmons	0.0005
SZ Lyn	58461.9277	168727	0.0340	V	N. Simmons	0.0006
SZ Lyn	58461.9285	168727	0.0348	V	G. Samolyk	0.0007
SZ Lyn	58468.7998	168784	0.0356	V	G. Samolyk	0.0009
RR Lyr	58299.7454	27126	–0.5288	V	G. Samolyk	0.0011
RR Lyr	58308.8115	27142	–0.5325	V	G. Samolyk	0.0011
RR Lyr	58314.4807	27152	–0.5320	V	T. Arranz	0.0009
RR Lyr	58348.4848	27212	–0.5400	V	T. Arranz	0.0010
RZ Lyr	58234.8227	33353	–0.0677	V	G. Samolyk	0.0009
RZ Lyr	58286.4704	33454	–0.0555	V	T. Arranz	0.0009
RZ Lyr	58300.7834	33482	–0.0573	V	K. Menzies	0.0012
RZ Lyr	58302.8272	33486	–0.0585	V	T. Polakis	0.0007
RZ Lyr	58331.4470	33542	–0.0682	V	T. Arranz	0.0008
RZ Lyr	58334.5150	33548	–0.0677	V	T. Arranz	0.0009
CX Lyr	58296.7139	40908	1.6373	V	T. Polakis	0.0011
CX Lyr	58299.8006	40913	1.6407	V	T. Polakis	0.0010
ST Oph	58301.6918	66337	–0.0271	V	G. Samolyk	0.0011
AV Peg	58360.4618	37323	0.1909	V	T. Arranz	0.0008
AV Peg	58362.4156	37328	0.1928	V	T. Arranz	0.0008
CV Peg	58301.8457	7933	–0.0038	V	T. Polakis	0.0013
GV Peg	58377.6763	23742	0.2439	V	K. Menzies	0.0016
FR Psc	58465.7085	10433	0.0060	V	R. Sabo	0.0009
DF Ser	58215.8166	65121	0.1066	V	G. Samolyk	0.0009
RV UMa	58135.9195	27903	0.1303	V	G. Samolyk	0.0017
RV UMa	58151.8389	27937	0.1357	V	G. Samolyk	0.0014
TU UMa	58230.6182	27614	–0.0638	V	K. Menzies	0.0009
AE UMa	58229.4031	263030	–0.0009	V	T. Arranz	0.0005
AE UMa	58231.3797	263053	–0.0027	V	T. Arranz	0.0003
GW UMa	58148.7747	30261	0.0017	V	G. Samolyk	0.0009
GW UMa	58148.9779	30262	0.0017	V	G. Samolyk	0.0010

Recent Minima of 242 Eclipsing Binary Stars

Gerard Samolyk
P.O. Box 20677, Greenfield, WI 53220; gsamolyk@wi.rr.com

Received February 15, 2019; accepted February 15, 2019

Abstract This paper continues the publication of times of minima for eclipsing binary stars from CCD observations reported to the AAVSO Eclipsing Binary section. Times of minima from observations received from September 2018 through January 2019 are presented.

1. Recent observations

The accompanying list contains times of minima calculated from recent CCD observations made by participants in the AAVSO's eclipsing binary program. This list will be web-archived and made available through the AAVSO ftp site at ftp://ftp.aavso.org/public/datasets/gsamj471eb242.txt. This list, along with the eclipsing binary data from earlier AAVSO publications, is also included in the Lichtenknecker database administrated by the Bundesdeutsche Arbeitsgemeinschaft für Veränderliche Sterne e. V. (BAV) at: http://www.bav-astro.de/LkDB/index.php?lang=en. These observations were reduced by the observers or the writer using the method of Kwee and van Worden (1956). The standard error is included when available. Column F indicates the filter used. A "C" indicates a clear filter.

The linear elements in the *General Catalogue of Variable Stars* (GCVS; Kholopov *et al.* 1985) were used to compute the O–C values for most stars. For a few exceptions where the GCVS elements are missing or are in significant error, light elements from another source are used: CD Cam (Baldwin and Samolyk 2007), AC CMi (Samolyk 2008), CW Cas (Samolyk 1992a), DV Cep (Frank and Lichtenknecker 1987), Z Dra (Danielkiewicz-Krosniak and Kurpinskw-Winiarska 1996), DF Hya (Samolyk 1992b), DK Hya (Samolyk 1990), EF Ori (Baldwin and Samolyk 2005), and GU Ori (Samolyk 1985).

The light elements used for QX And, V463 And, V599 Aur, DU Boo, HH Boo, AH Cnc, CZ CMi, V776 Cas, YY CrB, V772 Her, DE Lyn, BR Per, HX UMa, and KM UMa are from Kreiner (2004).

The light elements used for MU Aqr, AH Aur, XY Boo, DN Boo, CW CMi, V1261 Cyg, 2181 Cyg, V1065 Her, V1097 Her, V470 Hya, CE Leo, HI Leo, VW LMi, DI Lyn, HN Lyn. V502 Oph, V1853 Ori, KV Peg, VZ Psc, EQ UMa, II UMa, GR Vir, IR Vir, and NN Vir are from Paschke (2014).

The light elements used for V731 Cep and V337 Gem are from Nelson (2014).

The light elements used for V449 Aur, V972 Her, and V391 Vir are from the AAVSO VSX site (Watson *et al.* 2014). O–C values listed in this paper can be directly compared with values published in the AAVSO EB monographs.

References

Baldwin, M. E., and Samolyk, G. 2005, *Observed Minima Timings of Eclipsing Binaries No. 10*, AAVSO, Cambridge, MA.

Baldwin, M. E., and Samolyk, G. 2007, *Observed Minima Timings of Eclipsing Binaries No. 12*, AAVSO, Cambridge, MA.

Danielkiewicz-Krośniak, E., and Kurpińska-Winiarska, M., eds. 1996, *Rocznik Astron.* (SAC 68), **68**, 1.

Frank, P., and Lichtenknecker, D. 1987, *BAV Mitt.*, No. 47, 1.

Kholopov, P. N., *et al.* 1985, *General Catalogue of Variable Stars*, 4th ed., Moscow.

Kreiner, J. M. 2004, *Acta Astron.*, **54**, 207 (http://www.as.up.krakow.pl/ephem/).

Kwee, K. K., and van Woerden, H. 1956, *Bull. Astron. Inst. Netherlands*, **12**, 327.

Nelson, R. 2014, Eclipsing Binary O–C Files (http://www.aavso.org/bob-nelsons-o-c-files).

Paschke, A. 2014, "O–C Gateway" (http://var.astro.cz/ocgate/).

Samolyk, G. 1985, *J. Amer. Assoc. Var. Star Obs.*, **14**, 12.

Samolyk, G. 1990, *J. Amer. Assoc. Var. Star Obs.*, **19**, 5.

Samolyk, G. 1992a, *J. Amer. Assoc. Var. Star Obs.*, **21**, 34.

Samolyk, G. 1992b, *J. Amer. Assoc. Var. Star Obs.*, **21**, 111.

Samolyk, G. 2008, *J. Amer. Assoc. Var. Star Obs.*, **36**, 171.

Watson, C., Henden, A. A., and Price, C. A. 2014, AAVSO International Variable Star Index VSX (Watson+, 2006–2014; http://www.aavso.org/vsx).

Table 1. Recent times of minima of stars in the AAVSO eclipsing binary program.

Star	JD (min) Hel. 2400000+	Cycle	O–C (day)	F	Observer	Standard Error (day)
RT And	58372.6608	27397	–0.0101	V	G. Samolyk	0.0001
RT And	58386.4961	27419	–0.0112	V	T. Arranz	0.0001
RT And	58404.7338	27448	–0.0125	V	G. Samolyk	0.0001
TW And	58397.3841	4700	–0.0641	V	T. Arranz	0.0001
UU And	58375.7307	11253	0.1018	V	G. Samolyk	0.0002
UU And	58390.5921	11263	0.1003	V	T. Arranz	0.0002
UU And	58393.5663	11265	0.1019	V	T. Arranz	0.0002
UU And	58396.5385	11267	0.1015	V	T. Arranz	0.0001
WZ And	58385.5505	25175	0.0804	V	T. Arranz	0.0001
WZ And	58410.5952	25211	0.0814	V	G. Samolyk	0.0001
WZ And	58415.4656	25218	0.0822	V	T. Arranz	0.0001
WZ And	58436.3360	25248	0.0829	V	T. Arranz	0.0001
WZ And	58461.3801	25284	0.0833	V	T. Arranz	0.0001
XZ And	58374.8802	25343	0.1923	V	G. Samolyk	0.0001
XZ And	58392.5252	25356	0.1927	V	T. Arranz	0.0001
XZ And	58396.5971	25359	0.1928	V	T. Arranz	0.0001
XZ And	58415.6002	25373	0.1940	V	N. Simmons	0.0001
XZ And	58464.4630	25409	0.1948	V	T. Arranz	0.0001
XZ And	58479.3924	25420	0.1941	V	T. Arranz	0.0001
XZ And	58494.3227	25431	0.1944	V	T. Arranz	0.0001
AB And	58382.6482	67109.5	–0.0473	V	G. Samolyk	0.0001
AB And	58397.4190	67154	–0.0457	V	L. Corp	0.0001
AB And	58414.3451	67205	–0.0461	V	T. Arranz	0.0001
AB And	58414.5087	67205.5	–0.0485	V	T. Arranz	0.0001
AB And	58428.6164	67248	–0.0462	TG	G. Conrad	0.0001
AB And	58428.7805	67248.5	–0.0480	TG	G. Conrad	0.0001
AB And	58458.3198	67337.5	–0.0471	V	T. Arranz	0.0001
AB And	58489.5170	67431.5	–0.0478	V	G. Samolyk	0.0001
AB And	58492.3389	67440	–0.0470	V	T. Arranz	0.0001
AD And	58382.6338	19651.5	–0.0446	V	G. Samolyk	0.0002
AD And	58391.5082	19660.5	–0.0459	V	T. Arranz	0.0001
AD And	58415.6697	19685	–0.0462	V	G. Samolyk	0.0001
BD And	58371.8497	50571	0.0175	V	G. Samolyk	0.0002
BD And	58390.3650	50611	0.0167	V	T. Arranz	0.0001
BD And	58404.7127	50642	0.0144	V	G. Samolyk	0.0002
BD And	58414.4343	50663	0.0151	V	T. Arranz	0.0001
BX And	58387.5829	35827.5	–0.1021	V	T. Arranz	0.0003
BX And	58440.3586	35914	–0.1014	V	T. Arranz	0.0001
BX And	58462.3225	35950	–0.1017	V	T. Arranz	0.0001
BX And	58490.3864	35996	–0.1031	V	T. Arranz	0.0001
DS And	58409.6978	22035.5	0.0054	V	G. Samolyk	0.0001
QR And	58408.8113	34233	0.1570	V	R. Sabo	0.0002
QR And	58415.4224	34243	0.1635	V	T. Arranz	0.0003
QX And	58390.7908	14292	0.0046	V	G. Samolyk	0.0002
QX And	58409.5425	14337.5	0.0025	V	G. Samolyk	0.0003
QX And	58409.7500	14338	0.0039	V	G. Samolyk	0.0002
V463 And	58373.6504	14463	–0.0078	C	G. Frey	0.0002
RY Aqr	58407.5879	8941	–0.1431	V	G. Samolyk	0.0001
RY Aqr	58407.5882	8941	–0.1428	V	N. Simmons	0.0001
CX Aqr	58376.6621	39431	0.0152	C	G. Frey	0.0001
CX Aqr	58376.6629	39431	0.0160	V	G. Samolyk	0.0001
CX Aqr	58410.5792	39492	0.0172	V	G. Samolyk	0.0001
MU Aqr	58368.7150	25366	0.0222	C	G. Frey	0.0001
XZ Aql	58409.5619	7716	0.1803	V	G. Samolyk	0.0001
KO Aql	58424.6296	5774	0.1036	V	S. Cook	0.0003
KP Aql	58390.5983	5343.5	–0.0201	V	G. Samolyk	0.0002
OO Aql	58381.3376	39006.5	0.0706	V	L. Corp	0.0003
OO Aql	58384.6327	39013	0.0715	V	G. Samolyk	0.0001
OO Aql	58414.5333	39072	0.0716	V	G. Samolyk	0.0001
V343 Aql	58388.6592	16234	–0.0429	V	G. Samolyk	0.0001
V346 Aql	58374.4136	14874	–0.0137	V	T. Arranz	0.0001
V724 Aql	58377.6839	6501	–0.0211	C	G. Frey	0.0002
V805 Aql	58360.7051	2469	–0.0108	V	S. Cook	0.0006
RX Ari	58385.7302	19412	0.0606	V	G. Samolyk	0.0001
SS Ari	58376.8409	47658	–0.3971	V	G. Samolyk	0.0002
SS Ari	58389.6262	47689.5	–0.4006	V	T. Arranz	0.0002
SS Ari	58404.6523	47726.5	–0.3963	V	T. Arranz	0.0002
SS Ari	58416.6253	47756	–0.4001	V	T. Arranz	0.0001
SS Ari	58436.7231	47805.5	–0.3989	V	N. Simmons	0.0001
SS Ari	58459.6590	47862	–0.4017	V	G. Samolyk	0.0001
RY Aur	58413.8697	7384	0.0196	V	G. Samolyk	0.0002
TT Aur	58390.9043	27874	–0.0075	V	G. Samolyk	0.0001
WW Aur	58488.6350	10116	0.0013	V	N. Simmons	0.0001
WW Aur	58488.6357	10116	0.0020	V	G. Samolyk	0.0002
AH Aur	58461.4690	67113	–0.0141	V	T. Arranz	0.0001
AH Aur	58480.4917	67151.5	–0.0145	V	T. Arranz	0.0002
AP Aur	58404.8570	28033	1.7107	V	G. Samolyk	0.0002
AP Aur	58467.7779	28143.5	1.7226	V	G. Samolyk	0.0002
AP Aur	58498.5254	28197.5	1.7272	V	T. Arranz	0.0002
AR Aur	58409.8397	4839	–0.1326	V	G. Samolyk	0.0002
CL Aur	58414.7018	20450	0.1858	V	G. Samolyk	0.0002
CL Aur	58464.4770	20490	0.1864	V	T. Arranz	0.0001
CL Aur	58489.3631	20510	0.1852	V	T. Arranz	0.0002
CL Aur	58494.3416	20514	0.1862	V	T. Arranz	0.0001
EM Aur	58404.8240	15046	–1.1207	V	G. Samolyk	0.0003
EP Aur	58426.5485	54343	0.0188	V	T. Arranz	0.0001
EP Aur	58455.5097	54392	0.0206	V	T. Arranz	0.0001
EP Aur	58491.5612	54453	0.0206	V	T. Arranz	0.0002
EP Aur	58494.5171	54458	0.0214	V	T. Arranz	0.0002
EP Aur	58508.7014	54482	0.0215	TG	G. Conrad	0.0002
HP Aur	58388.8674	10987	0.0713	V	G. Samolyk	0.0001
HP Aur	58425.8590	11013	0.0697	V	G. Samolyk	0.0002
HP Aur	58470.6789	11044.5	0.0710	V	G. Samolyk	0.0003
HP Aur	58508.3843	11071	0.0719	V	T. Arranz	0.0002
IM Aur	58456.5207	14384	–0.1315	V	T. Arranz	0.0001
IM Aur	58466.4967	14392	–0.1338	V	T. Arranz	0.0001
IM Aur	58486.4535	14408	–0.1338	V	T. Arranz	0.0001
IM Aur	58491.4407	14412	–0.1358	V	T. Arranz	0.0002
IM Aur	58496.4312	14416	–0.1344	V	T. Arranz	0.0001
V449 Aur	57405.8179	12656.5	–0.1700	V	V. Petriew	0.0003
V449 Aur	57701.0105	13076	–0.1577	V	V. Petriew	0.0007
V449 Aur	57702.7693	13078.5	–0.1581	V	V. Petriew	0.0003
V449 Aur	57703.8280	13080	–0.1548	V	V. Petriew	0.0001
V449 Aur	57709.8076	13088.5	–0.1562	V	V. Petriew	0.0002
V599 Aur	57415.6940	15529.5	–0.0016	V	V. Petriew	0.0004
TU Boo	58486.9244	78389.5	–0.1597	V	G. Samolyk	0.0002
TU Boo	58498.9219	78426.5	–0.1608	V	K. Menzies	0.0001
TY Boo	58487.9365	75698	0.0649	V	G. Samolyk	0.0001
TZ Boo	58486.9403	63447	0.0611	V	G. Samolyk	0.0002
VW Boo	58486.9145	79789	–0.2849	V	G. Samolyk	0.0002
VW Boo	58499.0659	79824.5	–0.2860	V	R. Sabo	0.0001
XY Boo	58244.6952	49359	0.0147	C	G. Frey	0.0005
DN Boo	58272.6992	7536	0.0031	C	G. Frey	0.0003
DU Boo	57475.9347	4712.5	0.0228	V	V. Petriew	0.0005
GM Boo	58247.7039	17297	0.0286	C	G. Frey	0.0002
HH Boo	57531.7565	15789	–0.0027	V	V. Petriew	0.0002
SV Cam	58407.6983	26663	0.0590	V	G. Samolyk	0.0002
CD Cam	58470.7591	7469.5	–0.0138	V	G. Samolyk	0.0003
CD Cam	58493.6816	7499.5	–0.0169	V	G. Samolyk	0.0004
AH Cnc	58499.5737	16643.5	0.0208	V	T. Arranz	0.0002
R CMa	58488.7479	12500	0.1307	V	N. Simmons	0.0002
RT CMa	58487.7903	24628	–0.7735	V	G. Samolyk	0.0001
RT CMa	58512.3709	24647	–0.7746	V	T. Arranz	0.0001
SX CMa	58504.7882	18722	0.0228	V	G. Samolyk	0.0001
TZ CMa	58462.8004	16395	–0.2278	V	G. Samolyk	0.0002
UU CMa	58463.8395	6400	–0.0714	V	G. Samolyk	0.0001
AC CMi	58144.6650	7110	0.0039	C	G. Frey	0.0001
AC CMi	58462.9337	7477	0.0041	V	G. Samolyk	0.0002
AK CMi	58462.9345	27145	–0.0251	V	G. Samolyk	0.0001
AK CMi	58510.4715	27229	–0.0235	V	T. Arranz	0.0001

Table continued on following pages

Table 1. Recent times of minima of stars in the AAVSO eclipsing binary program, cont.

Star	JD (min) Hel. 2400000 +	Cycle	O–C (day)	F	Observer	Standard Error (day)
CW CMi	58152.6516	18921.5	–0.0500	C	G. Frey	0.0002
CZ CMi	58136.6738	13219	–0.0130	C	G. Frey	0.0001
RW Cap	58336.8686	4620	–0.7639	V	S. Cook	0.0007
TY Cap	58401.7264	9560	0.0943	V	S. Cook	0.0003
RZ Cas	58463.6905	12770	0.0800	V	G. Samolyk	0.0001
TV Cas	58394.4629	7609	–0.0304	V	T. Arranz	0.0001
TV Cas	58463.3435	7647	–0.0285	V	T. Arranz	0.0001
TW Cas	58415.5607	11487	0.0156	V	G. Samolyk	0.0002
ZZ Cas	58367.7498	20048	0.0255	V	S. Cook	0.0004
AB Cas	58385.8125	11465	0.1417	V	G. Samolyk	0.0001
AB Cas	58396.7468	11473	0.1410	V	S. Cook	0.0003
AB Cas	58470.5582	11527	0.1412	V	G. Samolyk	0.0001
CW Cas	58382.6331	52532	–0.1208	V	T. Arranz	0.0001
CW Cas	58385.3461	52540.5	–0.1181	V	T. Arranz	0.0001
CW Cas	58422.6519	52657.5	–0.1194	V	K. Menzies	0.0001
CW Cas	58456.6095	52764	–0.1209	TG	G. Conrad	0.0002
DO Cas	57653.9075	34655.5	0.0042	V	V. Petriew	0.0006
DO Cas	57698.7515	34721	0.0025	V	V. Petriew	0.0001
DZ Cas	58375.6030	38147	–0.2129	V	G. Samolyk	0.0003
DZ Cas	58394.4424	38171	–0.2109	V	T. Arranz	0.0001
GT Cas	58372.8063	10379	0.2056	V	G. Samolyk	0.0004
IR Cas	58412.5881	23577	0.0144	V	G. Samolyk	0.0001
IS Cas	58391.5355	16082	0.0701	V	T. Arranz	0.0001
IS Cas	58393.3788	16083	0.0719	V	T. Arranz	0.0001
IS Cas	58413.6357	16094	0.0721	V	N. Simmons	0.0001
IT Cas	58384.7470	7602	0.0709	V	G. Samolyk	0.0002
IV Cas	58415.5149	17587	–0.1325	V	G. Samolyk	0.0001
MM Cas	58377.5392	19833	0.1207	V	T. Arranz	0.0001
MM Cas	58399.5468	19852	0.1174	V	T. Arranz	0.0002
MM Cas	58413.4501	19864	0.1190	V	T. Arranz	0.0001
OR Cas	58376.5881	11372	–0.0332	V	T. Arranz	0.0001
OR Cas	58381.5705	11376	–0.0337	V	T. Arranz	0.0001
OR Cas	58416.4509	11404	–0.0332	V	T. Arranz	0.0001
OR Cas	58417.6966	11405	–0.0332	V	S. Cook	0.0008
OR Cas	58461.2977	11440	–0.0320	V	T. Arranz	0.0001
OX Cas	58463.6037	6907	0.0782	V	G. Samolyk	0.0002
OX Cas	58488.4917	6917	0.0727	V	G. Samolyk	0.0004
PV Cas	58375.4003	10367.5	–0.0004	V	T. Arranz	0.0001
V364 Cas	58387.3890	15588.5	–0.0245	V	T. Arranz	0.0001
V364 Cas	58413.6214	15605.5	–0.0243	V	G. Samolyk	0.0001
V364 Cas	58495.4048	15658.5	–0.0235	V	T. Arranz	0.0001
V375 Cas	58376.5343	16108.5	0.2681	V	T. Arranz	0.0002
V375 Cas	58390.5315	16118	0.2682	V	T. Arranz	0.0001
V375 Cas	58393.4784	16120	0.2683	V	T. Arranz	0.0001
V380 Cas	58461.5779	24178	–0.0737	V	N. Simmons	0.0002
V776 Cas	57659.6480	11714.5	–0.0034	V	V. Petriew	0.0006
V776 Cas	57659.8670	11715	–0.0046	V	V. Petriew	0.0005
V776 Cas	57677.7040	11755.5	–0.0045	V	V. Petriew	0.0007
V1261 Cas	58381.5540	13770	0.0080	V	T. Arranz	0.0002
V1261 Cas	58416.3756	13879.5	0.0081	V	T. Arranz	0.0002
U Cep	58375.7471	5549	0.2234	V	G. Samolyk	0.0002
U Cep	58385.7167	5553	0.2208	V	G. Samolyk	0.0001
SU Cep	58372.5339	35552.5	0.0063	V	T. Arranz	0.0001
WY Cep	58361.7326	26611	0.023	V	S. Cook	0.0003
XX Cep	58377.6210	5792	0.0231	V	T. Arranz	0.0001
XX Cep	58384.6340	5795	0.0242	V	N. Simmons	0.0002
ZZ Cep	58412.6721	14233	–0.0183	V	G. Samolyk	0.0004
CQ Cep	58358.7257	15782	–0.1341	V	S. Cook	0.0006
DV Cep	58415.6311	10028	–0.0052	V	G. Samolyk	0.0002
V731 Cep	58377.5895	377	–0.4082	V	T. Arranz	0.0001
SS Cet	58409.7539	5366	0.0697	V	G. Samolyk	0.0001
TT Cet	58372.8778	53147	–0.0823	V	G. Samolyk	0.0001
TW Cet	58388.7820	50545.5	–0.0337	V	G. Samolyk	0.0001
TX Cet	58467.6759	20767	0.0121	V	G. Samolyk	0.0001
RW Com	58487.9404	77800	0.0131	V	G. Samolyk	0.0001
RW Com	58496.0107	77834	0.0136	V	R. Sabo	0.0001
RZ Com	58488.8935	69870	0.0567	V	G. Samolyk	0.0001
SS Com	58507.9306	81165.5	0.9596	V	G. Samolyk	0.0002
CC Com	58493.9253	85915.5	–0.0298	V	G. Samolyk	0.0001
YY CrB	57485.7716	13240	0.0150	V	V. Petriew	0.0002
AE Cyg	58382.6046	14235	–0.0039	V	G. Samolyk	0.0002
AE Cyg	58386.4811	14239	–0.0042	V	T. Arranz	0.0001
AE Cyg	58388.4191	14241	–0.0045	V	T. Arranz	0.0001
AE Cyg	58414.5869	14268	–0.0048	V	G. Samolyk	0.0001
DK Cyg	58380.3763	43299.5	0.1270	V	T. Arranz	0.0001
KR Cyg	58414.6086	34678	0.0249	V	G. Samolyk	0.0001
V387 Cyg	58391.3903	47465	0.0212	V	T. Arranz	0.0001
V387 Cyg	58416.3734	47504	0.0210	V	T. Arranz	0.0001
V387 Cyg	58425.3419	47518	0.0212	V	T. Arranz	0.0001
V456 Cyg	58341.6724	15070	0.0529	V	N. Simmons	0.0002
V456 Cyg	58374.6464	15107	0.0528	V	G. Samolyk	0.0001
V466 Cyg	58371.6195	21269	0.0071	V	G. Samolyk	0.0001
V477 Cyg	58374.4358	6044	–0.0393	V	T. Arranz	0.0001
V704 Cyg	58371.5060	35801.5	0.0377	V	T. Arranz	0.0001
V836 Cyg	58380.4535	20702	0.0238	V	T. Arranz	0.0001
V1425 Cyg	57641.9449	13766.5	0.0153	V	V. Petriew	0.0001
V1425 Cyg	57646.9519	13770.5	0.0127	V	V. Petriew	0.0007
V1425 Cyg	57648.8310	13772	0.0132	V	V. Petriew	0.0002
V1425 Cyg	57692.6643	13807	0.0130	V	V. Petriew	0.0008
V1425 Cyg	57697.6738	13811	0.0129	V	V. Petriew	0.0003
V2181 Cyg	58414.6400	13354	–0.007	V	G. Samolyk	0.0002
W Del	58371.6598	3130	0.0173	V	G. Samolyk	0.0002
TT Del	58390.6432	4583	–0.1142	V	G. Samolyk	0.0002
TY Del	58384.6102	12950	0.0720	V	G. Samolyk	0.0001
YY Del	58371.6864	19433	0.0137	C	G. Frey	0.0002
YY Del	58375.6487	19438	0.0105	V	G. Samolyk	0.0001
DM Del	58376.7455	16427	–0.1352	TG	G. Conrad	0.0001
DM Del	58409.6802	16466	–0.1428	V	S. Cook	0.0005
FZ Del	58398.3955	34568	–0.0267	V	T. Arranz	0.0001
Z Dra	58470.9014	6355	–0.0035	V	G. Samolyk	0.0001
TZ Eri	58415.8534	6140	0.3495	V	G. Samolyk	0.0001
YY Eri	58467.7890	52523.5	0.1670	V	G. Samolyk	0.0001
YY Eri	58498.3306	52618.5	0.1666	V	T. Arranz	0.0001
RW Gem	58462.6010	14015	0.0027	V	T. Arranz	0.0001
RW Gem	58488.3906	14024	0.0029	V	T. Arranz	0.0001
SX Gem	58488.8555	28867	–0.0529	V	G. Samolyk	0.0001
WW Gem	58461.9653	26238	0.0233	V	G. Samolyk	0.0001
WW Gem	58486.7206	26258	0.0224	V	G. Samolyk	0.0001
AF Gem	58489.8231	25193	–0.0696	V	G. Samolyk	0.0001
AF Gem	58498.5283	25200	–0.0689	V	T. Arranz	0.0001
AL Gem	58486.7253	23116	0.0978	V	G. Samolyk	0.0001
EG Gem	58140.6633	24184	0.3162	C	G. Frey	0.0001
V337 Gem	58132.7069	2935	0.0026	C	G. Frey	0.0004
CC Her	58338.7008	10767	0.3184	C	G. Frey	0.0001
CT Her	58337.7198	8853	0.0117	V	S. Cook	0.0008
V772 Her	57545.8225	5737	–0.0026	V	V. Petriew	0.0004
V772 Her	57552.8581	5745	–0.0030	V	V. Petriew	0.0004
V772 Her	57559.8981	5753	0.0009	V	V. Petriew	0.0006
V772 Her	57560.7788	5754	0.0021	V	V. Petriew	0.0004
V972 Her	57515.9164	20347.5	–0.0028	V	V. Petriew	0.0012
V972 Her	57516.8078	20349.5	0.0024	V	V. Petriew	0.0009
V1065 Her	58293.7147	17588	–0.0122	C	G. Frey	0.0002
V1097 Her	58281.7294	16124	0.0092	C	G. Frey	0.0001
WY Hya	58508.7691	25052.5	0.0408	V	G. Samolyk	0.0001
AV Hya	58488.9489	31922	–0.1198	V	G. Samolyk	0.0001
DF Hya	58489.8799	47495	0.0084	V	K. Menzies	0.0005
DK Hya	58467.9361	29936	0.0000	V	G. Samolyk	0.0002
V470 Hya	58508.7770	14687	0.0094	V	G. Samolyk	0.0004
SW Lac	58374.6382	40843.5	–0.0736	V	G. Samolyk	0.0001
SW Lac	58383.4603	40871	–0.0713	V	T. Arranz	0.0001

Table continued on following pages

Table 1. Recent times of minima of stars in the AAVSO eclipsing binary program, cont.

Star	JD (min) Hel. 2400000+	Cycle	O–C (day)	F	Observer	Standard Error (day)
SW Lac	58413.2879	40964	–0.0707	V	L. Corp	0.0002
SW Lac	58413.6075	40965	–0.0719	V	G. Samolyk	0.0001
SW Lac	58458.3480	41104.5	–0.0719	V	T. Arranz	0.0001
VX Lac	58383.5000	12215	0.0873	V	T. Arranz	0.0001
VX Lac	58425.4053	12254	0.0874	V	T. Arranz	0.0001
VX Lac	58470.5336	12296	0.0869	V	G. Samolyk	0.0001
AW Lac	58371.6627	27779	0.2161	V	G. Samolyk	0.0002
CO Lac	58376.6901	19999	0.0095	V	G. Samolyk	0.0001
CO Lac	58387.4861	20006	0.0101	V	T. Arranz	0.0001
CO Lac	58410.6190	20021	0.0098	V	G. Samolyk	0.0001
CO Lac	58438.3792	20039	0.0103	V	T. Arranz	0.0001
GX Lac	58376.6122	2931	–0.0413	V	G. Samolyk	0.0002
Y Leo	58459.8357	7724	–0.0671	V	G. Samolyk	0.0001
UV Leo	58493.8052	33417	0.0458	V	G. Samolyk	0.0001
VZ Leo	58507.8320	25088	–0.0457	V	G. Samolyk	0.0003
CE Leo	58218.6640	34733	–0.0086	C	G. Frey	0.0001
HI Leo	58219.6799	17584	0.0189	C	G. Frey	0.0001
T LMi	58409.9194	4309	–0.1331	V	G. Samolyk	0.0001
VW LMi	57481.7411	18807.5	0.0183	V	V. Petriew	0.0002
Z Lep	58459.8174	31232	–0.2005	V	G. Samolyk	0.0001
RR Lep	58470.6861	30689	–0.0458	V	G. Samolyk	0.0002
RR Lep	58494.4894	30715	–0.0436	V	T. Arranz	0.0002
RR Lep	58495.4012	30716	–0.0472	V	T. Arranz	0.0001
VZ Lib	58335.6916	37813.5	–0.0893	V	S. Cook	0.0004
RY Lyn	58462.7855	10854	–0.0173	V	G. Samolyk	0.0002
DE Lyn	58162.7323	13851	–0.0204	V	K. Menzies	0.0001
DI Lyn	57477.6736	1716.5	0.0109	V	V. Petriew	0.0008
HN Lyn	57444.6862	11658.5	0.0029	V	V. Petriew	0.0015
Beta Lyr	58295.32	719	2.48	B	G. Samolyk	0.01
Beta Lyr	58295.33	719	2.48	V	G. Samolyk	0.01
Beta Lyr	58295.33	719	2.49	R	G. Samolyk	0.02
Beta Lyr	58301.75	719.5	2.43	V	G. Samolyk	0.02
Beta Lyr	58301.78	719.5	2.47	B	G. Samolyk	0.02
Beta Lyr	58301.79	719.5	2.47	R	G. Samolyk	0.02
AT Mon	58450.9512	15694	0.0113	V	G. Samolyk	0.0001
BB Mon	58488.7463	43411	–0.0041	V	G. Samolyk	0.0001
BO Mon	58507.7881	6741	–0.0122	V	G. Samolyk	0.0001
V456 Oph	58322.7061	16166.5	0.0166	C	G. Frey	0.0001
V456 Oph	58357.7681	16201	0.0266	V	S. Cook	0.0004
V501 Oph	58324.7090	28321	–0.0093	C	G. Frey	0.0001
V502 Oph	58357.6574	21742	–0.0017	V	S. Cook	0.0005
V508 Oph	58375.6316	38554	–0.0271	V	N. Rivard	0.0002
V508 Oph	58384.5969	38580	–0.0264	V	G. Samolyk	0.0001
V566 Oph	58372.6909	40368	0.2520	V	S. Cook	0.0004
V839 Oph	58371.7305	43822	0.3247	V	S. Cook	0.0005
EF Ori	58509.3619	3803	0.0114	V	T. Arranz	0.0001
EQ Ori	58497.3478	15497	–0.0405	V	T. Arranz	0.0001
ER Ori	57760.5799	38106	0.1340	V	N. Rivard	0.0002
ER Ori	58489.6827	39828	0.1443	V	G. Samolyk	0.0001
ER Ori	58496.4574	39844	0.1446	V	T. Arranz	0.0001
FL Ori	58468.7212	8460	0.0425	V	G. Samolyk	0.0001
FR Ori	58470.8361	34658	0.0426	V	G. Samolyk	0.0002
FT Ori	58461.7597	5432	0.0223	V	G. Samolyk	0.0001
FZ Ori	58154.7550	35327	–0.0299	C	G. Frey	0.0001
FZ Ori	58489.7458	36164.5	–0.0279	V	G. Samolyk	0.0001
FZ Ori	58509.3457	36213.5	–0.0273	V	T. Arranz	0.0002
FZ Ori	58510.3464	36216	–0.0266	V	T. Arranz	0.0001
GU Ori	58463.6938	32705.5	–0.0666	V	G. Samolyk	0.0002
GU Ori	58509.3497	32802.5	–0.0668	V	T. Arranz	0.0001
V1853 Ori	58133.6551	10619	0.0003	C	G. Frey	0.0001
U Peg	58369.5018	58322	–0.1695	V	L. Corp	0.0001
U Peg	58375.8725	58339	–0.1701	V	G. Samolyk	0.0001
U Peg	58410.7273	58432	–0.1700	V	G. Samolyk	0.0001
U Peg	58463.5710	58573	–0.1705	V	G. Samolyk	0.0001
TY Peg	58373.8410	5796	–0.4501	V	G. Samolyk	0.0001
UX Peg	58384.7357	11627	–0.0052	V	G. Samolyk	0.0001
AT Peg	58375.6965	11479	0.0293	C	G. Frey	0.0001
AW Peg	58368.7284	1377	0.0206	V	S. Cook	0.0009
BB Peg	58411.6443	40518	–0.0312	V	G. Samolyk	0.0001
BB Peg	58422.4884	40548	–0.0322	V	K. Menzies	0.0004
BB Peg	58459.5448	40650.5	–0.0297	V	G. Samolyk	0.0002
BG Peg	58390.7552	6586	–2.3767	V	G. Samolyk	0.0003
BO Peg	58370.6995	21981	–0.0575	C	G. Frey	0.0001
BX Peg	58385.6591	50604	–0.1322	V	G. Samolyk	0.0001
BX Peg	58412.5787	50700	–0.1330	V	G. Samolyk	0.0001
DI Peg	58405.6837	18557	0.0113	V	G. Samolyk	0.0001
GP Peg	58468.6591	17661	–0.0566	V	G. Samolyk	0.0001
KV Peg	58372.7166	23177	–0.0175	C	G. Frey	0.0003
KW Peg	58385.6326	12527	0.2227	V	G. Samolyk	0.0002
KW Peg	58412.5717	12560	0.2212	V	G. Samolyk	0.0002
Z Per	58504.5628	4203	–0.3388	V	G. Samolyk	0.0001
RT Per	58379.9687	29437	0.1132	V	R. Sabo	0.0001
RT Per	58488.6933	29565	0.1145	V	G. Samolyk	0.0001
ST Per	58461.7831	6051	0.3210	V	G. Samolyk	0.0001
XZ Per	58411.8518	12942	–0.0744	V	G. Samolyk	0.0001
XZ Per	58470.5854	12993	–0.0741	V	G. Samolyk	0.0001
BR Per	57392.7204	4551	0.0001	V	V. Petriew	0.0002
BR Per	57695.8817	4833	0.0006	V	V. Petriew	0.0001
IT Per	58414.7995	18929	–0.0420	V	G. Samolyk	0.0002
IU Per	58467.6783	15001	0.0088	V	G. Samolyk	0.0002
IU Per	58468.5347	15002	0.0081	V	N. Simmons	0.0001
KW Per	58468.5661	17250	0.0191	V	G. Samolyk	0.0001
V432 Per	58414.8687	70106.5	0.0611	V	G. Samolyk	0.0001
V432 Per	58468.5312	70273.5	0.0303	V	G. Samolyk	0.0002
V432 Per	58489.6146	70339	0.0543	V	G. Samolyk	0.0001
Beta Per	58409.7612	4453	0.1417	V	G. Samolyk	0.0001
Y Psc	58374.8054	3383	–0.0254	V	G. Samolyk	0.0001
RV Psc	58384.8570	61379	–0.0642	V	G. Samolyk	0.0002
RV Psc	58468.5081	61530	–0.0658	V	G. Samolyk	0.0002
RV Psc	58489.5606	61568	–0.0650	V	G. Samolyk	0.0001
VZ Psc	58363.4278	55620	0.0101	V	L. Corp	0.0003
VZ Psc	58461.5231	55995.5	0.0025	V	G. Samolyk	0.0002
VZ Psc	58461.6561	55996	0.0050	V	G. Samolyk	0.0002
VZ Psc	58480.3345	56067.5	0.0033	V	T. Arranz	0.0001
UZ Pup	58489.8017	17457.5	–0.0114	V	G. Samolyk	0.0001
AO Ser	58350.6799	27540	–0.0129	V	S. Cook	0.0002
RW Tau	58467.6286	4617	–0.2894	V	G. Samolyk	0.0001
RZ Tau	58413.8377	49888	0.0913	V	G. Samolyk	0.0001
RZ Tau	58463.7185	50008	0.0911	V	G. Samolyk	0.0001
RZ Tau	58468.7074	50020	0.0919	V	N. Simmons	0.0001
RZ Tau	58493.6481	50080	0.0921	V	G. Samolyk	0.0001
TY Tau	58409.9148	34545	0.2741	V	G. Samolyk	0.0001
TY Tau	58489.6407	34619	0.2756	V	G. Samolyk	0.0002
WY Tau	58478.8052	30424	0.0656	V	K. Menzies	0.0001
AC Tau	58411.8299	6252	0.1782	V	G. Samolyk	0.0002
AM Tau	58487.7616	6475	–0.0762	V	G. Samolyk	0.0001
AQ Tau	58385.8983	23629	0.5287	V	G. Samolyk	0.0003
CT Tau	58414.8156	19511	–0.0694	V	G. Samolyk	0.0001
EQ Tau	57697.8430	51222	–0.0338	V	V. Petriew	0.0001
EQ Tau	57698.0136	51222.5	–0.0339	V	V. Petriew	0.0003
EQ Tau	58406.8179	53299	–0.0397	V	R. Sabo	0.0001
EQ Tau	58425.7623	53354.5	–0.0402	V	G. Samolyk	0.0001
EQ Tau	58488.5709	53538.5	–0.0397	V	G. Samolyk	0.0001
V Tri	58409.7923	57989	–0.0060	V	G. Samolyk	0.0001
V Tri	58470.6536	58093	–0.0061	V	N. Simmons	0.0001
X Tri	58450.3725	16415	–0.0988	V	T. Arranz	0.0001
X Tri	58451.3441	16416	–0.0987	V	T. Arranz	0.0001
X Tri	58483.4042	16449	–0.0993	V	T. Arranz	0.0001
X Tri	58484.3757	16450	–0.0993	V	T. Arranz	0.0001
X Tri	58485.3473	16451	–0.0993	V	T. Arranz	0.0001

Table continued on next page

Table 1. Recent times of minima of stars in the AAVSO eclipsing binary program, cont.

Star	JD (min) Hel. 2400000 +	Cycle	O–C (day)	F	Observer	Standard Error (day)
X Tri	58486.3188	16452	–0.0993	V	T. Arranz	0.0001
RS Tri	58407.9074	10722	–0.0593	V	G. Samolyk	0.0001
RV Tri	58388.8748	16394	–0.0415	V	G. Samolyk	0.0001
RV Tri	58422.7896	16439	–0.0417	V	K. Menzies	0.0001
TY UMa	57831.6868	51613	0.3895	V	N. Rivard	0.0004
TY UMa	58463.8610	53396	0.4214	V	K. Menzies	0.0001
TY UMa	58504.8125	53511.5	0.4237	V	G. Samolyk	0.0002
UX UMa	57991.6549	104534	–0.0011	V	N. Rivard	0.0001
VV UMa	58425.9284	18346	–0.0816	V	G. Samolyk	0.0001
XZ UMa	58415.9234	10020	–0.1490	V	N. Simmons	0.0001
XZ UMa	58470.9261	10065	–0.1507	V	G. Samolyk	0.0001
EQ UMa	57400.7189	24274.5	0.0013	V	V. Petriew	0.0007
EQ UMa	57400.8996	24275	0.0029	V	V. Petriew	0.0007
HX UMa	57477.8377	13128	0.1641	V	V. Petriew	0.0009
II UMa	57478.9404	10880.5	0.0177	V	V. Petriew	0.0012
II UMa	57510.7117	10919	0.0178	V	V. Petriew	0.0027
KM UMa	57443.8348	14050	–0.0059	V	V. Petriew	0.0003
AG Vir	58223.7181	19904	–0.0170	C	G. Frey	0.0002
AH Vir	58496.9397	31120.5	0.2974	V	R. Sabo	0.0002
AW Vir	58242.6907	37345	0.0296	C	G. Frey	0.0001
AZ Vir	58246.6992	40811	–0.0236	C	G. Frey	0.0002
BF Vir	58255.7286	19022	0.1221	C	G. Frey	0.0001
GR Vir	58263.7133	37891.5	0.0199	C	G. Frey	0.0002
IR Vir	58226.6805	22743.5	–0.0075	C	G. Frey	0.0001
NN Vir	58251.7341	20286	0.0097	C	G. Frey	0.0002
V391 Vir	58232.7304	19680.5	0.0064	C	G. Frey	0.0002
RS Vul	58375.7324	5710	0.0169	V	S. Cook	0.0008
AW Vul	58388.6531	15008	–0.0347	V	G. Samolyk	0.0002
AY Vul	58409.5385	6518	–0.1667	V	G. Samolyk	0.0002
BE Vul	58411.6398	11791	0.1080	V	G. Samolyk	0.0001
BO Vul	58376.6403	11505	–0.0135	V	G. Samolyk	0.0001
BO Vul	58411.6661	11523	–0.0134	V	G. Samolyk	0.0001
BS Vul	58373.6428	31729	–0.0340	V	G. Samolyk	0.0001
BS Vul	58413.6236	31813	–0.0348	V	G. Samolyk	0.0004
BU Vul	58373.6587	43656	0.0173	V	S. Cook	0.0003
BU Vul	58385.6052	43677	0.0149	V	G. Samolyk	0.0001

Radial Velocities for Four δ Sct Variable Stars

Elizabeth J. Jeffery
Physics Department, California Polytechnic State University, San Luis Obispo, CA 93407; ejjeffer@calpoly.edu

Thomas G. Barnes, III
The University of Texas at Austin, McDonald Observatory, 1 University Station, C1402, Austin, TX 78712-0259; tgb@astro.as.utexas.edu

Ian Skillen
Isaac Newton Group, Apartado de Correos 321, 38700 Santa Cruz de La Palma, Canary Islands, Spain; wji@ing.iac.e

Thomas J. Montemayor
The University of Texas at Austin, McDonald Observatory, 1 University Station, C1402, Austin, TX 78712-0259; tjm@texas.net

Received February 27, 2019; revised April 23, 2019; accepted May 6, 2019

Abstract We report 124 radial velocities for the δ Sct variable stars VZ Cnc, AD CMi, DE Lac, and V474 Mon, covering full pulsation cycles. Although we also obtained data on KZ Hya, the signal-to-noise ratios of those spectra were too low to determine meaningful velocities. We confirm the multi-periodic behavior of both VZ Cnc and V474 Mon. However, our data do not cover a long enough baseline to address the multi-periodic nature of AD CMi claimed by Derekas *et al.* (2009). Our velocities, added to those in the literature for DE Lac, suggest discordant center-of-mass velocities, but the data are few.

1. Introduction

Delta Scuti variable stars are intrinsic pulsators of spectral type A to F, with short pulsation periods (usually less than 0.3 day). They lie on the H-R Diagram at the intersection of the Main Sequence and the classical instability strip. These stars can be either Population I or Population II, the latter also known as SX Phoenicis stars. Many δ Scuti stars are radial pulsators, while others pulsate nonradially. Breger (2000) points out that these stars represent a transition between the large amplitude radial pulsation of the classical instability strip (e.g., Cepheids) and the nonradial pulsations that occur in the hot side of the H-R Diagram.

Precise radial velocities are useful in several aspects of δ Scuti studies. For example, radial velocities are needed when employing the Baade-Wesselink method for determining distances, which has been applied to δ Scutis by Guiglion *et al.* (2013) and Burki and Meylan (1986), among others.

In this paper we present new radial velocity observations of four δ Scuti stars: VZ Cnc, AD CMi, DE Lac, and V474 Mon. We attempted velocities for the Pop II star KZ Hya, but were unsuccessful due to the required short integration time of the observations and the low metallicity of the star. For each star we have radial velocities for the complete pulsation cycle. Basic information about each star is given in Table 1, taken from the *General Catalogue of Variable Stars* (GCVS; Samus *et al.* 2017).

2. Observations

The observations were taken in 1996 (DE Lac) and 2003 (VZ Cnc, AD CMi, KZ Hya, V474 Mon, see Table 1) at the McDonald Observatory 2.1-meter Struve telescope at the f/13.5 Cassegrain focus with the Sandiford Échelle Spectrometer (McCarthy *et al.* 1993). All observations were taken at resolving power R = 60,000 per 2 pixels with a 1 arcsecond slit fixed in orientation. Two wavelength setups were used: the first, corresponding to observations taken 25 Jul 1996–26 Sep 1996, covered 5500–6900 Å in 27–28 orders (hereafter, the red region); the second set, taken 13 Dec 2003–18 Dec 2003, was observed using a bluer setup, 4280–4750 Å in 18 orders (hereafter, the blue region). Immediately following each stellar observation and prior to a telescope move, a Th-Ar emission spectrum was obtained for wavelength calibration. Internal flat field and bias frames were observed nightly for calibration of the Reticon 1200 × 400 CCD.

Each night we observed a spectrum of at least one radial velocity standard star chosen from the Geneva list of CORAVEL

Table 1. Observed δ Scuti stars.

Star	*R.A. (2000)* h m s	*Dec. (2000)* ° ′ ″	*V* (mag.)	*Spectral Class*	*Integration* (seconds)	*Principal Period* (days)
VZ Cnc	08 40 52.1	+09 49 27	7.18–7.91	A7 III-F2 III	240	0.17836356
AD CMi	07 52 47.2	+01 35 50	9.21–9.51	F0 III-F3 III	210	0.122974458
KZ Hya	10 50 54.1	–25 21 15	9.46–10.26	A0	120,180	0.059510388
DE Lac	22 10 07.8	+40 55 11	10.08–10.43	F5-F8	see Table 5	0.253692580
V474 Mon	05 59 01.1	–09 22 56	5.93–6.36	F2 IV	240	0.1361258

Table 2. Adopted radial velocity standard stars.

Star	R.A. (2000) h m s	Dec. (2000) ° ′ ″	V (mag.)	Spectral Class	Radial Velocity (km s^{-1})
HD65934	08 02 11	+26 38 16	7.70	G8 III	35.8
HD102870	11 50 42	+01 45 53	3.59	F8 V	4.3
HD136202	15 19 19	+01 45 55	5.04	F8 III-IV	54.3
HD187691	19 51 02	+10 24 57	5.12	F8 V	–0.0
HD222368	23 39 57	+05 37 35	4.13	F7 V	5.6

standard stars (http://obswww.unige.ch/~udry/std/stdcor.dat) to set the velocity zero point. The standard stars used for these observations are listed in Table 2.

3. Data reduction

The CCD frames were reduced using standard IRAF procedures (Tody 1993) (IRAF is distributed by the National Optical Astronomy Observatories, which are operated by the Association of Universities for Research in Astronomy, Inc., under cooperative agreement with the National Science Foundation.). The IRAF task FXCOR determines a velocity for every échelle order. Because of instrumental effects or telluric features, seven of the red region orders always produced erroneous velocities; these were rejected for all spectra. For each spectrum (blue and red regions) we rejected any anomalous velocities in the remaining orders using an iterative Chauvenet's criterion (Chauvenet 1864; Taylor 1997). To apply Chauvenet's criterion we computed the mean and standard deviation of the ensemble of velocities in a spectrum. We determined the distance of each velocity from the mean in units of the standard deviation and computed the probability that this deviation would occur for a sample of our size. If this probability is less than 50%, that value is rejected. The criterion is iterated until no rejections occur. The mean and standard deviation of the velocities from the remaining orders were then computed. For the red region (DE Lac only), the typical number of orders retained was 9, and for the blue region, 17. The red region had few lines for these hot stars, hence the paucity of orders retained.

A more extensive description of our reduction process from raw CCD frames to radial velocities is given in Barnes *et al.* (2005, Paper I).

The median standard-error-of-the-mean in an individual radial velocity is ± 0.05 km s^{-1} for the red region and ± 0.22 km s^{-1} for the blue region, computed from the scatter in the velocities from the multiple échelle orders. This internal scatter underestimates the actual uncertainty in the measured radial velocity of the star. The order-to-order scatter does not include uncertainty due to error in the adopted radial velocities of the standard stars, uncertainty due to motion of the image in the spectrograph slit, uncertainty in the wavelength scale, and uncertainty due to potential differences in the velocity scale between the δ Sct variable stars and the standards used for the velocity zero point.

As in Papers I and II (Jeffery *et al.* 2007) we inferred the likely uncertainty in our observations by looking at the scatter in the radial velocity pulsation curves. For this paper we used AD CMi as it has numerous observations and is not affected by multi-mode behavior. We fit Fourier Series of successively greater order to the velocities given in Table 4. The scatter in fits of order 4–6 was stable at the ± 0.30 km s^{-1} level. We interpret this as a reasonable estimate of the uncertainties in the radial velocities presented here. In Papers I and II we obtained estimated uncertainties of ± 0.40 km s^{-1} (Cepheids) and ± 0.50 km s^{-1} (RR Lyrae) from the same instrumentation and reduction process.

4. Radial velocity results

We list individual radial velocities in Tables 3–6 together with the mid-point Heliocentric Julian Dates and pulsation phases. The data we obtained for KZ Hya were too low in signal-to-noise, given the short integration time required for its pulsation period (Kim *et al.* 2007) and the low metallicity of the star ([Me/H] = -0.99 ± 0.1, Fu *et al.* 2008), to obtain radial velocities. McNamara and Budge (1985) and Przybylski and Bessell (1979) both observed spectra of KZ Hya, but note that no lines except those of H and Ca II H and K were visible. The latter are too blue for our observations. Although we see Hγ on some (but not all) of the spectra, the low SNR of our observations prevents reliable radial velocities from being measured. We therefore disregard this star in our remaining analysis.

With the exception of DE Lac (see below) integration times were set not to exceed 2% of the pulsation period to avoid smearing the velocities.

4.1. VZ Cnc

VZ Cnc is a double-mode δ Sct. Boonyarak *et al.* (2009) did a study of times of maximum for VZ Cnc to determine its period change, correcting for the double-mode behavior. They deduced, based on 194 times of maximum, that the period is increasing at the rate of 1.4×10^{-8} per year. We inverted their quadratic ephemeris of times of maximum light to compute the phases of our observations, where phase is the fractional part of the epoch, E:

$$HJD_{obs} = 2431550.7197 (\pm 0.002) + 0.17836356 (\pm 0.00000002)E + 0.7 (\pm 0.15) \times 10^{-8}E^2. \quad (1)$$

Our radial velocities are presented in Table 3 and shown in Figure 1. We have used different symbols to represent the different dates of observation: triangles, dots, and crosses represent (UT) 15 Dec, 16 Dec, and 17 Dec 2003, respectively. Its multimodal behavior is clearly seen in this figure. Uncertainties in the velocities are ± 0.30 km/s.

4.2. AD CMi

Boonyarak *et al.* (2011) analyzed times of maximum for AD CMi based on 100 maxima. They found a very small quadratic variation in the times of maximum, too small to be of significance across the dates of our observations, nonetheless we adopted their quadratic period variation for computation of the phases, as described above, to be consistent with previous studies:

$$HJD_{obs} = 2436601.8215 + 0.122974458E + 2.7 \times 10^{-13}E^2. \quad (2)$$

Table 3. Radial velocities of VZ Cnc.

UT Date	*HJD–2450000 (days)*	*Phase*	V_r *(km s⁻¹)*
15 Dec 2003	2988.8682	0.8688	22.93
	2988.8752	0.9077	27.00
	2988.8795	0.9316	29.20
16 Dec 2003	2989.8343	0.2349	30.44
	2989.8376	0.2532	22.65
	2989.8429	0.2827	11.69
	2989.8429	0.2827	11.69
	2989.8517	0.3315	1.89
	2989.8550	0.3499	1.10
	2989.8604	0.3799	0.52
	2989.8637	0.3982	1.16
16 Dec 2003	2989.8687	0.426	3.26
	2989.8721	0.4448	4.44
	2989.8779	0.4771	7.02
	2989.8823	0.5015	9.56
	2989.8873	0.5293	12.33
	2989.8910	0.5498	14.56
	2989.8963	0.5793	16.91
	2989.8996	0.5976	19.43
	2989.9059	0.6326	22.89
	2989.9376	0.8087	35.96
	2989.9432	0.8398	39.01
	2989.9466	0.8586	39.67
	2989.9525	0.8914	41.07
	2989.9561	0.9114	42.45
	2989.9615	0.9414	42.81
	2989.9649	0.9603	43.04
	2989.9706	0.9920	43.91
	2989.9741	0.0114	43.25
	2989.9791	0.0392	43.08
	2989.9825	0.0581	42.94
	2989.9877	0.0869	43.02
	2989.9912	0.1064	42.63
	2989.9964	0.1353	40.12
	2989.9998	0.1541	38.29
	2990.0049	0.1825	35.28
	2990.0083	0.2014	32.18
	2990.0133	0.2291	24.75
17 Dec 2003	2990.9403	0.3780	10.25
	2990.9457	0.4080	9.92
	2990.9492	0.4275	9.92
	2990.9549	0.4591	10.53
	2990.9583	0.4780	9.62
	2990.9641	0.5102	9.94
	2990.9678	0.5308	10.08
	2990.9732	0.5608	10.69
	2990.9765	0.5791	11.42
17 Dec 2003	2990.9821	0.6102	12.51
	2990.9854	0.6285	13.23
	2990.9907	0.6580	14.57
	2990.9940	0.6763	15.62

Note: Phases are calculated from the ephemeris HJD_{obs} *= 2431550.7197 + 0.17836356E + 0.7 × 10⁻⁸E².*

Table 4. Radial velocities of AD CMi.

UT Date	*HJD–2450000 (days)*	*Phase*	V_r *(km s⁻¹)*
14 Dec 2003	2987.8613	0.4862	42.33
	2987.8667	0.5301	44.47
	2987.8696	0.5537	45.81
	2987.8758	0.6042	48.28
	2987.8800	0.6383	49.34
	2987.8853	0.6814	50.36
	2987.8913	0.7302	50.18
	2987.8968	0.775	48.53
	2987.9001	0.8018	47.34
	2987.9076	0.8627	38.70
	2987.9106	0.8872	35.42
15 Dec 2003	2988.8902	0.853	39.93
	2988.8932	0.8774	36.93
	2988.8983	0.9189	31.38
	2988.9013	0.9433	28.49
	2988.9067	0.9872	25.91
	2988.9097	0.0116	25.67
	2988.9150	0.0547	25.78
	2988.9180	0.0791	26.25
	2988.9228	0.1181	27.32
	2988.9259	0.1433	28.70
	2988.9308	0.1832	30.28
	2988.9354	0.2206	31.34
	2988.9405	0.262	32.24
	2988.9435	0.2865	33.68
	2988.9486	0.3279	35.72
	2988.9516	0.3524	35.82
	2988.9568	0.3946	37.06
	2988.9599	0.4198	38.87
	2988.9648	0.4597	41.09
	2988.9678	0.484	41.96
	2988.9728	0.5247	45.41
	2988.9759	0.5499	45.20
17 Dec 2003	2990.9030	0.2207	31.52
	2990.9078	0.2597	33.03
	2990.9110	0.2857	33.87
	2990.9168	0.3329	35.21
	2990.9199	0.3581	35.80

Note: Phases are calculated from the ephemeris HJD_{obs} *= 2436601.8215 + 0.122974458E + 2.7 × 10⁻¹³E².*

Table 5. Radial Velocities of DE Lac.

UT Date	*HJD-2450000 (days)*	*Integration (seconds)*	*Phase*	*Vr (km s⁻¹)*
25 Jul 1996	289.8800	1200	0.8475	–0.46
29 Jul 1996	293.8169	1200	0.3658	–9.01
21 Aug 1996	316.7488	900	0.7575	9.12
22 Aug 1996	317.8427	900	0.0694	–23.15
26 Sep 1996	352.6953	1200	0.4494	–4.68

Note: Phases are calculated from the ephemeris HJD_{obs} *= 2428807.1385 + 0.253692580E + 1:29 × 10⁻¹¹E².*

Table 6. Radial velocities of V474 Mon

UT Date	HJD–2450000 (days)	Phase	V_r (km s^{-1})
13 Dec 2003	2986.8627	0.7105	21.49
	2986.8727	0.7839	21.56
	2986.8762	0.8096	21.67
	2986.8823	0.8544	21.44
	2986.8859	0.8809	20.51
	2986.8921	0.9264	23.23
	2986.8955	0.9514	22.56
	2986.9015	0.9955	19.70
	2986.9049	0.0205	16.49
	2986.9109	0.0645	15.91
	2986.9149	0.0939	13.84
	2986.9207	0.1365	11.51
	2986.9241	0.1615	10.91
	2986.9299	0.2041	10.50
	2986.9333	0.2291	9.99
	2986.9402	0.2798	9.23
	2986.9435	0.3040	13.48
	2986.9499	0.3510	12.78
	2986.9533	0.3760	12.14
	2986.9593	0.4201	15.05
	2986.9627	0.4451	15.15
	2986.9688	0.4899	16.74
	2986.9721	0.5141	19.52
14 Dec 2003	2987.9205	0.4812	18.08
	2987.9259	0.5209	18.96
	2987.9303	0.5532	20.16
	2987.9362	0.5966	20.96
	2987.9418	0.6377	21.61
	2987.9469	0.6752	22.10
	2987.9502	0.6994	22.25

Note: Phases are calculated from the ephemeris HJD_{obs} *= 2451500.000 + 0.1361258E.*

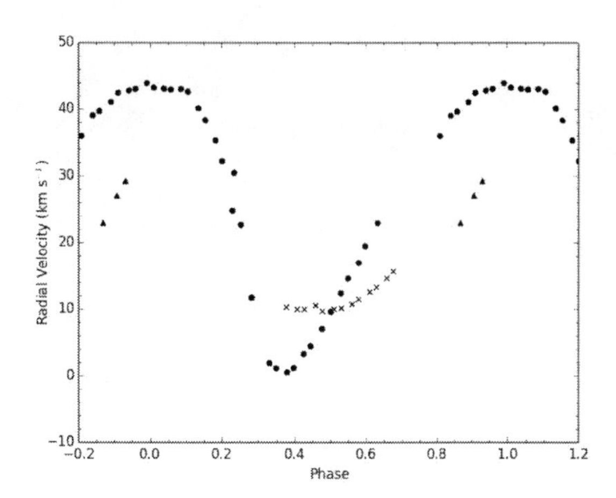

Figure 1. Phased radial velocity curve of VZ Cnc; phases are calculated from the ephemeris $HJD_{obs} = 2431550.7197 + 0.17836356E + 0.7 \times 10^{-8}E^2$. Triangles, dots, and crosses represent UT dates 15 Dec 2003, 16 Dec 2003, and 17 Dec 2003, respectively.

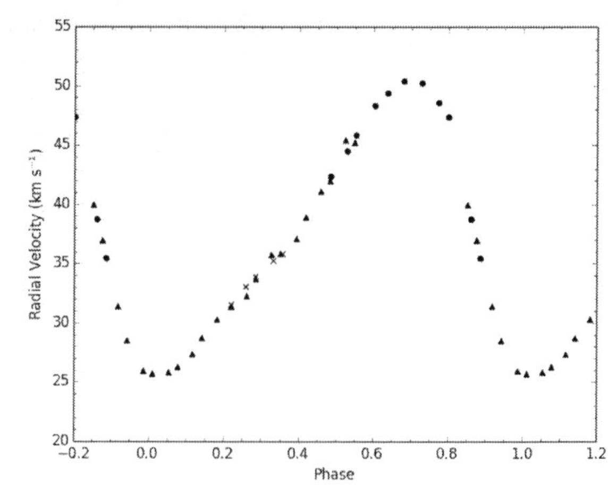

Figure 2. Phased radial velocity curve of AD CMi; phases are calculated from the ephemeris $HJD_{obs} = 2436601.8215 + 0.122974458E + 2.7 \times 10^{-13}E^2$. Dots, triangles, and crosses represent UT dates 14 Dec 2003, 15 Dec 2003, and 17 Dec 2003, respectively.

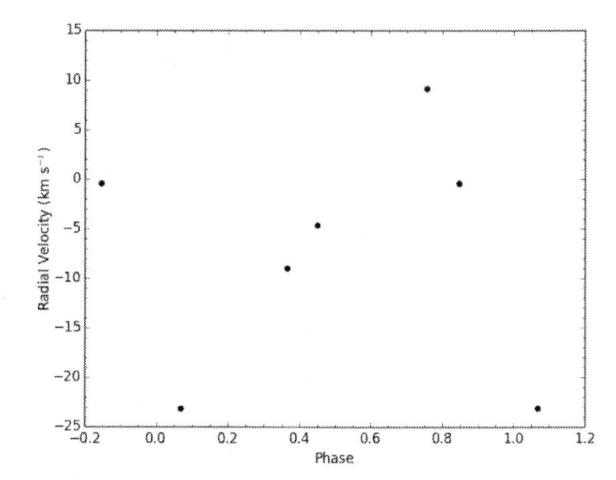

Figure 3. Phased radial velocity curve of DE Lac; phases are calculated from the ephemeris $HJD_{obs} = 2428807.1385 + 0.253692580E + 1.29\ 10^{-11}E^2$. Observations were taken on UT dates 25 Jul 1996, 29 Jul 1996, 21 Aug 1996, 22 Aug 1996, and 26 Sep 1996.

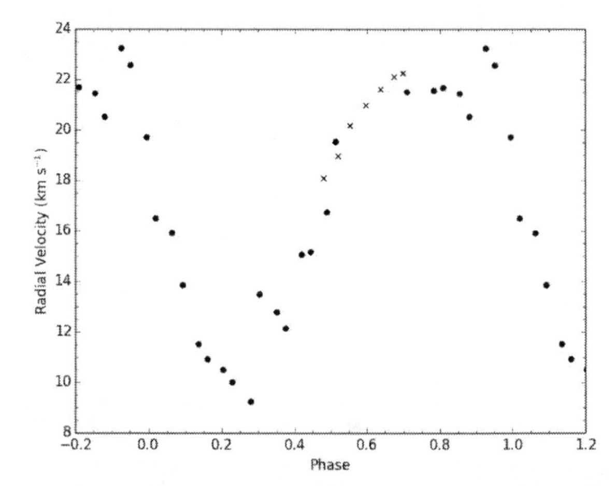

Figure 4. Phased radial velocity curve of V474 Mon; phases are calculated from the ephemeris $HJD_{obs} = 2451500.000 + 0.1361258E$. Dots and crosses represent (UT) 13 Dec 2003 and 14 Dec 2003, respectively.

They give no uncertainties for the ephemeris parameters. Our radial velocities are presented in Table 4 and shown in Figure 2. We have used different symbols to represent the different days of our observations: dots, triangles, and x's represent (UT) 14 Dec, 15 Dec, and 17 Dec, respectively. Uncertainties in the velocities are ±0.30 km/s.

The mean radial velocity in Table 4 is 37.4 km s^{-1} with an amplitude 24.7 km s^{-1}. This mean velocity agrees nicely with Balona and Stobie (1983), 38.8 km s^{-1}, and with the data in Figure 8 of Derekas *et al.* (2009), ~ 40 km s^{-1}. Abyankhar (1959) gives a value of 34.5 km s^{-1} based on 6 measurements with large scatter.

Derekas *et al.* (2009) find variation in the radial velocity curve ~ 4 km s^{-1} over 4 months that is not seen in our nearly contiguous nights of data. They attribute this to multi-periodic variation. Hurta *et al.* (2007) determined an orbit for the binary based on variation in times of maximum light. They give a period of 42.9 y with an estimated radial velocity amplitude of 2K = 2.2 km s^{-1}. The variation seen in the radial velocities of Derekas *et al.* (2009) greatly exceeds expectation for a 42.9 y orbital period.

4.3. DE Lac

We inadvertently observed DE Lac as part of our RR Lyrae program (Jeffery *et al.* 2007, Paper II) through a typo in our observing list. We should have observed it with an integration time no longer than 435 sec to adhere to our policy of limiting exposure times to 0.02 cycle, instead of the actual times of 900 sec (0.041 cycle) and 1,200 sec (0.055 cycle). Our radial velocities are presented in Table 5 and shown in Figure 3. Uncertainties in the velocities are ±0.30 km/s.

The effect of the exposure length, t_{exp}, is to smear the underlying signal through boxcar smoothing, so that the amplitude derived from the observations is less than the intrinsic amplitude. For a sinusoidal signal with period P, the reduction in amplitude is $\mathrm{sinc}(\pi t_{exp} / P)$ (Murphy 2012). For our 1,200 sec observations this corresponds to a factor of only 0.995.

Wang *et al.* (2014) analyzed times of maximum light and found a variable period. We used their quadratic solution to compute the phases for our observations as described above:

$$HJD_{obs} = 2428807.1385\ (\pm\ 0.0005) + 0.253692580\ (\pm\ 0.000000004)E + 1.29\ (\pm\ 0.16) \times 10^{-11}E^2. \quad (3)$$

Taking a straight mean of our five velocities we find –5.6 km s^{-1} with an amplitude 32.1 km s^{-1}. Woolley and Aly (1966) obtained +3.6 km s^{-1} and 24.6 km s^{-1}, respectively, based on eight velocities (taken from their Tables VI and VII, Rowland metal values). On the other hand, McNamara and Laney (1976) measured –16.5 km s^{-1} for the gamma velocity and 26.5 km s^{-1} for the amplitude based on 17 velocities. These gamma velocities are discordant with a spread ~ 20 km s^{-1}. Possibly the star is a binary with 2K ~ 20 km s^{-1}; however, analysis of 66 years of times of maximum light by Wang *et al.* (2014) made no mention of an orbital light travel time effect.

4.4. V474 Mon

V474 Mon is known to be a multi-periodic star with at least three periods (Balona *et al.* 2001). For the phases in Table 6 and Figure 4 we adopted the principal radial component from that source. We have used different symbols to represent the different days of our observations: dots and x's represent (UT) 13 Dec and 14 Dec, respectively. We inverted the following to compute phases for our observations:

$$\mathrm{HJDobs} = 2451500.000 + 0.1361258E. \quad (4)$$

The scatter in our velocities is more than twice the expected uncertainty (± 0.30 km/s), no doubt as a consequence of the multi-periodic variation. Our velocities compare well with those of Jones (1971) and with considerably smaller scatter. Szatmary (1990) lists V474 Mon as a binary with P = 15.492 days. Liakos and Niarchos (2017) list it as ambiguous for binarity.

5. Conclusions

We have presented 124 new radial velocities for the δ Sct variable stars VZ Cnc, AD CMi, DE Lac, and V474 Mon. All radial velocity curves cover complete pulsation cycles. Except for DE Lac integration times were limited to ≤ 2% of the pulsation cycle. The impact of the longer integration times for DE Lac is negligible. We have presented phased radial velocity curves for each star. Our radial velocities confirm the multi-periodic behavior of both VZ Cnc and V474 Mon. However, our data do not cover a long enough baseline to address the multi-periodic nature of AD CMi claimed by Derekas *et al.* (2009). Our velocities, added to those in the literature for DE Lac, suggest discordant center-of-mass velocities, but the data are few. Although we obtained data on KZ Hya, the signal-to-noise ratio of our spectra were too low to determine meaningful velocities for this star.

6. Acknowledgements

Financial support is gratefully acknowledged from National Science Foundation grant AST-9986817 (TGB). Travel support to obtain the observations is acknowledged from McDonald Observatory (EJJ, TGB, TJM). Travel support from the European Space Agency is also gratefully acknowledged (IS). We thank Dr. Michel Breger for helpful comments on the paper.

References

Abhyankar, K. D. 1959, *Astrophys. J.*, **130**, 834.

Balona, L. A., and Stobie, R. S. 1983, *South African Astron. Obs. Circ.*, **7**, 19.

Balona, L. A., *et al.* 2001, *Mon. Not. Roy. Astron. Soc.*, **321**, 239.

Barnes, T. G., Jeffery, E. J., Montemayor, T. J., and Skillen, I. 2005, *Astrophys. J., Supp. Ser.*, **156**, 227.

Boonyarak, C., Fu, J.-N., Khokhuntod, P., and Jiang, S.-Y. 2011, *Astrophys. Space Sci.*, **333**, 125.

Boonyarak, C., Khokhuntod, P., and Jiang, S.-Y. 2009, *Astrophys. Space Sci.*, **324**, 5.

Breger, M. 2000, in *Delta Scuti and Related Stars*, eds. M. Breger and M. H. Montgomery, ASP Conf. Ser. 210, Astronomical Society of the Pacific, San Francisco, 3.

Burki, G., and Meylan, G. 1986, *Astron. Astrophys.*, **159**, 261.

Chauvenet, W. 1864, *A Manual of Spherical and Practical Astronomy*, volume 2, 2nd ed., Lippincott, Philadelphia.

Derekas, A., *et al.* 2009, *Mon. Not. Roy. Astron. Soc.*, **394**, 995.

Fu, J. N., *et al.* 2008, *Astron. J.*, **135**, 1958.

Guiglion, G., *et al.* 2013, *Astron. Astrophys.*, **550**, 10.

Hurta, Zs., Pocs, M. D., and Szeidl, B. 2007, *Inf. Bull. Var. Stars*, No. 5774, 1.

Jeffery, E. J., Barnes, T. G., III, Skillen, I., and Montemayor, T. J. 2007, *Astrophys. J., Supp. Ser.*, **171**, 512.

Jones, D. H. P. 1971, *Roy. Obs. Bull. Greenwich-Cape*, No. 163, 239.

Kim, C., Kim, S.-L., Jeon, Y.-B., Kim, C.-H., and Gilmore, A. 2007, *Astrophys. Space Sci.*, **312**, 41.

Liakos, A., and Niarchos, P. 2017, *Mon. Not. Roy. Astron. Soc.*, **465**, 1181.

McCarthy, J. K., Sandiford, B. A., Boyd, D., and Booth, J. A. 1993, *Publ. Astron. Soc. Pacific*, **105**, 881.

McNamara, D. H., and Budge, K. G. 1985, *Publ. Astron. Soc. Pacific*, **97**, 322.

McNamara, D. H., and Laney, C. D. 1976, *Publ. Astron. Soc. Pacific*, **88**, 168.

Murphy, S.J. 2012, *Mon. Not. Roy. Astron. Soc.*, **422**, 665.

Przybylski, A., and Bessell, M. S. 1979, *Mon. Not. Roy. Astron. Soc.*, **189**, 377.

Samus N. N., Kazarovets E. V., Durlevich O. V., Kireeva N. N., and Pastukhova E. N. 2017, *General Catalogue of Variable Stars: Version GCVS 5.1*, *Astron. Rep.*, **61**, 80.

Szatmary, K. 1990, *J. Amer. Assoc. Var. Star Obs.*, **19**, 52.

Taylor, J. R. 1997, *Introduction to Error Analysis*, 2nd ed., University Science Books, Sausalito, CA.

Tody, D. 1993, in *Astronomical Data Analysis Software and Systems II*, eds. R. J. Hanisch, R. J. V. Brissenden, J. Barnes, ASP Conf. Ser. 52, Astronomical Society of the Pacific, San Francisco, 173.

Wang, S.-M., Qian, S.-B., Liao, W.-P., Zhang, J., Zhou, X., and Zhao, E.-G. 2014, *Bull. Astron. Soc. India*, **42**, 19.

Woolley, R., and Aly, K. 1966, *Roy. Obs. Bull. Greenwich-Cape*, No. 114, 259.

Discovery and Period Analysis of Seven Variable Stars

Tom Polakis
Command Module Observatory, 121 W. Alameda Drive, Tempe, AZ 85282; tpolakis@cox.net

Received May 3, 2019; revised May 20, 2019; accepted May 25, 2019

Abstract Tens of thousands of images have been acquired during the course of asteroid photometry observations at Command Module Observatory (MPC V02). Nightly sets of these images were analyzed using a software routine that locates field stars whose brightness shows anomalously high photometric scatter. A total of 32 variable stars were identified by this method, six of which are discoveries. An additional discovery was made serendipitously by noticing variability of a comparison star. Follow-up V-band observations were made for all of these stars, and combined with sparse survey data to perform period analysis and create light curves that are presented in this paper. All seven stars are eclipsing binary systems: four are of type W UMa, two are type β Lyr, and one is an Algol-type. Results have been entered into the International Variable Star Index.

1. Introduction

As of May 2019, the International Variable Star Index (VSX; Watson *et al.* 2014) contained data for more than 600,000 variable stars. Many recent discoveries have been made by data mining, taking advantage of the great wealth of available survey data. Despite the preponderance of these data and automated detection methods, a search for bright variable stars was undertaken with the expectation that they still await discovery.

In a typical night of asteroid photometric observations, two to three hundred images are obtained for several targets. After three years of such observing, nearly 100,000 images have been captured. This paper describes the instrumentation and methods that were employed to locate variable stars in these images. Also discussed are follow-up V-band imaging and analysis augmented by survey data, and the securing of precise periods and determination of variable star types.

2. Instrumentation and methods

2.1. Data acquisition

All of the CCD photometric observations were performed at V02 in Tempe, Arizona. Light pollution brightens the sky at this suburban site by roughly three magnitudes per square arcsecond. Images were taken between June 2017 and April 2019 using a 0.32-m f/6.7 Modified Dall-Kirkham telescope and an SBIG STXL-6303 CCD camera. Since asteroids do not vary significantly in color, a "clear" glass filter was used for all images. Exposure time for all the images was 2 minutes, which is a balance between minimizing photometric errors and providing adequate cadence. The image scale after 2 × 2 binning was 1.76 arcsec/pixel. Images were calibrated using a dozen bias, dark, and flat frames. Flat-field images were made using an electroluminescent panel. Image calibration and alignment was performed using MAXIM DL software (Diffraction Limited 2012).

Despite the nightly motion of asteroids, new field centers were required only every fourth night, since it takes a minor planet more than three nights to traverse the 45' × 30' field of the CCD. The relevant benefit to this work of maintaining the same field is that any variable stars that may be discovered could be followed up with images from the two subsequent nights.

2.2. Data reduction

MPO CANOPUS (Warner 2019) software was used for its features that are unique to performing photometric measurements of Solar System targets. The software also has a "Variable Star Search" feature, in which standard deviation vs. average magnitude for field stars is plotted using images for a particular field during a single night. Time-series plots of the magnitude of the ensemble of chosen comparison stars are also created to confirm that the standard deviations are not a result of local atmospheric effects. Figure 1 is an example of such a plot. The upward ramp for brighter stars at the left end is caused by the detector entering its non-linear range, while the higher errors at the right end are a result of reduced signal-to-noise ratio for fainter stars. Potential variable stars are found in the middle range of the plot.

False positives greatly outnumber actual variable stars in these plots. Stars too near the edge of the frame, companion stars near the edge of the measuring aperture, extended objects, and clouds are just some of the causes. Therefore, an "eye test" is performed by clicking on each data point, which brings up a time-series plot of that star's magnitude relative to the average of the ensemble of comparison star magnitudes. Those plots with a single outlying point or exhibiting a step change in brightness are quickly rejected. Legitimate variable stars show steady ramps, and curvature at regions of inflection, as shown in the example in Figure 2.

This method was employed for 170 fields, each typically containing between 60 and 80 images for a single night. These targeted searches turned up 32 variable stars, 26 of which were already catalogued in VSX.

One of the seven new variable stars was discovered by the luckier method of using it as a comparison star. CANOPUS provides a plot that compares the magnitude of each comparison star against the average of the other four in the ensemble. Short-period variables often show up as waviness in the plot that should ideally be flat, as shown in Figure 3. Typically, these variables are found to already be catalogued, but this particular case is a new discovery.

For the newly discovered variable stars, magnitudes from the Clear-filtered discovery images were transformed to Johnson V. For each new variable, magnitudes for an ensemble of five comparison stars were obtained from the APASS DR10 catalogue (Henden *et al.* 2019). Comparison stars are solar colored, with B–V ranging from 0.5 to 0.9. In all but one case, the target variable stars are also in this color range. Rudimentary light curves were created from the discovery images only to confirm periodic variability. Follow-up observing for each variable involved targeted imaging of these variables through a V filter, observing them with the appropriate number of nights and cadences.

3. Results

3.1. Light curve created from solely V02 data using V-band images

Figure 4 is the phased light curve for GSC 01845-00905 using data from only V02. The star was observed on nine nights, during which 634 images were obtained. While the minima are adequately defined, there remains a gap in the data near phase 0.8. The period spectrum is plotted in Figure 5. RMS error of the curve fit is plotted on the vertical axis against period solutions ranging from 2.14 to 2.17 days. Note that it shows multiple aliases, creating an ambiguous period solution. It becomes clear that augmentation by survey data would prove beneficial.

3.2. Period solutions and light curves including survey data

Of the many photometric surveys posting online data, ASAS-SN (Kochanek *et al.* 2017) proved to be the most fruitful for V-band measurements. For each target, hundreds of data points gathered between 2016 and the present were used to augment the V02 data. The zero-point difference between the two datasets typically amounted to 0.050 to 0.100 mag., so those points were shifted in the software by the appropriate amount. The longer time range of data was useful in eliminating any ambiguity in the period spectra, and in greatly increasing the precision of the periods.

The benefit of including ASAS-SN survey data for GSC 01845-00905 is illustrated in Figure 6, which shows the period spectrum with the inclusion of ASAS-SN data. Comparing this plot to Figure 5, it can be seen that the survey data's longer time baseline also results in an unambiguous period solution with greater precision.

Period determination was done using the MPO Canopus Fourier-type FALC fitting method (Harris *et al.* 1989). Pertinent data for all seven variable stars are presented in Table 1. Figures 7 through 13 are phased light curves. The smoothed curves are Fourier fits to the data. The first and last data points for each night are highlighted with circles and boxes, respectively. V02 and ASAS-SN data are identified in the legends.

3.3. Discussion of results

In this section, we discuss the columns appearing in Table 1. Where possible, the primary identification is from the Hubble Guide Star Catalog, Version 1.2 (Morrison *et al.* 2001). Otherwise, the UCAC4 (Zacharias *et al.* 2013) designation is adopted. Precise J2000 coordinates were acquired from Gaia DR2 (Gaia Collab. *et al.* 2016, 2018).

We can identify the types of these variables by simply observing the character of their light curves. The majority of the seven stars are W UMa (EW: Figures 7, 10, 12, 13), whose ellipsoidal components are similarly bright, so the primary and secondary minima are nearly equal. Two stars are β Lyr eclipsing binaries (EB: Figures 9 and 11), in which the components are of different brightness, and the two minima are significantly different. Finally, one star is an Algol-type eclipsing binary (EA: Figure 8). In this case, the light does not vary for a long portion of the period, during which the two stars are not eclipsing.

Period solutions are given in days to a precision of six decimal places, with errors indicated. Range is the difference between maximum and minimum light in the Fourier fit.

Most approaches for determining the epoch of minimum light require dense coverage surrounding the minimum with low data scatter. The ASAS-SN data have too much scatter for this method to be viable. Therefore, a pair of data points with known phase information were located on either side of phase = 0 on the Fourier-fitted curve, and their heliocentric Julian Dates were interpolated to arrive at a result that is estimated to 0.0001 day.

4. Conclusions

Despite the large quantity of catalogued variable stars and the volume of survey data, these discoveries with a small telescope at a suburban site demonstrate that many more bright variable stars await discovery. Follow-up observations done at the appropriate level coupled with photometric survey data yielded precise period solutions and light curves.

Most variable stars having large amplitudes have already been discovered. It is more likely that new discoveries will be of variables with amplitudes of several tenths of a magnitude. This is a common range for eclipsing binary stars, so they are likely candidates for discovery, as was the case for all seven stars in this work.

5. Acknowledgements

The effort was inspired by a presentation by Maurice Clark (Clark 2018), who discovered variable stars using a similar method. The author would like to thank Brian Skiff for his indispensable mentoring in data acquisition and reduction. Thanks also go out to Brian Warner for support of his MPO Canopus software package. Sebastián Otero was responsive and helpful with matters relating to VSX. This research was made possible through the use of the AAVSO Photometric All-Sky Survey (APASS), funded by the Robert Martin Ayers Sciences Fund and NSF AST-1412587.

References

Clark, M. 2018, in *The Society for Astronomical Sciences 37th Annual Symposium on Telescope Science*. Society for Astronomical Sciences, Rancho Cucamonga, CA, 105.

Diffraction Limited. 2012, MAXIMDL image processing software (http://www.cyanogen.com).

Gaia Collaboration, *et al.* 2016, *Astron. Astrophys.*, **595A**, 1.

Gaia Collaboration, *et al.* 2018, *Astron. Astrophys.*, **616A**, 1.

Harris, A. W., *et al.* 1989, *Icarus*, **77**, 171.

Henden, A. A., *et al.* 2019, AAVSO Photometric All-Sky Survey, data release 10 (http://www.aavso.org/apass).

Kochanek, C. S., *et al.* 2017, *Publ. Astron. Soc. Pacific*, **129**, 4502.

Morrison, J. E., Röser, S., McLean, B., Bucciarelli, B., and Lasker, B. 2001, *Astron. J.*, **121**, 1752.

Warner, B. D. 2019, MPO CANOPUS software (http://bdwpublishing.com).

Watson, C., Henden, A. A., and Price, C. A. 2014, AAVSO International Variable Star Index VSX (Watson+, 2006–2014; http://www.aavso.org/vsx).

Zacharias, N., Finch, C. T., Girard, T. M., Henden, A., Bartlett, J. L., Monet, D. G., and Zacharias, M. I. 2013, *Astron. J.*, **145**, 44.

Table 1. Photometry for recently discovered variable stars.

Ident.	*R.A. (J2000.0)* h m s	*Dec. (J2000.0)* ° ' "	*Type*	*Period (d)*	*Error (d)*	*Max. (V)*	*Min. (V)*	*Epoch of Min. (HJD)*	*Figure*
GSC 01224-00315	02 56 17.57	+16 35 26.1	EW	0.423042	0.000003	14.02	14.11	2458053.7046	7
GSC 01845-00905	05 05 31.19	+22 43 07.1	EA	2.152089	0.000003	14.37	14.92	2458107.8820	8
GSC 01299-01898	05 45 25.71	+15 22 08.5	EB	0.529882	0.000003	13.24	13.61	2458108.8332	9
GSC 01347-00934	07 26 52.98	+15 14 46.0	EW	0.492463	0.000002	13.70	13.82	2458488.9509	10
UCAC4 522-042119	07 40 07.27	+14 22 33.8	EB	0.713958	0.000002	14.12	14.40	2458158.6820	11
GSC 00790-00941	07 43 33.69	+14 23 49.5	EW	0.235043	0.000001	13.98	14.63	2458151.8873	12
GSC 04963-01164	13 06 41.20	–07 03 39.9	EW	0.445836	0.000001	12.31	12.43	2457542.8621	13

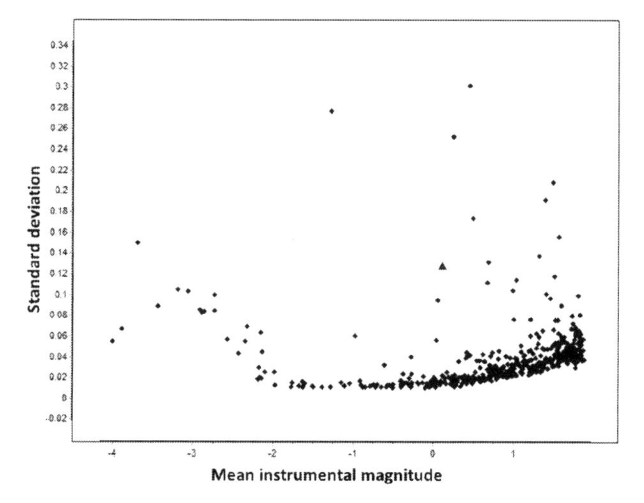

Figure 1. Standard deviation for each catalogued star from 70 images plotted against their mean magnitudes. The triangle identifies variable star GSC 01299-01898, while other outlying points are false positives.

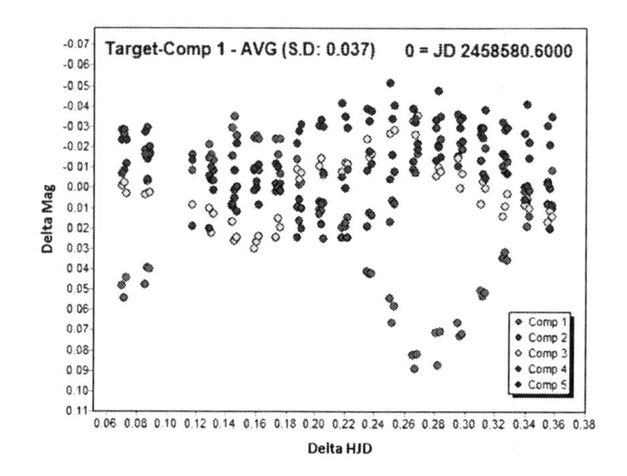

Figure 3. Variation in magnitude of each comparison star relative to the average of the other four. Star #1 (GSC 04963-01164) is clearly variable. Since it is used in the averaging, its variations cause smaller, mirror-image variations of the other four comparison stars.

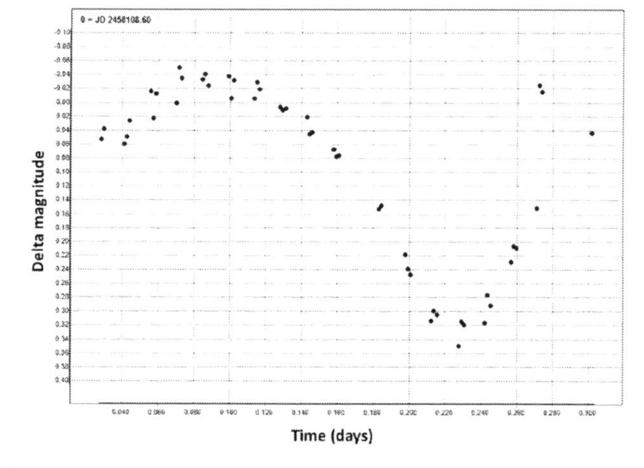

Figure 2. Variation in magnitude of GSC 01299-01898, which was identified in Figure 1

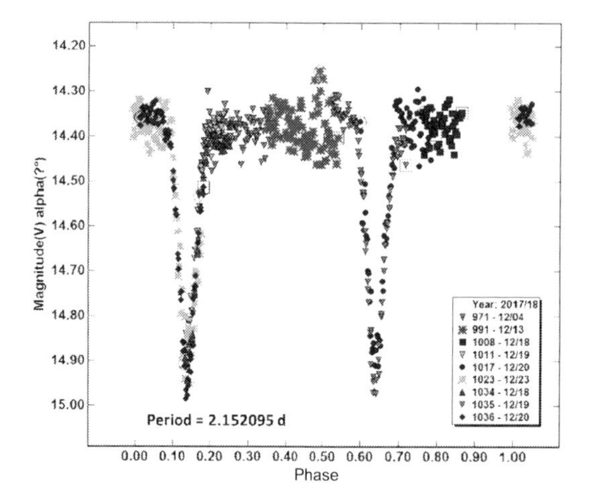

Figure 4. Phased light curve for GSC 01845-00905, using data only from V02.

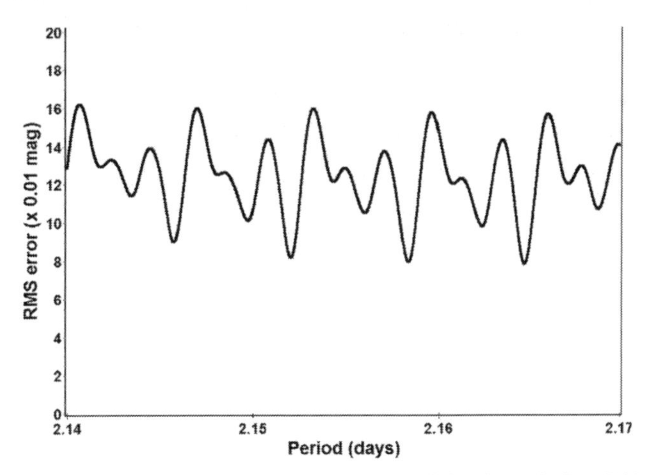

Figure 5. Period spectrum for GSC 01845-00905, using data only from V02. The short time baseline causes multiple aliases between 2.14 and 2.17 days.

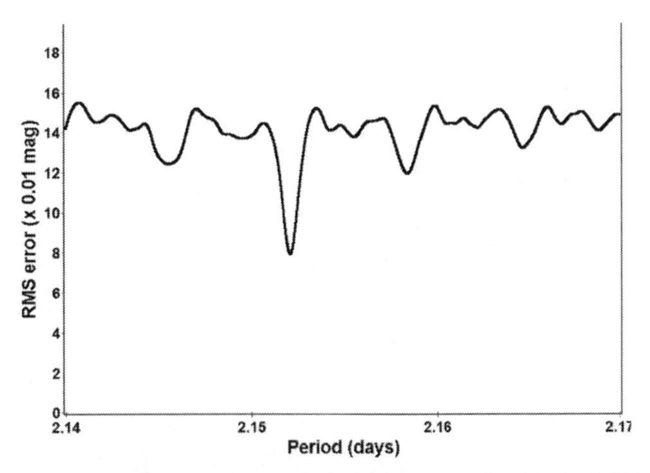

Figure 6. Period spectrum for GSC 01845-00905, using data from both V02 and ASAS-SN survey. Compare with Figure 5. The longer baseline clarifies the period solution.

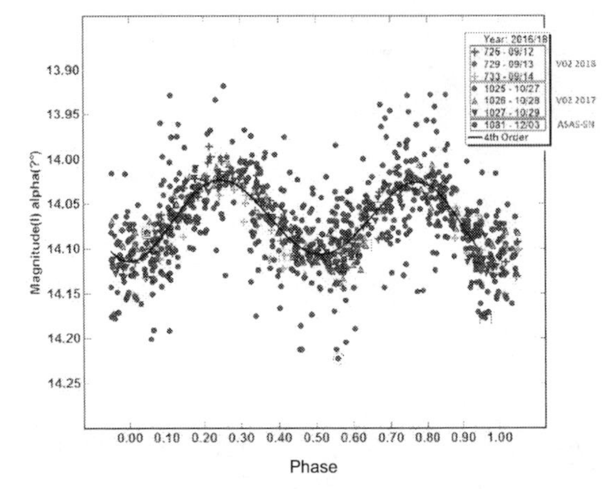

Figure 7. Phased light curve for GSC 01224-00315. Period = 0.423042 d. Epoch of primary minimum = 2458053.7046 HJD.

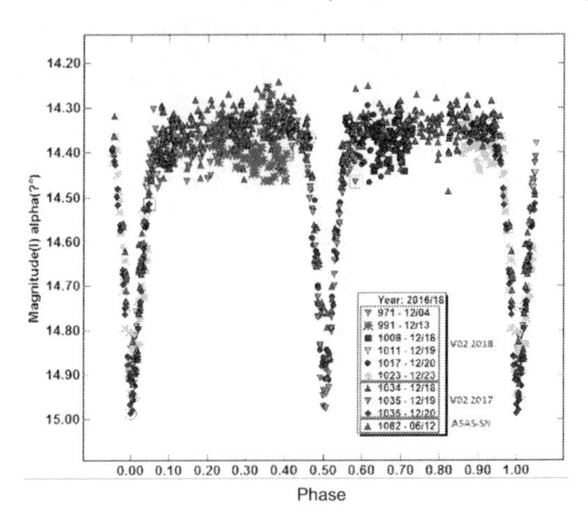

Figure 8. Phased light curve for GSC 01845-00905. Period = 2.152089 d. Epoch of primary minimum = 2458107.8820 HJD.

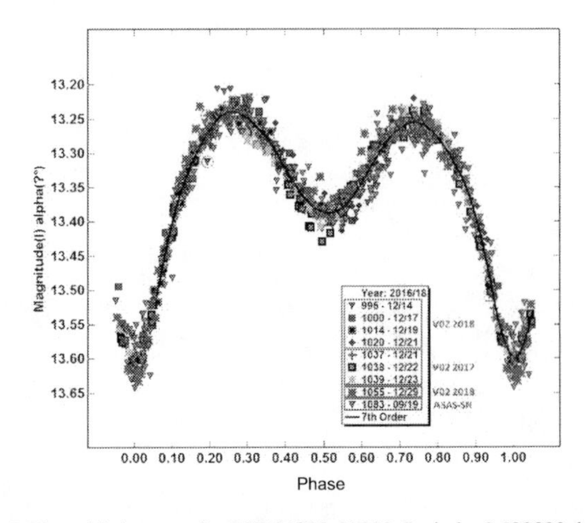

Figure 9. Phased light curve for GSC 01299-01898. Period = 0.529882 d. Epoch of primary minimum = 2458108.8332 HJD.

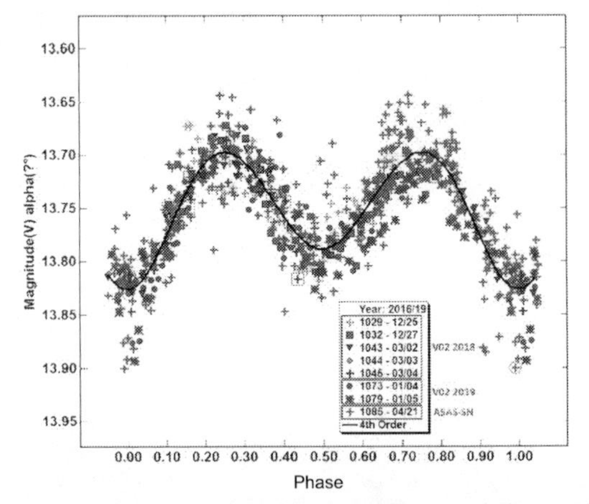

Figure 10. Phased light curve for GSC 01347-00934. Period = 0.492463 d. Epoch of primary minimum = 2458488.9509 HJD.

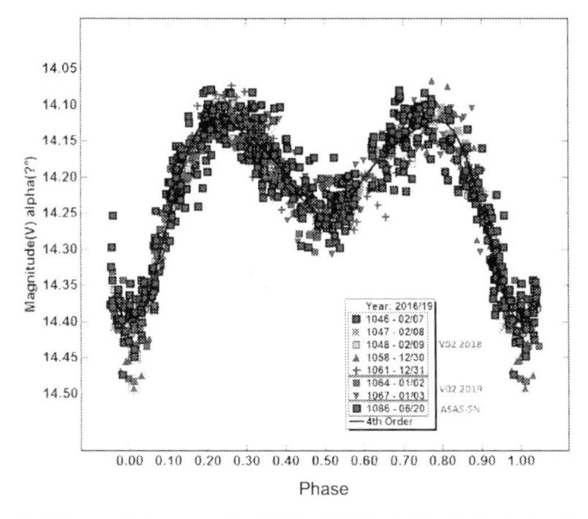

Figure 11. Phased light curve for UCAC4 522-042119. Period = 0.713958 d. Epoch of primary minimum = 2458158.6820 HJD.

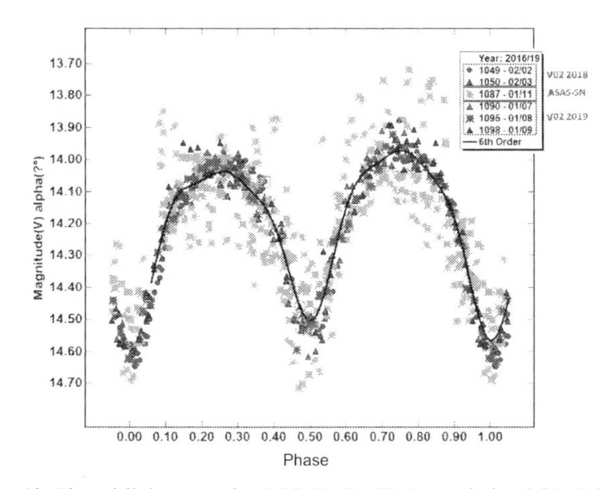

Figure 12. Phased light curve for GSC 00790-00941. Period = 0.235043 d. Epoch of primary minimum = 2458151.8873 HJD.

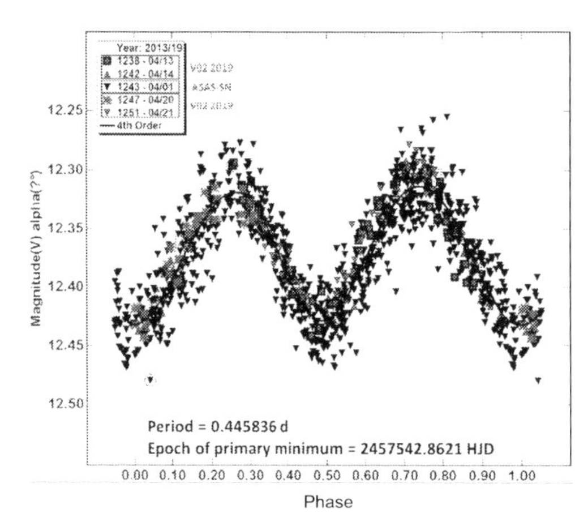

Figure 13. Phased light curve for GSC 04963-01164. Period = 0.445836 d. Epoch of primary minimum = 2457542.8621 HJD.

The History of AAVSO Charts, Part III: The Henden Era

Tim R. Crawford
79916 W. Beach Road, Arch Cape, OR 97102; tcarchcape@yahoo.com

Received December 6, 2018; revised December 31, 2018; accepted January 2, 2019

Abstract In this third paper covering the history of AAVSO charts I continue to describe not only the evolution of charts but the significant leap forward in all aspects of the chart production process that took place during the Henden Era, which started before Arne Henden became director and extends beyond his service as a Director of the AAVSO (2005–2015).

1. Introduction

When Arne Henden assumed the Directorship of the AAVSO in March of 2005, chart making had evolved to a computer-aided analog process producing charts which were also now available through the internet. While the chart production process still required manual labeling of the magnitudes and entering of the coordinates of comparison star data, the chart improvements were enabled by the use of planetarium software and art programs then being used by volunteer Michael A. Simonsen and staff member Aaron Price, as discussed in Malatesta *et al.* (2007).

Prior to Henden becoming director, Aaron Price, Vance Petriew (another volunteer), and Simonsen had already held numerous discussions on what tools and processes needed to be developed to meet chart needs going into the future. They recognized that ultimately some type of automated chart plotter needed to be developed and towards that goal they would have to establish a database of comparison stars to enable an automated chart plotter to function. The original concept for this database was actually one of the last contributions by Director Janet Mattei (Price 2018).

Price established and took charge of the Comparison Star Database Working Group to build a comparison star database in anticipation of an automated chart plotter, while Simonsen was put in charge of the AAVSO Chart Team to implement the guidelines, previously established by the International Chart Working Group, for the creation of new sequences and revisions of existing ones (Price *et al.* 2004a).

After a few months Price, during the AAVSO 2003 Spring Meeting, was able to persuade Petriew (during lunch on top of Kitt Peak), who was also a professional database manager, to take charge of the Comparison Star Database Working Group (Price 2018).

Shortly after assuming the Directorship, Henden informed everyone that he wanted the AAVSO out of the manual chart business by bringing the process fully into the digital age (Simonsen 2014). As Price noted:

> This is where he [Henden] was brilliant. Before this there was a little bit of residual grumbling and resistance-both among staff and membership. Arne's decisiveness put an end to it. At this point on, it was all hands on deck and everyone began working together (Price 2018).

The ultimate success of the process resulted in not only the original objective but a number of support elements as well. The whole of this endeavor involves a number of individual parts that were either in existence, being developed, or had yet to be created at the time Henden was appointed Director, and will be presented in a somewhat chronological order.

2. Chart error tracking tool (CHET)

A frustrating problem that had plagued chart making through the years was the tracking and fixing of errors on existing charts as reported by observers.

This problem was finally solved in 2003 when volunteer Chris Watson developed the web-based Chart Error Tracking Tool (CHET) that permitted observers to identify their problems and then have them tracked as sequence team members were able to respond.

The CHET tool underwent a significant revision in the fall of 2017 with the help of Phil Manno, who volunteered to convert the old code into the modern Django framework as used by other AAVSO tools; staff member Will McMain helped Manno get set up and started on the project. One of the more significant enhancements was an automatic response system for observers that would let them know when their reported errors were resolved.

3. Variable star database (VSD)

After assuming leadership in the Spring of 2003, Petriew established the guidelines for the Comparison Star Database Working Group's challenge of documenting into a digital database (called CompDB) all the comparison stars and variables that then existed.

Group members included Aaron Price, Vance Petriew, Rick Merriman, Keith Graham, Dan Taylor, Brian Skiff, Tim Hager, Carlo Gualdoni, Mike Simenson, Bob Stein, Roy Axelsen, Mark Munkacsy, Christopher Watson, Curt Schneider, Jim Bedient, Radu Corlan, Joe Maffei, Arno Van Werven, Pedro Pastor, and Dolores Sharples (Price et al. 2004b).

The CompDB project was completed in July 2006 and documented 4,184 charts with 35,820 unique stars, and consumed an estimated 10,000+ hours of volunteer effort by twenty volunteers (Petriew *et al.* 2007).

CompDB contained both identified and suspected variable stars (later moved to the VSX database) as well as comparison stars (estimated at ~30,500).

During October of 2006 Henden updated the photometry of about 22,000 of those comparison stars with the then best-known data.

The goal was to create a database that would then enable some form of automated chart production. To accomplish this, CompDB was converted into a relational database (the Variable Star Database, or VSD) which would contain only comparison stars and their associated information: coordinates, magnitude values and their errors, photometry source, and remarks.

To build VSD each comparison star in the existing CompDB was then assigned an AAVSO Unique Identifier (AUID) and migrated in.

AUIDs were created in the form of three triplets: ###-XXX-### without vowels being allowed for the XXX (alpha) portion. It was estimated that this Alpha-numeric format would allow for ~10 billion permutations (Petriew *et al.* 2007).

4. Variable star plotter (VSP)

With the development of VSD Henden felt quite confident that an automated chart plotter could now be developed and awarded a contract to Michael Koppelman and his company (Clockwork Active Media Systems (CAMS)) to build an Automated Chart Plotter (ACP) in 2006 with assistance from Chris Watson and Aaron Price.

The design goals were to create an online accessible system whereby the observer could enter a target name, a radius size of the field of view, and a magnitude limit which would generate an onscreen chart, of the specified size, showing visual magnitude values. Screen access to a photometry table (magnitude values for various filters, if available) for the target chart was included. As Price notes:

> Chris Watson designed the interface, I coded the interface software, and CAMS did the chart generation engine. I think a key development here was the development of the unique chart ID. That allowed anyone to recreate the exact details of any chart at any time (Price 2018).

In December of 2007, Henden announced:

> The Beta release of the Variable Star Plotter (VSP) was made just prior to the Spring Meeting. The VSP plotting engine has been ready for quite some time, but we needed to populate its internal database with all of the variable stars and sequence stars that were present on the existing charts before it was functional enough for the membership to use.... Due to its increased functionality, [it] should be used in preference to the chart archive (Henden 2007).

What a tremendous leap forward in chart making. The AAVSO no longer had the burden of producing paper charts as they had been for close to 100 years, thus freeing up considerable staff time.

Observers, through the internet, could now configure how they wanted a chart to be presented for any field of view (current configuration shown in Figure 1), then printed out on their home printer. In addition, they could also request a photometry table which was important to the growing field of CCD observing (Figure 2).

If there was an Achilles Heel to this incredible improvement, at the time, it was simply a lack of calibrated photometry to satisfy the rapidly increasing demands of the many observers who were now using the new tool.

Another major feature was that the observers could continue to choose the older black dot charts—the dots being in the correct position and sized in proportion to their V magnitude, or they could now select a digital sky image of the field of view showing all the stars captured by the camera, which was of great benefit to CCD observers.

5. International Variable Star Index (VSX)

During 2005, because of inadequacies of catalogs identifying variable stars, AAVSO Council Member Lew Cook suggested that an ad hoc group be formed to solve this problem. With David B. Williams as Chair, the members of this group included Aaron Price, Mike Simonsen, Vance Petriew, John Greaves, Brian Skiff, Bill Gray, and Arne Henden (Williams and Saladyga 2011).

Simonsen then enlisted amateur astronomer Christopher Watson (CHET creator), who conceptualized and outlined a database of variable star information that could easily be updated. After installing his system on a private site, Watson invited Henden to explore and critique it; that action quickly drew in the rest of the ad hoc group to explore this new tool (Williams and Saladyga 2011).

The final product was the VSX database, which could also qualify as a global clearing house database for all the up-to-date information available on variable stars and suspected variable stars, and insure that naming conflicts were either avoided or simply included by reference as an alternate name. Additionally, this database could assign the previously described (section 3) AUIDs to known and suspected variables.

While a number of volunteers have served on the VSX team through the years, current active members are Patrick Wils and Sebastián Otero (both of whom were also original participants in the ad hoc group) as well as Klaus Bernhard and Patrick Schmeer. This team keeps VSX continuously updated with newly discovered variables as well as new information for existing database variable stars (Otero 2018).

As of the first part of September 2018, VSX contained data on 542,610 variable stars.

VSX also plays another important role in that neither SEQPLOT (see section 7) nor the VSP will recognize a variable star name unless it exists in VSX and has been assigned an AUID. In addition, observers are not able to report observations unless the target is both in VSX and has been assigned an AUID.

6. Sequence team

With the development of VSP (renamed from ACP), there was no longer any need to continue the Chart Team, as VSP could produce a chart for any coordinates entered. What was now required was a group of volunteers to create sequences

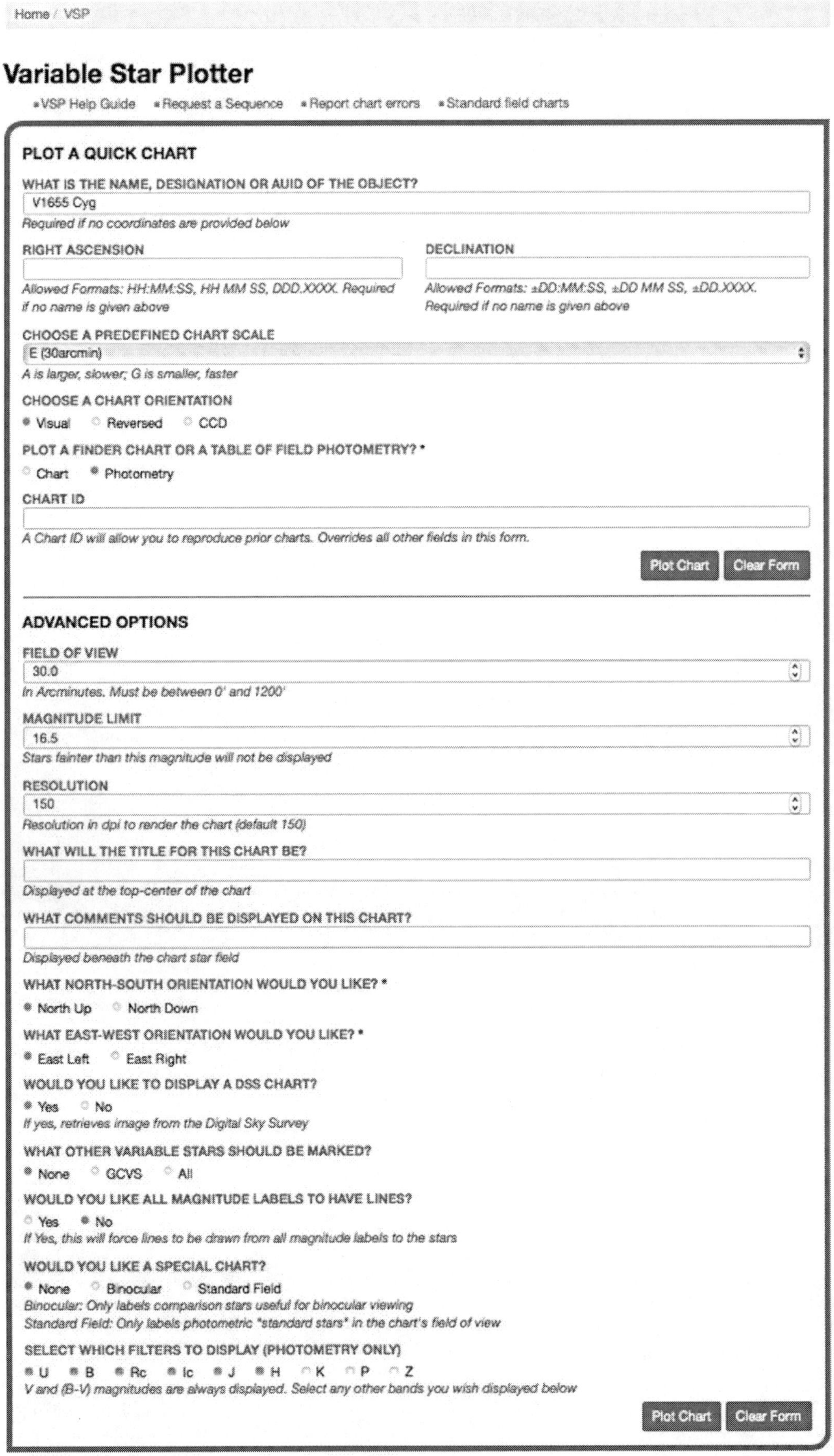

Home / VSP

Variable Star Plotter

- VSP Help Guide
- Request a Sequence
- Report chart errors
- Standard field charts

PLOT A QUICK CHART

WHAT IS THE NAME, DESIGNATION OR AUID OF THE OBJECT?

V1655 Cyg

Required if no coordinates are provided below

RIGHT ASCENSION

Allowed Formats: HH:MM:SS, HH MM SS, DDD.XXXX. Required if no name is given above

DECLINATION

Allowed Formats: ±DD:MM:SS, ±DD MM SS, ±DD.XXXX. Required if no name is given above

CHOOSE A PREDEFINED CHART SCALE

E (30arcmin)

A is larger, slower; G is smaller, faster

CHOOSE A CHART ORIENTATION

Visual Reversed CCD

PLOT A FINDER CHART OR A TABLE OF FIELD PHOTOMETRY? *

Chart Photometry

CHART ID

A Chart ID will allow you to reproduce prior charts. Overrides all other fields in this form.

Plot Chart Clear Form

ADVANCED OPTIONS

FIELD OF VIEW

30.0

In Arcminutes. Must be between 0' and 1200'

MAGNITUDE LIMIT

16.5

Stars fainter than this magnitude will not be displayed

RESOLUTION

150

Resolution in dpi to render the chart (default 150)

WHAT WILL THE TITLE FOR THIS CHART BE?

Displayed at the top-center of the chart

WHAT COMMENTS SHOULD BE DISPLAYED ON THIS CHART?

Displayed beneath the chart star field

WHAT NORTH-SOUTH ORIENTATION WOULD YOU LIKE? *

North Up North Down

WHAT EAST-WEST ORIENTATION WOULD YOU LIKE? *

East Left East Right

WOULD YOU LIKE TO DISPLAY A DSS CHART?

Yes No

If yes, retrieves image from the Digital Sky Survey

WHAT OTHER VARIABLE STARS SHOULD BE MARKED?

None GCVS All

WOULD YOU LIKE ALL MAGNITUDE LABELS TO HAVE LINES?

Yes No

If Yes, this will force lines to be drawn from all magnitude labels to the stars

WOULD YOU LIKE A SPECIAL CHART?

None Binocular Standard Field

Binocular: Only labels comparison stars useful for binocular viewing

Standard Field: Only labels photometric "standard stars" in the chart's field of view

SELECT WHICH FILTERS TO DISPLAY (PHOTOMETRY ONLY)

U B Rc Ic J H K P Z

V and (B-V) magnitudes are always displayed. Select any other bands you wish displayed below

Plot Chart Clear Form

Figure 1. VSP options.

Variable Star Plotter

▪ Plot Another Chart ▪ Star Chart for this Table

Field photometry **for V1655 Cyg** from the AAVSO Variable Star Database

Data includes all comparison stars within 0.25° of RA: **20:25:24.02 [306.35008°]** & Dec: **38:42:35.6 [38.70989°]**

Report this sequence as **X23135PR** in the chart field of your observation report.

AUID	RA	Dec	Label	B	V	B-V	Rc	Ic
000-BMS-452	20:25:18.30 [306.32626343°]	38:37:44.3 [38.6289711°]	**109**	11.454 (0.041)[29]	10.894 (0.022)[29]	0.560 (0.047)	10.506 (0.028)[29]	10.142 (0.033)[29]
000-BMS-455	20:26:04.72 [306.51965332°]	38:43:29.9 [38.72497177°]	**111**	11.614 (0.071)[29]	11.106 (0.044)[29]	0.508 (0.084)	10.738 (0.051)[29]	10.392 (0.056)[29]
000-BMS-456	20:24:57.68 [306.24032593°]	38:49:50.5 [38.83069611°]	**116**	12.558 (0.077)[29]	11.603 (0.048)[29]	0.955 (0.091)	11.044 (0.052)[29]	10.523 (0.056)[29]
000-BMS-457	20:25:35.33 [306.3972168°]	38:44:16.8 [38.73799896°]	**122**	12.876 (0.073)[29]	12.196 (0.043)[29]	0.680 (0.085)	11.767 (0.055)[29]	11.365 (0.065)[29]
000-BMS-453	20:25:31.57 [306.38153076°]	38:42:24.6 [38.70683289°]	**124**	13.015 (0.061)[29]	12.419 (0.037)[29]	0.596 (0.071)	12.037 (0.046)[29]	11.678 (0.054)[29]
000-BMS-460	20:25:36.18 [306.40075684°]	38:46:33.8 [38.77605438°]	**126**	13.222 (0.076)[29]	12.641 (0.050)[29]	0.581 (0.091)	12.283 (0.056)[29]	11.946 (0.062)[29]
000-BMS-454	20:25:07.05 [306.27938843°]	38:45:01.8 [38.75049973°]	**129**	13.489 (0.078)[29]	12.938 (0.048)[29]	0.551 (0.092)	12.547 (0.050)[29]	12.180 (0.052)[29]
000-BMS-461	20:25:56.55 [306.48562622°]	38:43:56.2 [38.73227692°]	**133**	14.169 (0.075)[29]	13.342 (0.047)[29]	0.827 (0.089)	12.840 (0.061)[29]	12.371 (0.072)[29]
000-BMS-459	20:25:11.21 [306.29672241°]	38:43:40.8 [38.72800064°]	**135**	14.138 (0.069)[29]	13.529 (0.046)[29]	0.609 (0.083)	13.130 (0.047)[29]	12.756 (0.047)[29]

Figure 2. Photometry table from a VSP request.

(photometry values) for any field of view for which an observer would have an interest.

Therefore, the old Chart Team was finally disbanded (their last batch of new paper charts was released in March of 2006), and a new reconfigured team was created by Henden and Simonsen (Chair) in 2008 known as the Sequence Team.

The primary purpose of the Sequence Team was not to create charts, as such, but to create calibrated sequences which would cover specific field of view sizes so that observers could evaluate any target's magnitude that would fall within a given specific field of view using their own chart created and printed using the VSP.

Since the completion of the initial population of the VSD database, all new sequences have subsequently been loaded into VSD by Sequence Team members, eventually primarily using the SEQPLOT tool (section 7) with a heavy dependency upon APASS (AAVSO Photometric All-Sky Survey ~10–16.5V) calibrations.

Currently, in addition, when appropriate, calibration data from BSM (Bright Star Monitor ~6.3–13V), GCPD (General Catalogue of Photometric Data ~0–6.5V), NOFS (Naval Observatory Flagstaff aka Henden ~13–18.7V), SRO (Sonoita Research Observatory ~10.3–17V), and Tycho-2 (named after Tycho Brahe ~6.5–10.2V) survey data are also occasionally used within the SEQPLOT tool by team members.

Outside of SEQPLOT, team members will sometimes source calibration data from the following surveys, if needed: CMC15 (Carlsberg Meridian Catalog ~10V–14.4V), SDSS (Sloan Digital Sky Survey ~14.8–18.7V), and Pan-STARRS1 (Panoramic Survey Telescope & Rapid Response System ~14.5–20V).

In 2012 Simonsen had the team, including Otero, working on creating sequences specifically for the new AAVSO Binocular Program. A total of 153 bright semiregular and Mira stars in both hemispheres were selected, with specific comparison stars annotated for these specific variables' sequences so that only those comparison stars would appear in the larger field of views encompassed by binoculars when the Binocular chart option in the VSP was selected (Simonsen 2014).

While some twenty individuals have been involved with the Sequence Team efforts since its inception, current active team members are Tom Bretl, Tim Crawford, Robert Fidrich, Jim Jones, Mike Poxon, and Brad Walter. Bretl, Crawford, Fidrich, and Jones have served on the team since at least 2009.

Since record keeping started in late 2008 (through September 8, 2018) the Sequence Team has created 3,592 new sequences and revised/extended some 2,129 existing sequences, for a total of 5,721 new or revised sequences. Bretl, Crawford, and Jones together have accounted for 84.4% of all sequence work during this period.

The comparison star database (VSD) has now grown to ~86,000 stars (unique AUIDs), through the end of August 2018, up from the ~30,500 that existed in 2007.

Team member Bretl was appointed to Chair the Sequence Team in 2015 by Simonsen who needed more time for other responsibilities. Bretl continues to serve in that capacity today.

7. SEQPLOT

During March of 2007, Aaron Price wrote a script so that sequence team members could load sequences directly into the VSD database; sequences were required to be presented in a text file for uploading, in a specified format (listed in its current configuration):

#Comp,RA,m,s,Dec,m,s,V,Verr,B-V,B-Verr,U-B,U-Berr,V-R,V-Rerr,R-I,R-Ierr,V-I,V-Ierr,Source,#Comments

This was somewhat of a labor-intensive process for the sequence team volunteers as they would have to locate a suitable survey either by possessing documents previously assembled, i.e., Henden USNO data, published papers, or an internet source of a specific survey or surveys, such as Tycho-2 and ASAS. The next step was to generate a listing of potential comparison stars from a specific survey, then choose a potential candidate and check it via a digital sky image (VSP or Aladin) to avoid doubles, and then once selected, to convert and transcribe the available data into the above format following previously established rules regarding how comps were to be selected. As Henden recalls:

> I developed a FORTRAN program called hfind in the 1990s, that charted all of the then-available NOFS calibration data and created star charts, along with an interactive cursor to extract photometric data for each plotted star. I demonstrated this program to Janet [Mattei] around 1998, as she was concerned over the large programming task to change from hand-drawn charts with an embedded sequence table. She then assigned her postdoc [staff member] (George Hawkins) to create an IDL chart-plotter, but had not made any progress in working with sequences before her passing (Henden 2018).
>
> Towards the end of 2008, I decided that we needed an easier method of creating sequences, now that the chart generator was working so well. [AAVSO staff member] Sara Beck had been learning JAVA with [staff member] Kate Davis, and when Sara returned from Tahiti in early 2009, I assigned her the task of essentially porting my hfind program into JAVA. She built a database using my calibrated star tables, and then created the query and plotting program to read that database and plot calibrated stars in a fashion similar to hfind (Henden 2018).

Sara Beck notes:

> Henden sent me an additional email in mid-February 2009 describing how the size of the star dots should be computed and the R.A/Dec. should be converted to the tangent plane. The database was set up a week later, and I created the download page on February 27th. Henden then announced it to the Sequence Team the same day (Beck 2018).

While the sequence catalog (Seqcat) was initially populated with only a few calibrated surveys of limited sky coverage (GCPD, Henden, SRO, and Tycho-2, for example), this was a major labor-saving tool; these surveys were shortly followed by SRO, BSM, and APASS (all described in section 6).

The sequence team member is presented with a list of calibrated data surveys by Name and ID from which they choose which survey they want to check (the current configuration is shown in Figure 3).

After entering the variable star name or its coordinates, search radius, and limiting magnitude, the sequence team member is presented with a screen image showing the relative distribution of potential comp stars within the requested field of view. The variable stars are displayed as yellow, and then the potential comparison stars as either blue, green, or red, depending upon their B–V values. When a sequence team member clicks on a potential comparison star they are presented with the coordinates and magnitude values (V and B–V) as well as the number of calibrated observations and the source ID (Figure 4). To add a comparison star to a sequence, the team member has a button to click for any selected star (Send to File) that will then add all the data to a text file, in the correct format as shown earlier, for uploading to VSD.

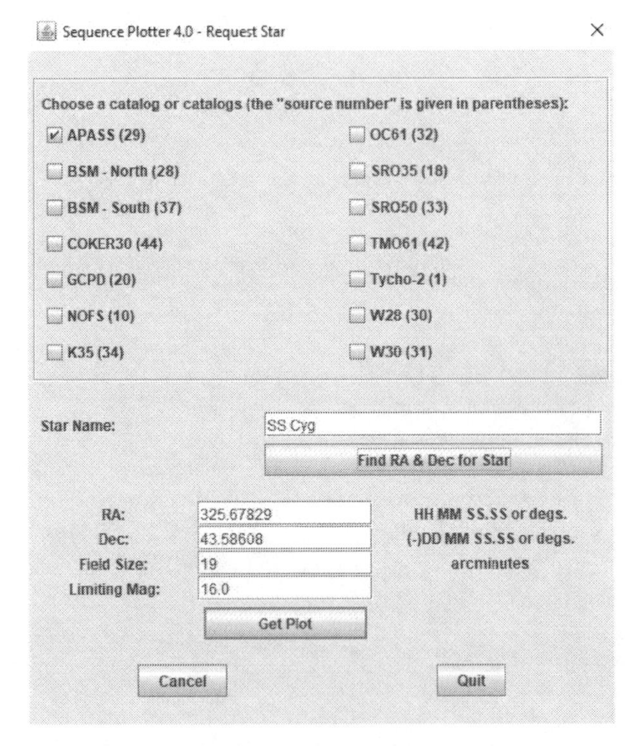

Figure 3. Current SEQPLOT Initial Entry Screen.

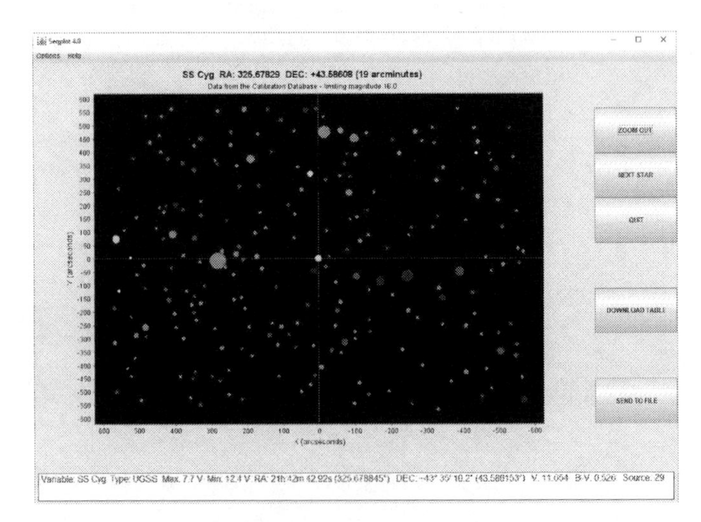

Figure 4. Current SEQPLOT Sequence Building Screen.

8. All-sky photometry

With the tools now well established for observer printing of charts and creation of sequences, the greatest deficiency was the absence of all-sky calibrated data needed for the Sequence Team to be able to create new sequences as were increasingly being requested by observers.

Henden had been a major contributor of calibrated data, as time permitted (starting in the 1990s), while employed at the Naval Observatory in Flagstaff, Arizona. With his appointment as Director of the AAVSO in 2005, this source ceased to exist. Henden describes the turn of events:

> However, the AAVSO became a member of the Sonoita Research Observatory (SRO) consortium in 2005. I used the Celestron C14 telescope for field calibrations for the next few years, primarily for CV and Mira fields. In 2008, two telescopes became available from the estate of Paul Wright. Sited near Cloudcroft, NM, at Tom Krajci's observatory, these telescopes and SRO formed the basis of the AAVSOnet robotic telescope network. While time was made available to the membership, all three telescopes also provided calibration photometry for AAVSO campaigns.
>
> In November 2009, the AAVSO submitted a proposal to NSF to cover the eclipse of epsilon Aurigae, a very long-period binary system. At 3rd magnitude, it was too bright to monitor with the AAVSOnet telescopes. I designed a new, inexpensive system consisting of a wide-field astrographic refractor, SBIG CCD camera and Celestron GOTO mount, and obtained permission from Krajci to site the system at his observatory. This "Bright Star Monitor" (BSM) was specified to be capable of imaging Polaris at V = 2, with a primary emphasis of monitoring the ε Aur eclipse. During the Fall 2009 AAVSO Council meeting, I presented the concept to the Council and got permission to proceed. Donations were obtained from D. Starkey, D. Welch, J. Bedient, M. Simonsen, G. Walker, K. Mogul. and D. Sworin for the equipment purchase. It saw first light October 12, 2009.
>
> The original BSM system was followed by others in Australia, Massachusetts, and Hawaii, all funded by donations from individuals. This BSM network was heavily used to provide field calibrations for the brighter stars in the AAVSO program, as well as monitoring bright variables (and acquiring thousands of data points during the eps Aur eclipse).
>
> Since many nights were calibrated using Landolt standards, and every star in every frame was measured by an automated pipeline, a database of all calibrated stars in all nightly fields was created. This file eventually contained about one million bright stars across the sky in isolated 3-square-degree patches.
>
> Similar databases were created for the other AAVSOnet telescopes, such as SRO. These individual databases were gathered together into the Seqcat database used by seqplot by the Sequence Team Members.
>
> I wrote an article about the AAVSO for *Sky & Telescope* for their November 2009 issue. Robert Ayers, an amateur in California, read the article and contacted me about possible financial contribution towards pro-am collaboration. After much discussion, a proposal was submitted to his Robert Martin Ayers Sciences Fund to create the AAVSO Photometric All-Sky Survey (APASS). This was a long-time dream of mine, to provide calibrated photometry in every part of the sky, so that the tedious process of all-sky calibration of individual FOVs could be avoided. An APASS site in both the northern and southern hemispheres were envisioned, to cover the entire sky from $10<V<17$ mag in multiple passbands. The northern site saw first light on November 6, 2009, and was sited at Tom Smith's observatory in Weed, NM. The southern site saw first light November 2, 2010 at the PROMPT complex at CTIO (La Serena, Chile). These sites typically calibrated 500 square degrees per photometric night in five passbands (Johnson B,V, and Sloan g,r,i).
>
> A team of professional collaborators (D. Welch, D. Terrell, S. Levine, A. Henden), software vendors (such as B. Denny and D. George) and volunteer amateurs (such as T. Smith, A. Sliski, and J. Gross) helped set up the system and monitor its progress. Multiple attempts at obtaining NSF funding for the project were made, with a grant made in 2014.
>
> The first data release from APASS occurred in September, 2010 and included about 4 million northern stars. Since then, 9 other releases have been issued, covering 99% of the sky.
>
> APASS was incorporated into Seqcat, starting with DR1, and forms the majority of the database at this time (Henden 2018).

APASS Data release 10 was installed in October of 2018, to be followed by DR 11 which will have about 100 additional nights of data with continued analysis improvement. It is anticipated that in about another year the final APASS data release, DR 12, will take place, which will include some missing Northern Hemisphere areas which are currently being covered as well as improved data analysis.

9. Request comparison stars for variable star charts

During 2010, sequence team member Tim Crawford realized that observers needed a way to make requests (as staff and team members were already able to do) for sequences for targets of interest that lacked comparison stars, and subsequently created a proposal to accomplish this. Team leader Simonsen and the AAVSO Director Henden readily approved this proposal and the staff time for creation of a web page for observers to make their requests.

The first sequence request was received on December 16, 2010 from Denis Denisenko (Russian Federation), and the 1,000th on November 25, 2015 from Stephen Hovell (New Zealand).

Through the end of August 2018, the Sequence Team has received some 1,963 requests from observers for new sequences. This total does not include staff requests for new sequences nor those self-generated by Sequence Team members.

10. Conclusions

Foremost, it needs to be recognized that while the important development of the various databases and tools and the creation of the Variable Star Plotter are significant for both current observers and future observers for producing charts and photometry tables, they would all have a limited and confining purpose were it not for the availability of calibrated data, principally APASS, that allows for the overwhelming majority of the sky in both hemispheres to be able to have sequences available in the ~10–16.5V range.

The APASS (as well as BSM) effort was a visionary triumph for Henden and quite an amazing feat given the limited resources available and the overworked team of both staff members and volunteers that aided Henden.

With the increasing availability of the fainter Pan-STARRS1 data (~14.5–20V) and the final release of APASS data, future generations of observers should never have a problem securing a sequence for any existing variable star or future variable star discovery, in either hemisphere.

Without the appointment of Arne Henden as the Director of the AAVSO, in 2005, I am skeptical that the charting/sequence process would have advanced to become as sophisticated and reliable as it is today. We are all indebted to Henden for his many past and continued tireless contributions to the process.

11. Acknowledgements

At the AAVSO Fall Meeting in 2014 Simonsen presented a PowerPoint presentation titled "History of AAVSO Charts III: the Henden Era." As future events unfolded, Simonsen was unable to advance this outline to a paper, which prompted me to undertake the project. I am indebted to Mike for both his initial outline as well as for bringing me on the chart team in 2007 and showing great patience as a mentor.

I owe a very large thank you to Arne Henden for his review of this paper and for allowing me to directly quote a number of his contributions. In addition, Aaron Price is to be acknowledged and thanked for also reviewing this paper and for allowing me to directly quote him several times.

I apologize if I have overlooked any of the many staff and volunteer contributors to the various portions of this process. I have attempted in good faith to recognize as many of you as I could identify.

References

Beck, S. J. 2018, private communication.

Henden, A. A. 2007, *AAVSO Newsletter*, No. 35 (December 2007), 1.

Henden, A. A. 2018, private communications.

Malatesta, K. H., Simsonen, M. A., and Scovil, C. E. 2007, *J. Amer. Assoc. Var. Star Obs.*, **35**, 377.

Otero, S. 2018, private communication.

Petriew, V., and Koppelman, M. 2007, 93rd AAVSO Spring Meeting presentation, Calgary, AB, Canada.

Price, A., Petriew, V., and Simonsen, M. 2004a, *J. Amer. Assoc. Var. Star Obs.*, **33**, 59.

Price, A., Petriew, V., and Simonsen, M. 2004b, "Charts and Comparison Stars: A Roadmap to the Future," 93rd AAVSO Spring Meeting presentation, Berkeley, CA.

Price, A. 2018, private communication.

Simonsen, M. 2014, "History of AAVSO Charts III: The Henden Era," 103rd AAVSO Annual Meeting presentation, Woburn, MA.

Williams, T. R., and Saladyga, M. 2011, *Advancing Variable Star Astronomy: The Centennial History of the American Association of Variable Star Observers*, Cambridge Univ. Press, Cambridge and New York, 322.

Abstracts of Papers and Posters Presented at the 107th Annual Meeting the AAVSO, Held in Flagstaff, Arizona, November 15–17, 2018

Discoveries for δ Scuti Variable Stars in the NASA Kepler 2 Mission

Joyce A. Guzik
Jorge Garcia
Jason Jackiewicz
address correspondence to J. A. Guzik, 432 Pruitt Avenue, White Rock, NM 87547; jguzik@mindspring.com

Abstract The NASA Kepler spacecraft launched nearly ten years ago has been observing fields along the ecliptic plane for about 90 days each to detect planets and monitor stellar variability. We analyzed the light curves of thousands of main-sequence stars observed as part of the Kepler Guest Observer program. Here we summarize the statistics of discovery and properties of the pulsation amplitude spectra for about 250 δ Scuti variable stars found in Kepler2 Campaigns 4 through 17. These stars are about twice as massive as the Sun, pulsating in many simultaneous radial and nonradial pulsation modes, with periods of about two hours. We discuss the potential and challenges for these stars of using pulsations to constrain stellar interior properties.

Stepping Stones to TFOP: Experience of the Saint Mary's College Geissberger Observatory

Ariana Hofelmann
Saint Mary's College of California, Department of Physics and Astronomy, 1928 Saint Mary's Road, Moraga, CA 94575; Ariana.hofelmann@gmail.com

Brian Hill
Saint Mary's College of California, Department of Physics and Astronomy, 1928 Saint Mary's Road, Moraga, CA 94575; brh3@stmarys-ca.edu

Abstract We upgraded our college's Meade 0.4-m telescope, including its software, mount, and imaging train, in order to perform exoplanet photometry. We used a variety of resources, beginning with Dennis Conti's Exoplanet CHOICE course. We will summarize the data-taking processes and analysis software typically used for this type of photometry. After having been accepted into the TESS Followup Observing Program Sub Group 1 (TFOP SG1), we have been going through a process of imaging and submitting false positives that conform to their data submission requirements. We hope our experience and the encouragement of the TFOP program leaders will inspire other AAVSO members to go through similar steps.

Small Observatory Operations: 2018 Highlights from the West Mountain Observatory

Michael Joner
Brigham Young University, Department of Physics and Astronomy, N-488 ESC, Provo, UT 84602 ; xxcygni@gmail.com

Abstract The West Mountain Observatory (WMO) is an off-campus astronomical observatory operated by Brigham Young University. WMO is located about 23 km southwest of the main BYU campus in Provo, Utah, at an elevation of 2,120 m. Observations are done for a variety of student and faculty projects using the three small telescopes (0.3 m, 0.5 m, and 0.9 m) housed at WMO. I will present a summary of recent upgrades and improvements at WMO, along with observational highlights from 2018 that include targets ranging from solar system objects out to active galaxies with a lot of ground to cover in between.

Comparison of North-South Hemisphere Data from AAVSO Visual Observers and the SDO Satellite Computer-Generated Wolf Numbers

Rodney Howe
3343 Riva Ridge Drive, Fort Collins, CO 80526; ahowe@frii.com

Abstract Rolling correlations and rolling covariance analysis are used for two different type data submitted to the AAVSO solar database. In this paper we look at rolling correlations from 35 visual solar observers and their Wolf numbers for north and south solar hemispheres, and compare those data with SDO (Solar Dynamics Observatory) satellite Wolf numbers calculated from HMI CCD images of north-south magnetograms and visual intensity CCD images (http://hmi.stanford.edu). The SDO computer generated group, sunspot, and Wolf numbers from HMI images show symmetric volatility in the plots when compared to the AAVSO solar observers who count group, sunspot, and Wolf numbers from observatories on Earth. Rolling correlation can be used to examine how correlative relationships between the two solar hemisphere Wolf numbers change over time. A value of 1 means both hemispheres are synchronized with each other. A value of –1 means that if one hemisphere's Wolf numbers decline, the other hemisphere's numbers rise. A correlation of zero means no correlation relationship exists.

…on Analysis of the Eccentric Eclipsing … Cassiopeiae

. Box 263, Rockyford, Alberta T0J 2R0, Canada; obs681@gmail.com

Abstract The Algol-type eclipsing binary system V1103 Cas was discovered by Otero *et al.* (2006), and identified as an eccentric system with period 6.1772 days. I observed it on multiple nights from 2012 to 2017, and found the primary and secondary eclipses to be of unequal depth and duration. The secondary eclipse is displaced from phase 0.5, and that displacement is slowly varying. Differential V-filtered lightcurves were modeled (using BINARYMAKER3) to determine the eccentricity (0.27) and inclination (87.5 degrees) of the system. These parameters, and 10 times of minima, were used to determine the apsidal rotation period (748 years), using the method described by Lacy (1992). The presentation will include material showing how the eclipse widths and timing will vary through the apsidal period.

Variable Stars and Cultural Astronomy

Kristine Larsen
Central Connecticut State University, 1615 Stanley Street, New Britain, CT 06050; larsen@ccsu.edu

Abstract Cultural astronomy encompasses the interdisciplinary, international, and multicultural fields of ethnoastronomy (the study of the astronomical knowledge and practices of current cultures) and archaeoastronomy (the study of the astronomical knowledge and practices of ancient cultures). Numerous universities across the United States (and beyond) have developed cultural astronomy courses in recent decades in recognition of the sophisticated astronomical knowledge developed across the globe without the use of modern technology (i.e. the telescope or imaging technology). Cultural astronomy provides a lens through which to study how individuals and cultures interacted with the heavens in personal and meaningful ways. While calendars, creation myths, celestially aligned structures, and navigation are usually the most common examples touted, what we now know to be variable stars have also played a role in the astronomical observations and mythology of numerous cultures. These include observations of supernovae, sunspots visible to the unaided eye, and possibly even Algol and other naked eye variables. Stellar variability has also been suggested as the reason why the Pleiades are widely known as a group of seven individuals (persons or animals) in mythologies from across the world despite the fact that only six are easily visible. This poster surveys these examples through a multicultural lens and suggests strategies for incorporating them in cultural astronomy courses and outreach programs.

Cold War Spy in the Sky now Provides an Eye on the Cosmos

Ken Steiner
10102 Mountain Apple Drive, Mint Hill, NC 28227; ksteiner30@gmail.com

Abstract The 12-meter satellite communications dish at Pisgah Astronomical Research Institute (PARI) was updated and converted from an electronic spy role to an astronomical radio telescope after being dormant for 20 years. The presentation will compare radio and optical telescopes and look at the cold war mission of this instrument when PARI was formerly a Department of Defense facility (Rosman Station).

The story of the update and conversion to a radio telescope, after the dormancy of 20 years, will be illustrated along with the current student involvement at PARI with the newly commissioned instrument. Finally, we will look at the variable star observation on August 21, 2017, to our knowledge the first time a solar eclipse was observed by a large radio telescope.

APASS DR10 Has Arrived!

Arne A. Henden
106 Hawkins Pond Road, Center Harbor, NH 03226; ahenden@gmail.com

Abstract The AAVSO Photometric All-Sky Survey (APASS) has reached a new milestone with its Data Release 10 (DR10). Approximately 128 million stars have been calibrated from $7 < V < 17$ and in passbands B,V,u',g',r',i',z'. This dataset has been made public on the AAVSO website.

APASS is designed to provide calibrated photometry everywhere in the sky and for nearly all CCD images. Much like the catalogs such as UCAC or GAIA do for astrometry, APASS gives the ability to photometrically calibrate your data without having to resort to all-sky photometry.

With DR10, the sky coverage is about 99%. There are still a few missing fields, primarily in the northern sky above declination 20 degrees. The upcoming DR11 will cover most of those fields, with about 200 new nights of data. It should be released very close to the time of the Annual meeting.

APASS is a volunteer-driven project. There are many opportunities for people at all skill levels to contribute to the project and help it reach its conclusion!

The Faint Cataclysmic Variable Star V677 Andromedae

Lewis M. Cook
1730 Helix Court, Concord, CA 94518; lew.cook@gmail.com

Enrique de Miguel
Departamento Fisica Aplicada, Facultad de Ciencias Experimentales, Universidad de Huelva, Huelva, 21071, Spain; demiguel@uhu.es

Geoffrey Stone
44325 Alder heights Road, Auberry, CA 93602; geofstone@earthlink.net

Gary E. Walker
114 Cove Road, West Dennis, NH 02670; bailyhill14@gmail.com

Abstract More than 5000 CCD/CMOS photometric observations of the cataclysmic variable star V677 Andromedae were made in the years 2015, 2016, and 2018 and the light curves are compared. The light curves are found to differ slightly from year to year for an unknown reason and there is also a dependence on the level of activity in this binary pair. In its active state, the light curve exhibits a double-humped shape, while in the declining state, the double hump in large part disappears. Our sparse photometry in this state nonetheless strongly suggests it is replaced by a single-hump shaped light curve. The star was too faint for us to observe in the inactive state except for stacked images covering entire cycles. A refined estimate of the period from our 2018 data was found to be 105 minutes, 18.6 seconds, however it is not firmly established if this is the orbital period or a "superhump" period.

The Fun of Processing a Stellar Spectrum—the Hard Way

Stanley A. Gorodenski
9440 E. Newtown Avenue, Dewey, AZ 86327; stanlep@commspeed.net

Abstract Freeware exists for processing spectra: VSPEC, IRIS, ISIS, AUDELA, BASS, and MIDAS, to name some. One feature in common with all is that they can, to some extent, be viewed as black boxes, and they are limited to doing certain very specific functions. Not only may some users feel it would be too daunting to attempt to write computer programs to do some of the same things these packages do, many of the packages are limited to the extent a spectrum can be explored in a statistical and graphics sense. There are a huge number of gifted computer people in astronomy. This talk is directed toward those, like the author, who is not so gifted but found it very interesting and exciting to be able to write programs to explore a spectrum. Although there are many things one can do, such as computing equivalent width, computing radial velocity, and estimating a continuum, this talk will only focus on two things. It will show the fun of developing a Gaussian curve and using it to identify large deviations from an ideal Gaussian distribution at the pixel column level of a spectrum. It discusses some of the difficulties of doing a dark sky subtraction and it describes an exploratory method for doing one that, although it may or may not be better than the method these other packages use and actually is very computer intensive, demonstrates what can be done if you can computer program and have a good statistical package with good programming capabilities.

Is sCMOS Really sCMAS?

Gary Walker
114 Cove Road, West Dennis, NH 02670; bailyhill14@gmail.com

Abstract The world of Astro Imaging has seen several technology changes. The author has experienced Tri-X film, push processing, Fuji 400, hyper sensitizing, CCD monochrome, colored filters, and now sCMOS. Many CCD chip manufacturers have shut down their factories—many to make space for new CMOS fabrication lines. Leveraging from the computer chip industry fabrication technology, CMOS chips offer small pixels, high speed, low noise, high dynamic range, and most important, lower cost. While this works well for DSLRs, cell phone cameras, security cameras, and machine vision applications, how does this affect Astro Imaging? At the 2016 NEAIC, the word from vendors was that for the point and stare application of long exposures common to astronomy, the CCD was still the detector of choice. The evolution of the CMOS technology may have closed the gap. The author investigates how CMOS can best be used for the point and stare applications that Astro Imagers need.

β Cepheid and Mira Variable Stars: A Spectral Analysis

Jesse D'Shawn Harris
22792 U.S. Highway 23 N, Duffield, VA 24244; jdh3uf@uvawise.edu

Lucian Undreiu
121 Beverly Avenue, Wise, VA, 24293; lundreiu@hotmail.com

Abstract The purpose of this project is to investigate and compare the spectra of several variable stars belonging to two classes, β Cephei and Mira, while trying to correlate our observations with the photometric survey maintained by AAVSO. For this study we have used a SBIG STT 8300 CCD camera, in conjunction with an LHIRES III Spectrograph that was attached to a 16-inch Meade SCT telescope.

The short pulsation periods (0.1–0.3 day) of β Cephei variables enabled us to follow their evolution through the entire cycle of pulsation. The identification of H Balmer and neutral He lines, which are originating in the upper atmospheric layers, confirms the spectral class (early B-type). The β Cephei intricate dynamics of the pulsation mechanism are revealed by the changes in the width of strong spectral lines, as well as by the line doubling. BW Vulpelculae is of particular interest, due to its large amplitude of pulsation, reflected in the variability in its spectrum with progression through its period.

Mira variables are cool, red giants, pulsating slowly (>100 days), while having large fluctuations in brightness. We collected and compared spectra of several Mira variables, identifying TiO and ZrO bands, typical for their spectral class (late M/S). Studying this class of variable stars is definitely relevant, as they offer us a way to see the future evolution of stars similar to our own Sun.

New Intense Multiband Photometric Observations of the Hot Carbon Star V348 Sagittarii

Franz-Josef Hambsch
Oude Bleken 12, Mol 2400, Belgium; hambsch@telenet.be

Christopher S. Jeffery
Armagh Observatory, College Hill, Armagh BT61 9DG, United Kingdom; csj@arm.ac.uk

Abstract V348 Sgr is one of four hot carbon-rich and hydrogen-deficient stars. It is also the central star of a planetary nebula with a strong stellar wind, an infrared dust excess, and a circumstellar dust shell. Since July 2014, near daily multi-band photometric observations have been obtained at the Remote Observatory Atacama Desert (ROAD) close to San Pedro de Atacama, Chile. Strong variations of the brightness of V348 Sgr have been observed, ranging from magnitude 19 to 11.2 in V band. No clear periodicity is discernible in the data. The observed light curve shows much more variation and on a much shorter time scale than that of R CrB, the prototype hydrogen deficient, carbon- and helium-rich star. The star becomes markedly redder during extinction phases as a consequence of obscuring dust. The particular challenge in this case is to understand what triggers the production of dust.

Camera Characterization and First Observation after Upgrade of Feder Observatory

Isobel Snellenberger
Adam Kline
1236 Belsly Boulevard, Moorhead, MN 56560; snellenbis@mnstate.edu

Abstract We prepared the Paul P. Feder Observatory at the Minnesota State University Moorhead Regional Science Center to observe exoplanet transits after recent upgrades of the camera and control system. We characterized the camera by measuring linearity, gain, read noise, and dark current. We also discuss how we minimize tracking error without a guide camera. We observed a transit of the exoplanet Kelt 16b, the first exoplanet transit observed with the new system. The goal is to observe exoplanet candidates identified by the Transiting Exoplanet Survey Satellite (TESS).

Bright Star Monitor Network

Michael Nicholas
6635 West Hill Lane, Glendale, AZ 85310; mnicholas5cox@gmail.com

Abstract The AAVSO Bright Star Monitor telescopes are a subset of the larger AAVSOnet network located at sites around the world. Each site is equipped with a small telescope, a high grade astronomical camera, and standard photometric filters. They are operated robotically, and are locally supported by AAVSO member volunteers. Each telescope is capable of performing precise CCD photometric measurements on the sky's relatively bright stars, those in the range of 3.0 to 13.0 V magnitudes. It is available free to all AAVSO members.

Solar System Objects and the AAVSO Photometric All-Sky Survey (APASS)

Stephen Levine
Lowell Observatory, 1400 West Mars Hill Road, Flagstaff, Arizona 86001; sel@lowell.edu

Arne Henden
106 Hawkins Pond Road, Center Harbor, NH 03226; ahenden@gmail.com

Dirk Terrell
4932 Peakview Street, Erie, CO 80516; terrell@boulder.swri.edu

Doug Welch
100 Melville Street, Dundas, ON L9h 2A3, Canada; welch@physics.mcmaster.ca

Brian Kloppenborg
3450 Miller Drive, Unit 1116, Atlanta, GA 30341; brian@kloppenborg.net

Abstract The AAVSO Photometric All-Sky Survey, data release 10 (APASS DR10) can be used for photometric calibration of observations of moving objects. Because APASS provides calibrated photometry over the whole sky, it makes it much simpler to tie together observations of objects, like asteroids and comets, that move appreciable distances over the time they are observed. Because the photometric standards are in each image, it will also be possible to recover photometry at the few percent level from non-photometric nights. In addition to providing calibration for new observations, the original APASS data comprise over 500,000 images, each 7.8 square degrees in size, taken over the course of more than nine years. We have searched those images for known Solar System bodies, and present the initial results of this search. For many of the objects found, we have simultaneous five color (B,V, g', r', and i') photometry. APASS provides photometric standards in at least five colors over the magnitude range 7 to 17, which makes it a good match for calibration for telescopes ranging from a few inches in size up to several meters.

Conducting the Einstein Gravitational Deflection Experiment

Richard L. Berry
22614 N. Santiam Highway, Lyons, OR 97358; rberry@wvi.com

Abstract In this presentation, I describe our experiment and results from the Einstein Eclipse Experiment carried out at Alpaca Meadows Observatory at the August 21, 2017, solar eclipse. For those intending to carry out a similar experiment at future solar eclipses, I describe the pitfalls we encountered and our successes in pre-eclipse preparation, image acquisition, data extraction, and data reduction. Thanks to PCC Sylvania's Dr. Toby Dittrich, students Andrew Jozwiak, Steve Pinkston, Abraham Salazar, and Jacob Sharkansky, participant Jeremy Britton, and donor David Vernier, and special thanks to Donald Bruns for his generous help and advice.

Erratum: Recent Minima of 266 Eclipsing Binary Stars

Gerard Samolyk
P.O. Box 20677, Greenfield, WI 53220; gsamolyk@wi.rr.com

In the article "Recent Minima of 266 Eclipsing Binary Stars" (*JAAVSO*, 2018, **46**, 184–188), the TOM listed for RS Sct on JD 58327 (page 187, right-hand column) should be listed as RS Ser.

Made in United States
Troutdale, OR
09/05/2023